Lecture Notes in Biomathematics

Managing Editor: S. Levin

W0106013

66

Nonlinear Oscillations in Biology and Chemistry

Proceedings of a meeting held at
the University of Utah, May 9–11, 1985

Edited by H. G. Othmer

Springer-Verlag
Berlin Heidelberg New York Tokyo

Editor

Hans G. Othmer
Department of Mathematics, University of Utah
Salt Lake City, UT 84112, USA

Mathematics Subject Classification (1980): 34-06, 92-06

ISBN-13: 978-3-540-16481-4 e-ISBN-13: 978-3-642-93318-9
DOI: 10.1007/978-3-642-93318-9

2146/3140-543210

Preface

This volume contains the proceedings of a meeting entitled 'Nonlinear Oscillations in Biology and Chemistry', which was held at the University of Utah May 9-11, 1985.

The papers fall into four major categories: (i) those that deal with biological problems, particularly problems arising in cell biology, (ii) those that deal with chemical systems, (iii) those that treat problems which arise in neurophysiology, and (iv), those whose primary emphasis is on more general models and the mathematical techniques involved in their analysis. Except for the paper by Auchmuty, all are based on talks given at the meeting. The diversity of papers gives some indication of the scope of the meeting, but the printed word conveys neither the degree of interaction between the participants nor the intellectual sparks generated by that interaction.

The meeting was made possible by the financial support of the Department of Mathematics of the University of Utah. I am indebted to Ms. Toni Bunker of the Department of Mathematics for her very able assistance on all manner of details associated with the organization of the meeting. Finally, a word of thanks to all participants for their contributions to the success of the meeting, and to the contributors to this volume for their efforts in preparing their manuscripts.

Hans G. Othmer
Salt Lake City, Utah
January 3, 1986

CONTENTS

BIOLOGICAL SYSTEMS

OSCILLATIONS AND CHAOS IN THE PANCREATIC β-CELL

Teresa Ree Chay
Department of Biological Sciences
University of Pittsburgh
Pittsburgh, PA 15260

1. INTRODUCTION

The voltage across the plasma membrane of the pancreatic β-cell displays burst activity when glucose between ~8.5 mM and 16.5 mM is added to the perfusion medium (Dean and Mathews, 1970; Ribalet and Beigleman, 1979; Atwater et al., 1980). This consists of an active phase during which spikes are generated and a silent phase during which the membrane is hyperpolarized. In high glucose concentrations above 16.6 mM, the burst pattern disappears and spike activity becomes continuous. The active phase has been associated with the stimulus for the release of insulin (Meissner, 1976; Scott et al., 1981).

A characteristic feature of the β-cell is that the duration and frequency of each burst can be altered by the glucose concentration. Increasing glucose increases the relative duration of the active phase (Beigelman et al. 1977; Meissner and Pressler, 1980; Atwater et al., 1980). The experiment of Meissner and Schmelz (1974) shows that, just above the onset of bursting, the period is longer than throughout the intermediate range of glucose concentrations. The period, moreover, lengthens again at concentrations just below the onset of continuous spiking. Other experiments (Beigelman et al., 1977) also show that when the glucose concentration is increased slowly but steadily from zero to 22.2 mM, the burst frequency increases at first but later decreases with an increase in the absolute burst duration.

The Ca^{2+}-activated K^+ channel is believed to play a crucial role for β-cell bursting; thus, it has been regarded as the regulatory link between $[Ca]_i$, insulin release, and Ca^{2+} influx (Atwater et al., 1983; Petersen and Maruyama, 1984). In cultured rat β-cells, there is another type of K^+-channel in which glucose is directly involved (Ashcroft et al., 1984). This so-called G-channel is completely inhibited by 20 mM glucose. The experiment on 10 mM glucose, however, suggests that the conductance of this channel is partially active at this glucose concentration. A partially active G-channel may contribute significantly to the burst activity.

In this paper, we first formulate a mathematical model for the bursting activity of β-cell using a barrier kinetic approach (Glasstone et al., 1941). With it we study the bursting behavior of the β-cell in response to a variation of the G-channel permeability (i.e., in response to the variation of glucose). We use a barrier kinetic approach instead of an equivalent electric circuit approach (Hodgkin and Huxley, 1952; Chay and Keizer, 1983) because the barrier kinetic model has proven to be particularly suited for studying ion transport across the excitable cell membrane when there is a large ionic variation of extracellular and/or intracellular ions in the β-cell electrical activity (Chay, 1980, 1983).

2. MODEL

In our model, there are three types of potassium channels: (i) the Ca^{2+} sensitive K^+ channel which is activated by intracellular Ca^{2+}, (ii) the G-channel which is inactivated by one (or a few) of the glucose metabolites, and (iii) the voltage gated channel which behaves like the Hodgkin-Huxley K^+ channel.

According to the barrier kinetic model (Chay, 1980, 1983; Lee et al 1983) the current carried by K^+ ions through these K^+ channels is expressed by

$$I_{K,j} = P_{K,j} \{[K^+]_o \exp(-FV/2RT) - [K]_i \exp(FV/2RT)\} \qquad (1)$$

where $[K^+]_o$ and $[K^+]_i$ are, respectively, the potassium concentration of the exterior and interior media, RT the gas constant times the absolute temperature, F the Faraday constant, and $P_{K,j}$ the permeability coefficient of the K^+ ions passing through the j-th type channel. Here, the subscript j refers to V, Ca, Glu, and thus $I_{K,V}$, $I_{K,Ca}$, and $I_{K,Glu}$ stand for the currents carried by K^+ ions passing through the voltage-, Ca_i-, and glucose-sensitive channels, respectively.

Note that in the equivalent electric circuit version of Hodgkin and Huxley (1952), the right side of eq. (1) takes $g_{K,j}(V_K - V)$, where $g_{K,j}$ is the conductance of the jth K^+-channel and V_K is the reversal potential for K^+ ions. Note also that a linearization around $F(V-V_K)/2RT$ in eq. (1) yields the equivalent circuit expression if $g_{K,j}$ is defined as $p_{K,j}$ $[K]_i \exp(FV/2RT)/RT$ and V_K is defined as $(RT/F)\ \ln([K]_o/[K]_i)$. Furthermore, if $p_{K,j}$ is proportional to $(FV/RT)/\{\exp(FV/2RT)-\exp(-FV/2RT)\}$, eq. (1) becomes the expression derived using a constant field approximation (Goldman, 1943). Since the membrane potential of the β-cell oscillates between -55 mV and -20 mV, eq. (1) gives nearly identical results as those obtained using the constant field approximation. Equation (1) does not differ significantly from a H-H type equivalent circuit model, as long as the external and/or internal potassium ion concentrations are held fixed.

The permeability coefficient of glucose sensitive K^+ channels takes a constant value, P_{glu}; however, the permeability constant of voltage-gated K^+-channels depends not only on membrane potential but also on time. We assume that the kinetics of this channel follow the Hodgkin-Huxley K^+ gating mechanism. Then, the permeability coefficient is expressed by

$$P_{K,V} = \bar{P}_{K,V} n^4 \tag{2}$$

where $\bar{P}_{K,V}$ is the maximal permeability, and n is the probability of opening of this channel:

$$dn/dt = [n_\infty - n]/\tau_n \tag{3}$$

where n_∞ and τ_n are given in Appendix I.

The permeability coefficient of Ca-sensitive K^+-channels is assumed to take the following simple binding form:

$$P_{K,Ca} = \bar{P}_{K,Ca}/(1+K_{K,Ca}/[Ca]_i) \tag{4}$$

where $[Ca]_i$ is the free intracellular calcium ionic concentration, $K_{K,Ca}$ is the effective dissociation constant of Ca^{2+} from the Ca_i-sensitive receptor protein.

Our model also contains the voltage-gated Ca^{2+}-channel. The current carried by the Ca^{2+} ions through this channel, I_{Ca}, is given by

$$I_{Ca} = P_{Ca} \{[Ca]_o \exp(-FV/RT) - [Ca]_i \exp(FV/RT)\} \tag{5}$$

where $[Ca]_o$ is the extracellular calcium concentration, and P_{Ca} is the permeability coefficient of calcium ions. We assume that P_{Ca} follows the kinetics which describe the Na^+ channel of the Hodgkin and Huxley:

$$P_{Ca} = \bar{P}_{Ca} m_\infty{}^3 h_\infty \tag{6}$$

where $\bar{p}_{Ca}$ is the maximal Ca^{2+} permeability coefficient, and m_∞ and h_∞ are the activation and inactivation probabilities of the Ca^{2+}-channel at the steady state. Note that we have replaced the time dependence of m and h of the Hodgkin-Huxley gating variables by their respective steady state values. This is because (i) the relaxation time for both m and h is on the order of ms (i.e., m and h follow very fast kinetics), (ii) we are not concerned with the exact shape of a spike, and (iii) this substitution greatly reduces the time required for computation while essentially giving identical results (Chay, 1985). The expression for m_∞ and h_∞ are also given in Appendix I.

The leak current in our model consists of the current carried by "leaking" Na^+ and K^+ ions and other currents, such as the electrogenic Na/K pump activity. Thus,

$$I_L = P_{L,o}\exp(-FV/2RT) - P_{L,i}\exp(FV/2RT) \quad (7)$$

where $P_{L,o}$ and $P_{L,i}$ are the leak permeability coefficients of extracellular and intracellular leaking ions, respectively.

The intracellular calcium concentration $[Ca]_i$ in our model is treated as a dynamic variable. Thus, the rate change of $[Ca]_i$ is due to the influx of extracellular calcium ions into the cell through the voltage sensitive calcium channel and the efflux of free intracellular Ca^{2+} ions from the cell to the extracellular medium by the action of Ca-ATPase pump activity. We thus have

$$f^{-1}\, d[Ca]_i/dt = 3I_{Ca}/4\pi r^3 F - k_{Ca}/(1+K_{Ca,p}/[Ca]_i) \quad (8)$$

where r is the radius of the β-cell, k_{Ca} the rate constant for the efflux of Ca_i, and $K_{Ca,p}$ the dissociation constant of $[Ca]_i$ from the Ca-ATPase. In the above equation, f is related to the free calcium and the total calcium concentrations in the cytoplasm (Ferreira and Lew, 1976).

In addition to the currents mentioned above, there is a capacitance current due to the charge on the inner and outer surfaces of the membrane. Thus, in the absence of an external current, the charge neutrality condition ensures us that

$$0 = I_{mem} + 2I_{Ca} + I_{K,Ca} + I_{K,V} + I_{K,Glu} + I_L \quad (9a)$$

where

$$I_{mem} = -4\pi r^2 C_m\, dV/dt \quad (9b)$$

and C_m is the membrane capacitance. Note that our present model contains a term $I_{K,Glu}$ in addition to the four other currents which appeared in our earlier model (Chay and Keizer, 1983, 1985).

In recapitulating, the present model contains three dynamic variables: (i) the membrane potential, V, whose variation with time is expressed as the sum of ionic currents carried by K^+ ions through three functionally different K^+ channels and by Ca^{2+} ions through the voltage sensitive Ca^{2+} channels, and by leakage ions through the time independent channel, (ii) the probability of opening of the voltage-dependent K^+ gate, n; and (iii) the intracellular Ca^{2+} ionic concentration. The three differential equations (3), (8) and (9) govern the time dependence of these variables.

3. RESULTS AND DISCUSSION

The differential equations representing the three dynamic variables were solved numerically on a DEC-10 computer (Digital Equipment Corp., Marlboro, MA). A Gear algorithm (Hindmarsh, 1974) has been used to solve

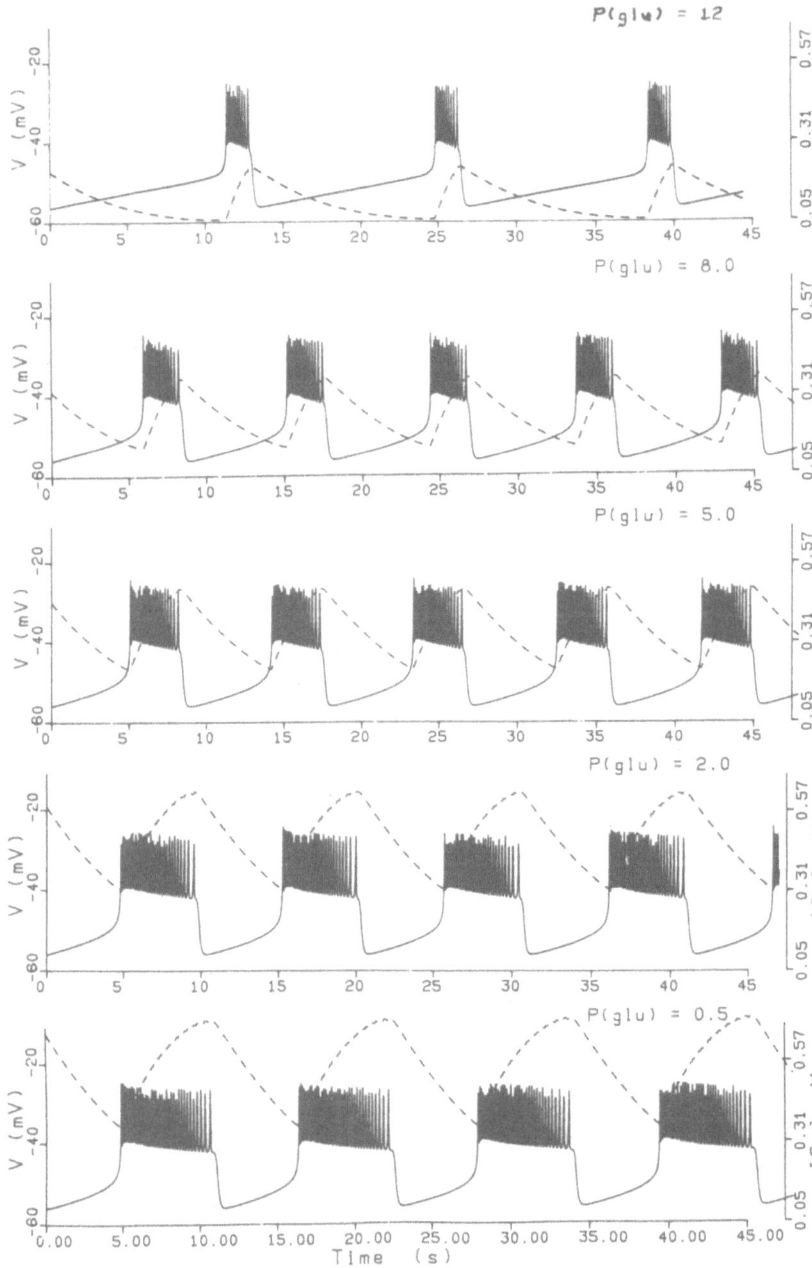

Figure 1: The effect of decreasing the permeability constant P_{glu}. The solid line shows a bursting behavior of the membrane potential, and the dashed line shows the associated changes in the concentration of intracellular Ca^{2+} ions. Here, the values of P_{glu}'s used for the computation are, from the top panel to the bottom, 12.0 pA M^{-1}, 8.0 pA M^{-1}, 5.0 pA M^{-1}, 2.0 pA M^{-1}, 0.5 pA M^{-1}, respectively. Other parametric values are listed in Table I.

the three equations in double precision, where we set the absolute and relative error tolerances at 10^{-8}. This numerical ordinary differential equation solver is very well suited to solve stiff equations like ours. The parameters used for the computation are listed in Table I. Most of the values in this table (e.g., the radius r, the capacitance C_m, the extracellular and intracellular potassium concentrations, the extracellular calcium concentration) came from experiments on β-cells. The values of $\bar{p}_{K,Ca}$ and P_{glu} were estimated from the work of Atwater et al. (1983) and Ashcroft et al. (1984) on the β-cell, respectively (using the coversion factor 1 $A \cdot M^{-1}$ = 11 S). Other parametric values were obtained from other cells such as the mammalian erythrocyte (e.g., $K_{Ca,p}$ value from the work of Ferreira and Lew, 1976).

Figure 1 illustrates the numerical solutions of the dynamics of P_{glu}-induced β-cell activity obtained from our model for five different values of P_{glu}. As can be seen in the figure, the bursting mode in each panel has the following characteristics: When the glucose concentration is raised (i.e., a decrease in P_{glu}) beyond a certain value, the membrane potential exhibits a typical pattern of a burst, which includes a silent phase of repolarization followed by a rapid depolarization and continuous spike activity. Oscillations in the membrane potential (solid line) result from a limit cycle oscillation which involves the intracellular Ca^{2+} concentrations (dashed line). The period of intracellular Ca^{2+} oscillations is identical to the burst period of membrane potential, although the shapes are quite different. At the beginning of a burst the Ca^{2+} concentration rises rapidly, peaking just after the burst ends. This increase is caused by the action potential spikes, during which Ca^{2+} ions flow into the cell from the perfusion medium. When $[Ca]_i$ approaches near the maximum level, the membrane potential falls abruptly to the minimum level of -56 mV. Note that an increase in glucose produces an increase not only in the amplitude but also in the magnitude of the intracellular Ca^{2+} concentration.

In the bursting regime, we observe that the shape of membrane potential depends very little on glucose: i.e., the maximum repolarization potential V_r is around -56 mV; the plateau potential V_p is around -40 mV (at the beginning of the burst -39 mV and at the end of the burst around -42 mV); the amplitude of spike potential is about 15 mV. Although the shapes of V_m oscillations look the same for all P_{glu} values, the duration of the silent phase becomes shorter and the active period becomes longer as the glucose concentration increases. This is consistent with the observed membrane potential oscillations of the β-cells on the glucose response (Beigelman et al. 1977; Ribalet and Beigelman, 1979).

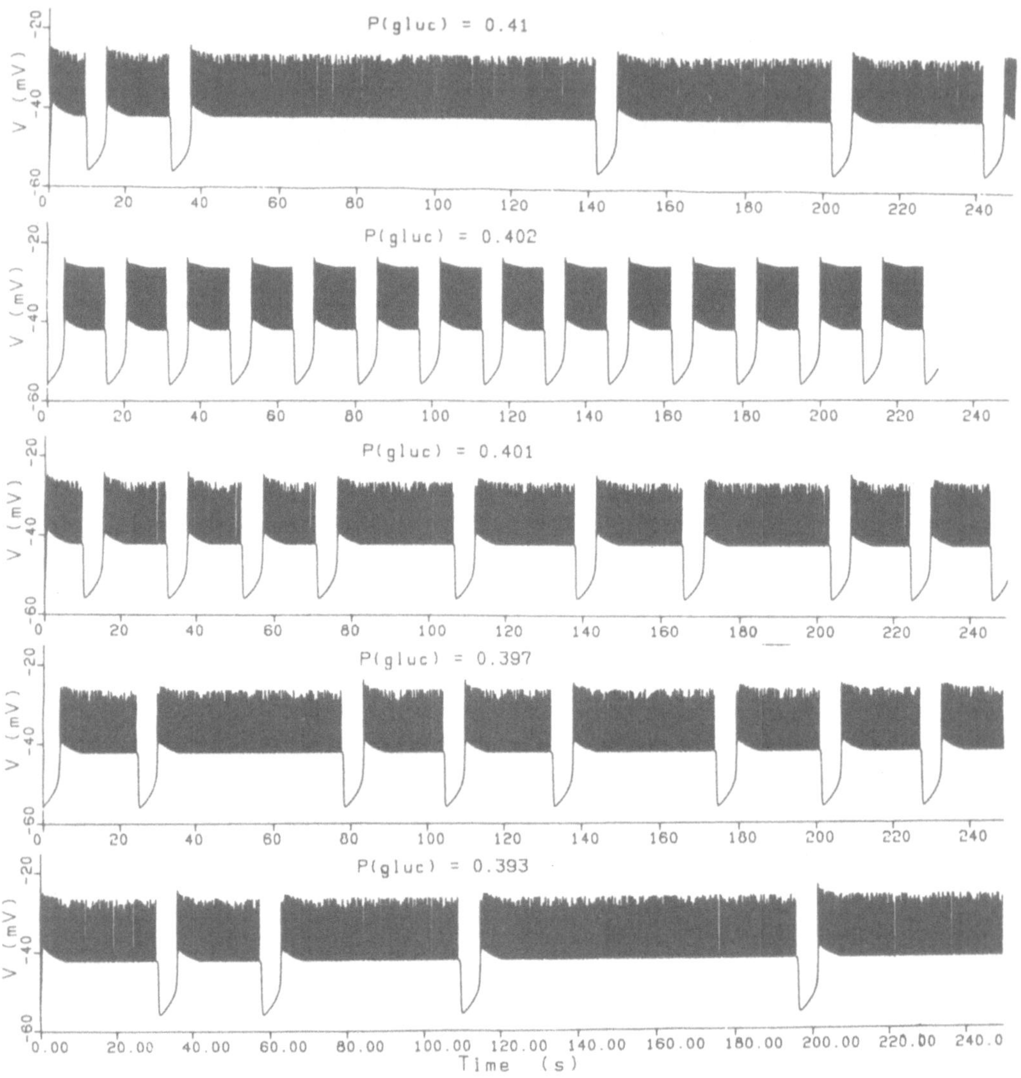

Figure 2: Chaotic bursting which occurs in a transition from a bursting mode to a continuous spiking mode. The P_{glu} values used for this computation are, from the top panel to the bottom, 0.41 pA M^{-1}, 0.402 pA M^{-1}, 0.401 pA M^{-1}, 0.397 pA M^{-1}, and 0.393 pA M^{-1}, respectively. The initial condition of the top trace is different from the rest.

Increasing glucose also results in a decrease in the burst periodicity. But, as the glucose concentration becomes saturated the burst period increases. Above a certain glucose concentration, the bursts disappear entirely and only spikes remain. As shown in Fig. 2, the period near the onset of continuous spiking (i.e., high intermediate glucose) is longer than that near the subthreshold glucose concentration (i.e., low glucose concentration). Note in Fig. 2 that aperiodicity occurs during the transition from the bursting mode to continuous spiking (i.e., in the ranges from ~0.45 to ~0.37 $pA \cdot M^{-1}$). Aperiodicity similar to those in Fig. 2 has been observed in experiments on the β-cells (Beigelman et al., 1977; Henquin et al., 1982).

Chaos that appears in the transition between the bursting and continuous spiking modes are very sensitive to the initial condition: Near the upper chaotic regime (i.e., above 0.4 $pA \cdot M^{-1}$), continuous regular bursting may appear, depending on certain initial conditions (e.g. see the second trace); similarly, near the lower bound of the chaotic regime (i.e., below 0.39 $pA \cdot M^{-1}$), continuous regular spiking may appear again, depending on the initial condition. As is apparent in Fig. 2, the initial conditions of the top trace differs from that of the bottom four traces. When the same initial condition is used (as shown in the bottom four traces) the average length of active phases increases with increasing P_{glu}. Such aperiodic behavior for a non-random system is known as deterministic chaos and is quite common in the excitable cells (Chay, 1984; Chay, 1985; Chay and Rinzel, 1985).

A useful quantitative tool for studying such deterministic chaotic behavior is a discrete time one-variable representation of the dynamic quantities such as V_m and $[Ca]_i$. The map representation allows a compact and simplified way to display the chaotic behavior. In Fig. 3, we have demonstrated how a deterministic nature of chaos appears in the transition region, by constructing a one-variable model that relates the spike-to-spike intervals (top left) and also that relates the recordings of V_m taken at every 50 ms intervals (bottom left). Here, $V_m(n)$ at time t is described by the difference equation, $V_m(n+1) = F(V_m(n))$. The map on the top left reveals intermittency in which periods of relative regularity (i.e., spikes) are abruptly and irregularly interrupted by bursts of quite different activity. The development of chaos can be seen more clearly from the limit cycle formed by V_m and $[Ca]_i$ in the top right and the phase portrait in the bottom right panel. This phase portrait was constructed by placing the calcium concentration at time t on the x-axis and its value after a delay of 100 ms on the y-axis.

Figures 4a and 4b show a hierarchy of doublet patterns (period-doubling cascade) as P_{glu} is increased from 0.0 to 0.31 $pA \cdot M^{-1}$. We show,

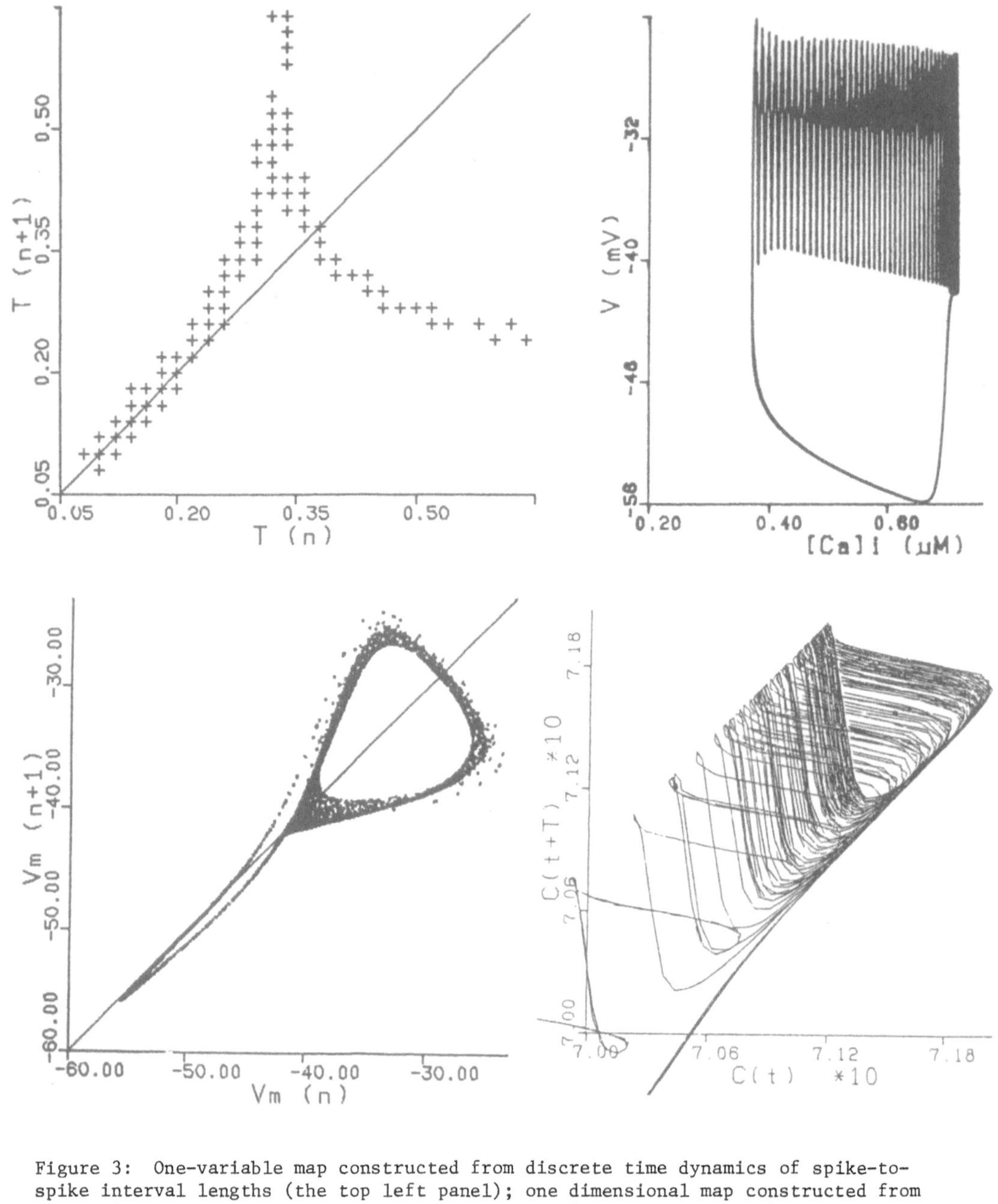

Figure 3: One-variable map constructed from discrete time dynamics of spike-to-spike interval lengths (the top left panel); one dimensional map constructed from the readings on V_m taken at every 50 ms interval (bottom left panel); the limit cycle formed by V_m and $[Ca]_i$ (top right panel); the trajectory of $[Ca]_i$ at time t vs. that at time t + 100 ms (the bottom right panel).

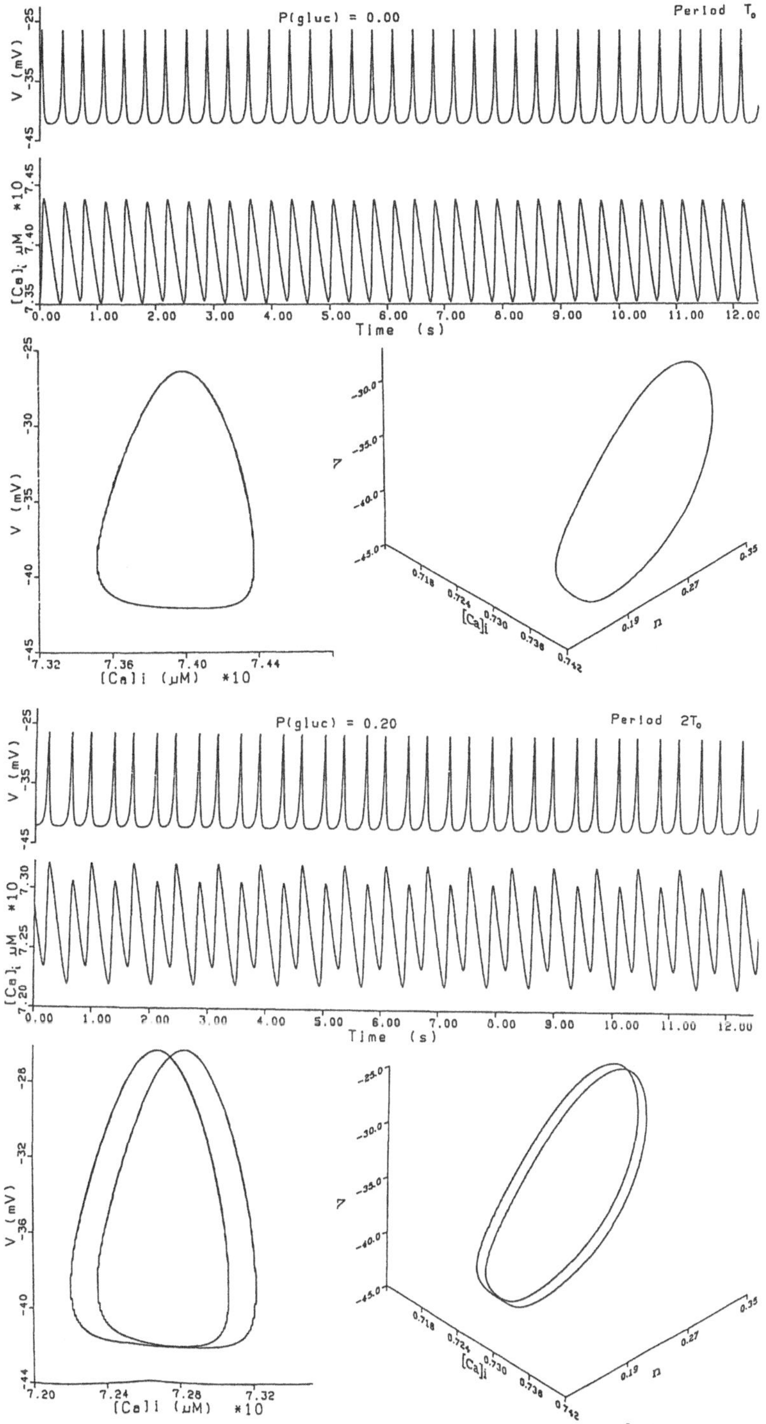

Figure 4a: Period-doubling as P_{glu} is increased from 0.0 pA M^{-1} (the top panel) to 0.20 pA M^{-1} (the bottom panel). The top two traces of each panel show the time course of V_m and $[Ca]_i$, respectively. In the bottom trace shows the limit cycles formed by V_m and $[Ca]_i$ (left) and by the 3 dynamic variables (right).

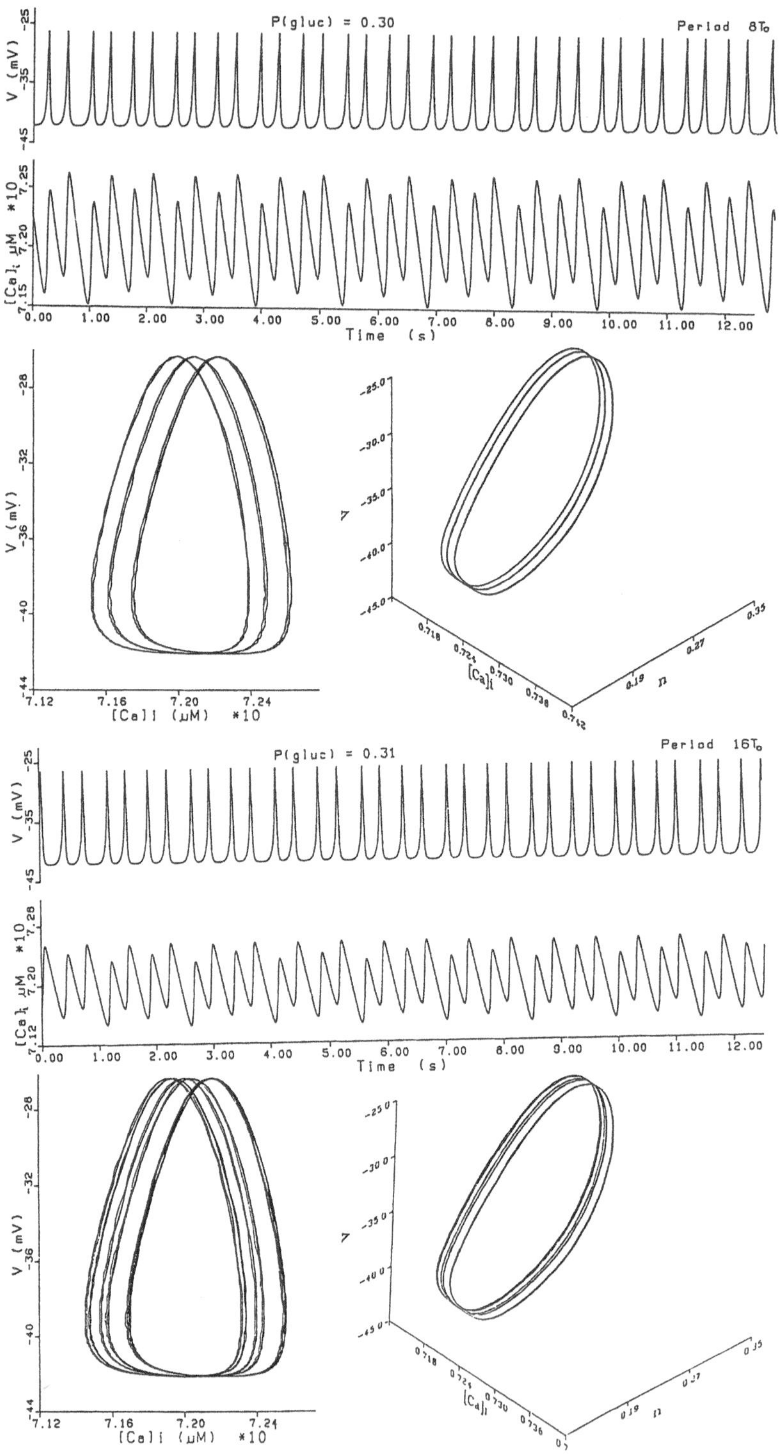

Figure 4b: The period of $8T_o$ and $16T_o$ as P_{glu} was increased further to 0.30 pA M^{-1} (the top panel) and 0.31 pA M^{-1} (bottom).

in the first two traces of each panel, the dynamics of membrane potentials and intracellular calcium concentrations. Note that when P_{glu} = 0.0 the cell fires repetitive action potentials at a regular interval (top panel of Fig. 4a). As P_{glu} increases, we observe a cascade of period doublings; the approximate periods are T_0 $2T_0$, $8T_0$, $16T_0$, where T_0 is the interspike interval of the basic spiking pattern in the upper panel of Fig. 4a. Although the voltage time courses show little variation in the doublet patterns from one case to the next, the calcium records (middle trace of each panel) reveal the fine structure of the successive period doublings. The Ca-V projects (bottom left) and the three dimensional limit cycles formed by the three dynamic variables (the bottom right) show clearly how a closed orbit bifurcates and is succeeded by a closed orbit with twice as many loops. At bifurcation, the period of the observed response exactly doubles. The bifurcation phenomenon is caused by a loss of stability in a periodic solution. This period-doubling cascade is evidently the route that leads to the chaotic bursting pattern of Fig. 2.

To see how the periodic and aperiodic states develop from the stable steady state and then disappear as P_{glu} changes, we have constructed a bifurcation diagram, and this is presented in Fig. 5. We used AUTO, a program for automatic bifurcation analysis developed by Doedel (1981, 1985), to construct the diagram. This analyzer computes the regions of the periodic branches, steady state branches, and Hopf bifurcations. It also gives the amplitude and period at a given value of the bifurcation parameter. The top trace of this figure shows the square root of the sum of the squares of the three dynamic variables. The second and third traces show the maxima of the membrane potential and $[Ca]_i$, respectively. The bottom trace reveals the period. Note that our model predicts that the maxima of the stable periodic $[Ca]_i$ must decrease with increasing P_{glu}.

In Fig. 5, negative values of permeability coefficient have no physical significance. Note that there are two Hopf bifurcation points (HB); one at 12.419 pA M^{-1} and -35.652 pA M^{-1}. The bifurcating branch of the periodic solution, which originates from the right bifurcation, terminates in a homoclinic orbit (i.e., orbit of infinite period containing a saddle point) at 12.437 pA M^{-1}, and one which originates from the left bifurcation terminates in another homoclinic orbit at 9.512 pA M^{-1}. The region AUTO was unable to connect is where the burst develops and bifurcates, as shown in the top two traces of Fig. 1. Note that near the two HBs, our differential equations have a stable oscillatory state (SPS)coexistent with a stable steady state (SSS). In these regions, instantaneous perturbations such as a brief current pulse may send stable rhythmic

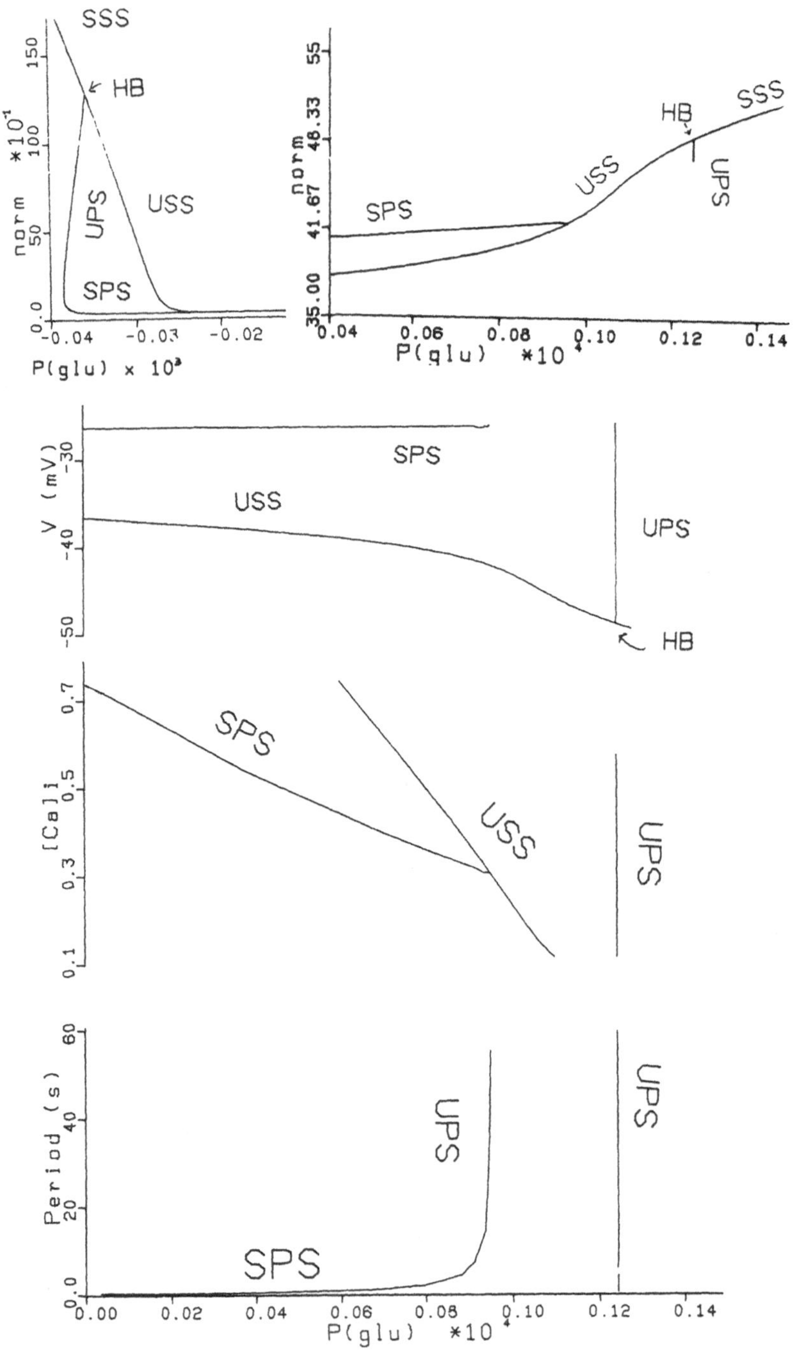

Figure 5: Bifurcation diagram revealing many different types of modes when P_{glu} is used as a bifurcation parameter. Here, HB stands for the Hopf bifurcation point, SSS for the stable steady state, USS for the unstable steady state, SPS for the stable periodic state, and UPS for the unstable periodic state. Here, the unit of P_{glu} is μA M^{-1}.

oscillation into the steady state.

3. CONCLUSION

We have presented a theoretical model for the bursting electrical activity of insulin-secreting pancreatic β-cells. Our treatment exploits the crucial role of G-channel on the electrical activity of the β-cell. As in the Chay-Keizer model (1983), our present model supports the belief that the bursting is, in fact, a consequence of the existence of Ca^{2+}-activated K^+ permeability, which triggers an increased influx of Ca^{2+} through voltage-gated Ca^{2+} channels. Thus, the intracellular Ca^{2+} concentration oscillates in synchrony with the membrane potential. Unlike the Chay-Keizer model, however, the dependence of glucose is due to the G-channel activity which controls the periodicity of bursting. The inactivation of the G-channel activity makes the magnitude and amplitude of $[Ca]_i$ increase with increasing glucose. The physiological implication of the increase in $[Ca]_i$ is that, as the glucose concentration increases, the cell secretes a larger amount of insulin. The controlling influence of P_{glu} on the magnitude of $[Ca]_i$ makes it a particularly important parameter, since glucose is known to inactivate G-channel activity (Ashcroft, 1984).

In the transition regime between the bursting and spiking modes of activity, our three dynamic variable model of the β-cell exhibits aperiodicity. The chaotic behavior was further studied numerically and analyzed via phase space concepts and in terms of one-variable discrete maps. As in many studies of determinitic chaos, this aperidicity is very sensitive to initial conditions. The oscillatory systems which are sensitive to an initial condition oftentimes exhibit aperiodic behavior in the region of transition behavior (Chay, 1984). This sensitivity makes the bursting cell transform to the spiking one (and vise versa) even with an infinitesimal perturbation.

Perhaps it should be mentioned that our model is one of a few biophysically realistic models which exhibit endogenous aperiodic chaos and that our numerical results look very similar to experimental results in the β-cells (Beigelman et al., 1976; Henquin et al., 1983). It appears from our simulations that the three dynamical models presented in this paper provides a fairly accurate quantitative description of limit cycle bursting in the pancreatic β-cell in response to glucose.

APPENDIX I

In this appendix we provide explicit forms for the steady state probability functions, m_∞, h_∞, and n_∞ and the relaxation time τ_n.

We assume that the voltage dependencies of these quantities take the same expressions as the original Hodgkin-Huxley equations (Hodgkin and Huxley, 1952) but V is shifted along voltage axis by 48, 48, and 28 mV, respectively. Thus,

$$y_\infty = \alpha_y/(\alpha_y + \beta_y) \qquad \text{(A1)}$$

where y_∞ stands for m_∞, h_∞, and n_∞, and

$$\alpha_m = 0.1(23+V)/(1=e^{-0.1V-2.3}) \qquad \text{(A2)}$$

$$\beta_m = 4e^{-(V+48)/18} \qquad \text{(A3)}$$

$$\alpha_h = 0.07e^{-0.05V-2.4} \qquad \text{(A4)}$$

$$\beta_h = 1/(1+e^{-0.1V-1.8}) \qquad \text{(A5)}$$

$$\alpha_n = 0.01(18+V)/(1-e^{-0.1V-1.8}) \qquad \text{(A6)}$$

$$\beta_n = 0.125e^{-(V+28)/80} \qquad \text{(A7)}$$

We also assume that the relaxation time τ_n in eq. (3) follows the expression similar to that of Hodgkin and Huxley (1952):

$$\tau_n = 0.003/(\alpha_n + \beta_n) \qquad \text{(A8)}$$

APPENDIX II

Parameter Values of the Model

Parameter	Unit	Numerical Value
C_m	$\mu F \cdot cm^{-2}$	1
$\bar{P}_{Ca}$	$\mu A \cdot M^{-1}$	0.125
$\bar{P}_{k,V}$	$\mu A \cdot M^{-1}$	1.2×10^{-2}
$\bar{P}_{K,Ca}$	$\mu A \cdot M^{-1}$	5.0×10^{-5}
$P_{L,o}$	μA	4.0×10^{-4}
$P_{L,i}$	μA	1.5×10^{-3}
$[K^+]_o$	mM	5
$[K^+]_i$	mM	130
$[Ca^{2+}]_o$	mM	2
r	μm	6
f		4.0×10^{-3}
T	^oK	310
k_{Ca}	$\mu M \cdot s^{-1}$	25

$K_{K,Ca}$	μM	1
$K_{Ca,p}$	μM	1

ACKNOWLEDGMENT: I would like to thank Dr. Illani Atwater at the National Institutes of Health and Dr. Bernard Ribalet at the University of California at Los Angeles for valuable discussion during this work. This work was supported by NSF PCM82 15583.

REFERENCES

Ashcroft, F.M., Harrison, D.E., Ashcroft, S.J.H.: Glucose induces closure of single potassium channels in isolated rat pancreatic β-cells. Nature 312, 446-448 (1984).

Atwater, I., Dawson, C.M., Scott, A., Eddlestone, G., Rojas, E.: The nature of the oscillatory behavior in electrical activity for pancreatic β-cell. In Biochemistry Biophysics of the Pancreatic β-cell. Georg Thieme Verlag, New York, 100-107 (1980).

Atwater, I., Rosario, L., Rojas, E.: Properties of the Ca-activated K^+ channel in pancreatic β-cells. Cell Calcium 4, 451-461 (1983).

Beigelman, P.M., Ribalet, B., Atwater, I.: Electrical activity of mouse pancreatic beta-cells. J. Physiol. Paris, 73, 201-217 (1977).

Chay, T.R.: Proton transport across charged membrane and pH oscillations. Biophys. J., 30, 99-118 (1980).

Chay, T.R.: Eyring rate theory in excitable membranes: application to neuronal oscillations. J. Phys. Chem., 87, 2935-2940 (1983).

Chay, T.R.: Abnormal discharges and chaos in a neuronal model system. Biol. Cybern., 50, 301-311 (1984).

Chay, T.R.: Chaos in a three-variable excitable cell model. Physica D (in press).

Chay, T.R., Keizer, J.: Minimal model for membrane oscillations in the pancreatic β-cell. Biophys. J., 42, 181-190 (1983).

Chay, T.R., Keizer, J.: Theory of the effect of extracellular potassium on oscillations in the pancreatic β-cell. Biophys. J. (in press)

Chay, T.R., Rinzel, J.: Bursting, beating, and chaos in an excitable membrane model. Biophys. J., 47, 357-366 (1985).

Dean, P.M., Mathews, E.K.: Glucose-induced electrical activity in pancreatic islet cells. J. Physiol., 210, 255-264 (1970).

Doedel, E.J.: AUTO: A program for the automatic bifurcation analysis of autonomous systems, (Proc. 10th Manitobe Conf. on Num. Math. Comput. Winnipeg, Canada), Cong. Num. 30, 265-284 (1981).

Doedel, E.J.: This volume (1985).

Ferreira, H.G., Lew, V.L.: Use of ionophore A23187 to measure cytoplasmic Ca buffering and activation of the Ca pump by internal Ca. Nature, 259, 47-49 (1976).

Glasstone, S.K., Laidler, J., Eyring, H.: "The Theory of Rate Processes" McGrow-Hill, New York, Chapter X (1941).

Goldman, D.E.: Potential, impedance, and rectification in membranes. J. Gen. Physiol., 27, 37 (1943).

Henquin, J.C., Meissner, H.P., Schmeer, W.: Cyclic variations of glucose-induced electrical activity in pancreatic β-cells. Pflugers Arch

393, 322-327 (1982).

Hindmarsh, A.C.: Ordinary Differential Equations Systems Solver. Lawrence Livermore Laboratory. Livermore, CA. Report UCID-30001 (1974).

Hodgkin, A., Huxley, A.F.: A quantitative description of membrane current and application to conduction and excitation in nerve. J. Physiol. (London) 117, 500-544 (1952).

Lee, Y.S., Chay, T.R., Ree, T.: On the mechanism of spiking and bursting in excitable cells. Biophys. Chem., 18, 25-34 (1983).

Meissner, H.P.: Electrical characteristics of the beta-cells in pancreatic islets. J. Physiol., Paris, 72, 757-767 (1976).

Meissner, H.P., Preissler, M.: Ionic mechanisms of the glucose-induced membrane potential changes in β-cells. In Biochemistry Biophysics of the Pancreatic-β-cell. Georg Thieme Verlag, New York, 91-99 (1980).

Meissner, H.P., Schmelz, H.: Membrane potential of beta-cells in pancreatic islets. Pflugers Arch., 351, 195-206 (1974).

Petersen, O.H., Maruyama, Y.: Calcium-activated potassium channels and their role in secretion. Nature, 307, 693-696 (1984).

Ribalet, B., Beigelman, P.M.: Cyclic variation of K^+ conductance in pancreatic β-cells: Ca^{2+} and voltage dependence. Am. J. Physiol., 237, C137-C146 (1979).

Scott, A.M., Atwater, I., Rojas, E.: A method for the simultaneous measurement of insulin release and B cell membrane potential in single mouse islets of Langerhans. Diabetologia, 21, 470-475 (1981).

ON DIFFERENT MECHANISMS FOR MEMBRANE POTENTIAL BURSTING

John Rinzel and Young Seek Lee
Mathematical Research Branch, NIADDK
National Institutes of Health
Bethesada, MD 20205

1. Introduction.

A number of mathematical models have been proposed to describe the electrical bursting activity of biological excitable membrane systems. Many of these models have been formulated for specific applications [4,5,14]. One of our goals has been to understand the basic underlying qualitative structure of these models and to distinguish, possibly different, classes of models for bursting. In this paper we contrast two examples which illustrate different mathematical mechanisms.

For these cases, and for many other bursting models, one can identify fast and slow subsystems with the following features. The fast system (time scale, msec) exhibits the fundamental features of excitability with spike-like pulses (triggerable from a stable steady state) or repetitive firing of pulses (corresponding to a limit cycle) depending upon the values of the slow variables. The slow dynamics (time scale, sec), with feedback from the fast subsystem, is then responsible for sweeping the slow variables into and out of, or back and forth through, a range of values so that the fast system is either oscillating or sitting near steady state. Hence one observes alternating active or silent phases of the bursting activity (Fig. 1).

To be more specific, we let X and Y denote the (vectors of) fast and slow variables respectively which solve a system of the general form:

$$\dot{X} = F(X,Y) \tag{1.1}$$

$$\dot{Y} = \varepsilon G(X,Y), \tag{1.2}$$

Here, $0<\varepsilon<<1$ to underscore the slowness of (1.2). The X-variables will be akin to those in a Hodgkin-Huxley-like description [11] for nerve membrane potential and conductances.

In one case which we describe (in Section 2), Y is a scalar variable, intracellular free calcium concentration, and the fast subsystem exhibits bistablity. Over a certain interval J of Y-values, the solution of (1.1) may be either a stable steady state X_{ss}, i.e. $0=F(X_{ss},Y)$, of low membrane potential or a stable oscillation, i.e. $\dot{X}_{osc}=F(X_{osc},Y)$, $X_{osc}(t+T)=X_{osc}(t)$, of high mean membrane potential. The slow dynamics are such that Y slowly decreases through J when $X=X_{ss}$. When X is in the oscillatory mode then Y increases through J by accumulating small increases during the successive cycles of the fast oscillation. This type of bursting depends on the hysteresis behavior of the fast subsystem and in some sense the fast subsystem drives the slow oscillations of Y. Figure 1A shows bursting of this general type; the underlying mean potential of the burst appears of relaxation type.

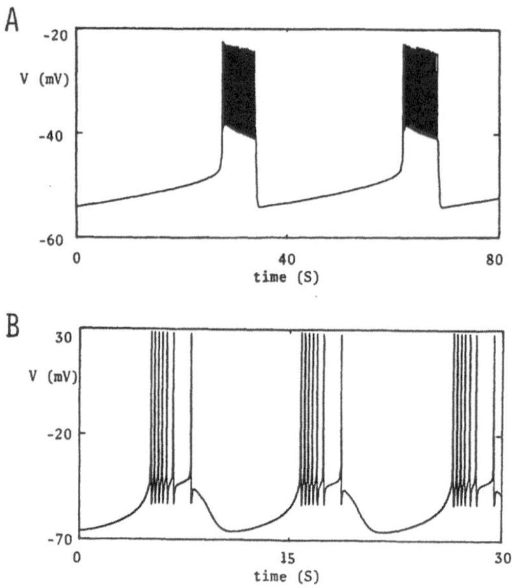

Fig. 1 Periodic bursting oscillation, membrane potential versus time, for two different models (by numerical integration using Gear's method): (A) simplified Chay-Keizer model [5, 6], equations (2.1)-(2.3), for electrical acitivity of pancreatic β-cells. Model has one slow variable Ca and, here, k_{Ca}=.00513; (B) Plant's model [15] for bursting behavior of R-15 neuron of Aplysia, equations (3.1)-(3.5). Two slow variables, Ca and x.

Our dissection (in Section 3) of a different example demonstrates that hysteresis in the fast subsystem is not necessary for burst generation. In this case, there are disjoint (adjoining) regions J_{ss} and J_{osc} in Y-space for which the fast subsystem is either in the steady state/excitable mode or in the oscillation/repetitive firing mode, respectively. Furthermore, the slow system has two or more components so that it is may exhibit an identifiable slow oscillation in spite of feedback from the X system. We show how the parameters of (1.2) may be tuned indepenently of (1.1) so that the slow oscillation is restricted to remain in the region J_{ss} of Y-space, i.e. subthreshold to the region of repetitive firing. Then, as the parameter values of (1.2) are varied, the slow oscillation trajectory spills over the separatrix boundary and for part of the cycle it resides in the region J_{osc}; this represents the active phase of a burst. The burst shown in Fig. 1B is from a model which exhibits this mathematical mechanism. Here, the underlying slow wave looks smooth and more sinusoidal. For this example there are two slow variables: intracellular free calcium concentration and a membrane ionic channel with a slow conductance for calcium.

To understand these two examples, we follow Rinzel [17] and employ numerical bifurcation and branch tracking methods (AUTO[7]). First, we describe the repertoire of behavior of the fast subsystem as a function of the slow variables when the latter are treated as parameters. The stable steady states and oscillatory solutions of (1.1) form the attracting branches of the slow manifold over which the dynamics of the Y system are then considered. Our procedure might be considered the numerical analog to an analytic singular perturbation approach. It is carried out numerically since we wish to consider the models as formulated biophysically and we do not have analytic representations for the fast subsystem behavior.

2. Bursts Driven by Spikes.

Insulin-secreting β-cells of the mammalian pancreas exhibit bursting electrical activity in response to glucose [3]. In an isolated islet (the functional unit of the pancreas), the release of insulin has been shown to be correlated with such electrical activity [18]. Atwater and coworkers [2] formulated a descriptive biophysical model to explain the observed behavior. This led to a five-variable mathematical model, proposed and investigated numerically by Chay and Keizer [5], which accounts for various aspects of the experimental data such as glucose-dependent effects on the burst duration, and appropriate voltage ranges and time scales of the response. Rinzel's qualitative analysis of this model [17] showed that the model, which has one slow variable, has a fast subsystem which exhibits hysteresis behavior. Here we demonstrate that a simplified (three-variable) Chay-Keizer model (see [6]) can be analyzed similarly. In this model, the fast subsystem has two variables so that it may be represented graphically and understood entirely by phase plane methods. The model, for a representative β-cell, takes the form

$$C_m\dot{V} = -I_{Ca}(V) - [\bar{g}_K n^4 + \bar{g}_{K-Ca}\, Ca/(1+Ca)](V-V_K) - \bar{g}_L(V-V_L) \qquad (2.1)$$

$$\dot{n} = \lambda[n_\infty(V)-n]/\tau_n(V) \qquad (2.2)$$

$$\dot{Ca} = f[-\alpha I_{Ca}(V) - k_{Ca}\, Ca] \;, \qquad (2.3)$$

where $I_{Ca}(V) = \bar{g}_{Ca} m_\infty^3(V) h_\infty(V)(V-V_{Ca})$.

The first two equations are a modified and reduced Hodgkin-Huxley [11] description for membrane potential V and ionic conductances of an excitable membrane. This model differs from the HH description for squid axon membrane in essentially the following ways: inward current is due primarily to calcium rather than sodium, the functions m_∞, h_∞, n_∞, and τ_n have been translated along the V-axis, m and h (activation and inactivation for the inward conductance) act instantaneously (rather than satisfy differential equations like (2.2)), the parameters V_{Ca}, V_K, V_L, $\bar{g}_{Ca}$, $\bar{g}_K$, $\bar{g}_L$ have been readjusted. As in [5] we treat V_K as a constant. Also, there is an additional channel type: a non-voltage dependent potassium channel which is activated by free intracellular calcium (concentration, Ca). This channel provides the feedback from the regulation of cytoplasmic calcium to the membrane dynamics. The third equation (2.3) says that calcium enters the cytoplasmic compartment by flux through the

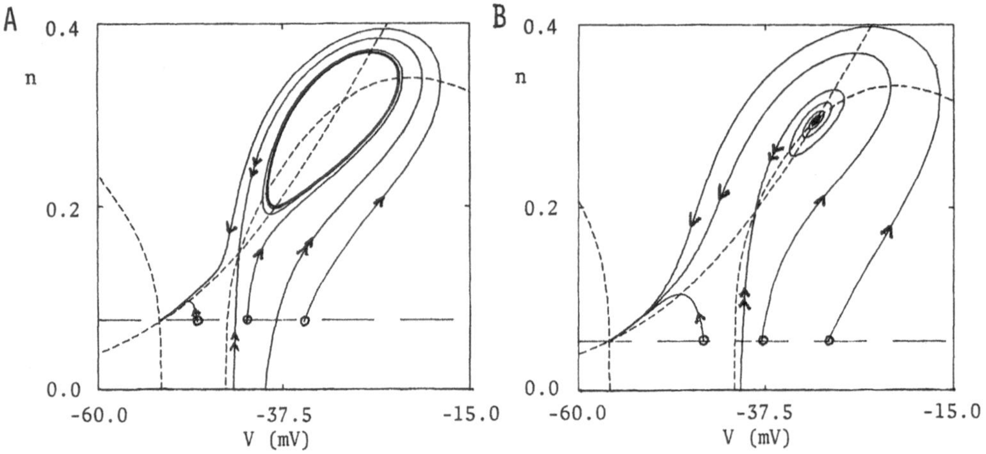

Fig. 2 Phase plane portrait for fast subsystem, (2.1)-(2.2), of Chay-Keizer model for two different fixed values of Ca: 0.7 μM for (A) and 1.1 μM for (B). V and n-nullclines are shown dashed. Three singular points in each case. Double arrowheads denote stable manifolds of saddle point. Responses for three different initial conditions (open circles) show: bistability in (A) with stable lower steady state and stable oscillation surrounding unstable upper steady state; in (B), limit cycle is absent and lower steady state is globally attracting.

membrane and leaves via first-order intracellular uptake (e.g., by mitochrondia or by endoplasmic reticulum) or extrusion through the membrane. The constant α involves Faraday's constant and the cell's surface area to volume ratio. The parameter f represents the ratio of free to total intracellular calcium. It is small since most of the intracellular calcium is involved in reversible, high-affinity binding to various cytoplasmic components [5]. Thus, Ca is the slow variable and equations (2.1)-(2.2) constitute the fast subsystem. Here, we use parameters from [6] with the exceptions: λ=0.3, f=0.0058, $\bar{g}_{K-Ca}$=47.5pS, $\bar{g}_{Ca}$=8.14nS, $\bar{g}_K$=7.68nS, $\bar{g}_L$=31.6pS, C_m=4.52pF.

Consider first the V-n subsystem. For high Ca, the conductance of the Ca-gated potassium channel is fully activated; the membrane has a unique stable steady state potential near V_K. For low Ca, the K-Ca channel is not activated and the system has a unique steady state with V between V_K (=-75 mV) and V_{Ca} (=100 mV). For intermediate Ca, there are three steady states. For some parameter values, both the upper and lower states are stable. In some other cases, the upper state is unstable and is surrounded by a stable oscillation (see Fig. 2A for which Ca=0.7). The responses to three different voltage perturbations from the lower state indicate the different domains of attraction. Both large and small displacements from the rest state can lead to trajectories which return to rest. For this value of Ca, the saddle's unstable manifold (not shown here), which exits upward and to the right from the saddle point, winds onto the limit cycle and remains "inside" the stable manifold which enters the saddle from above. As Ca increases, so does the size of the oscillation. At a critical value of Ca, Ca_{HC}, the oscillation makes contact with the saddle; the period is infinite and we have a homoclinic orbit. In this case, the stable and unstable manifolds coincide. As Ca increases beyond Ca_{HC} the oscillation is lost; the unstable manifold has "passed through" the stable manifold and is now "outside" of the latter (see Fig. 2B for which Ca=1.1). Now, the rest state is globally attracting.

The dependence of these phase plane dynamics upon Ca is summarized in the bifurcation diagram (Fig. 3A) of the fast subsystem. The Z-shaped curve represents the steady state branches (dashed indicates unstable). The stable oscillation around the upper steady state is represented in two ways: as the maximum and minimum values of V over one period (shown solid) and as the average of V over a period (dashed). This branch of periodics emerges via Hopf bifurcation at $Ca=Ca_{HB}$ and terminates with finite amplitude and infinite period at $Ca=Ca_{HC}$.

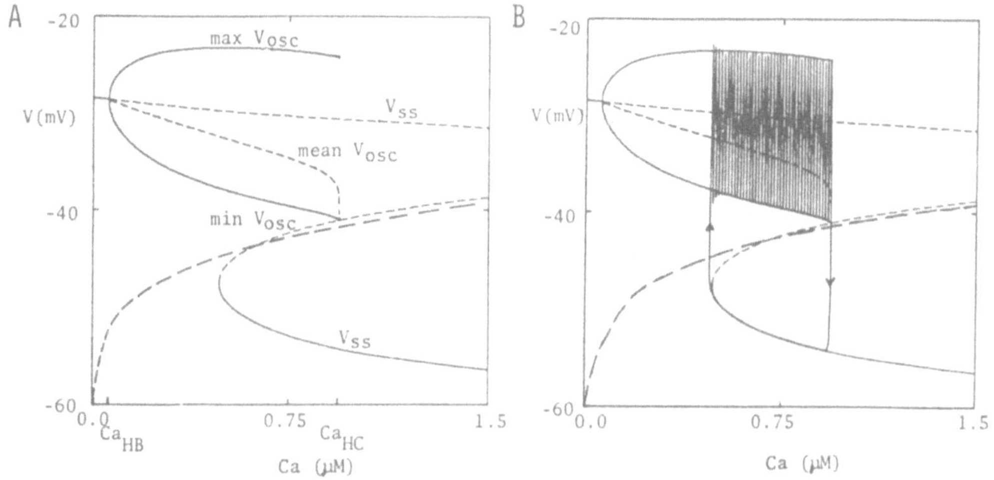

Fig. 3 Bifurcation diagram (computed with AUTO [7]) for fast subsystem, (2.1)-(2.2), of Chay-Keizer model. Dependence upon Ca of steady state ("ss") and periodic ("osc") solutions. Short dashes on triple-branched steady state curve denote instability. Branch of periodic solutions, for which max V_{osc}, min V_{osc}, and time-average of V_{osc} over a period are shown, begins with Hopf bifurcation at Ca_{HB} and terminates with homoclinic connection to saddle at Ca_{HC}. Long dashed curve is Ca-nullcline for slow equation, (2.3). The Ca-V projection of bursting pattern from Fig. 1A is shown in (B). The stable solutions of fast subsystem form attracting slow manifold for full system.

Next we consider the slow dynamics of Ca and continue with our graphical representation. The nullcline for Eqn (2.3) is shown (long dashes) in Fig. 3A. The intersection of the Ca-nullcline with the middle branch of the Z-curve corresponds to a steady state of the full three-variable system which, for this value of k_{Ca}, is unstable. There is an attracting periodic orbit which "surrounds" this steady state (Fig. 3B). It corresponds to the burst pattern shown in Fig. 1A. During the silent phase, the trajectory tracks the lower (stable) steady state of the fast subsystem. Then the trajectory is below the Ca-nullcline so Ca slowly decreases (by the uptake-mechanism). As the K-Ca conductance diminishes, V rises. Eventually, this low-V state coalesces with the falling saddle point and then the trajectory tends upward toward the only remaining attractor of the V-n subsystem - the upper stable oscillation. With this upward movement the trajectory crosses above the Ca-nullcline. Then, during the active phase of spiking, dCa/dt is positive and the trajectory moves rightward as it sweeps along the oscillation branch until it passes the homoclinic connection and then descends to reenter the silent phase.

The burst trajectory in the Ca-V plane is somewhat remniscent of a relaxation oscillation except that here the upper branch of the slow manifold does not represent a steady state solution of the fast subsystem but rather a spike-like oscillatory solution. Each spike during the active phase produces a net increment in Ca. Because f is small, the increment is small and the time course of Ca is smooth. This observation suggests that we might account for the dynamics of these small increments by the difference in the mean rates of Ca-influx and Ca-efflux over the duration of each spike. We seek to replace (2.3) by an "averaged" equation. First, we integrate (2.3) from t_n to t_{n+1}, the times of successive spike upstrokes, and then divide by T_n $(=t_{n+1}-t_n)$, the interspike interval, to obtain:

$$\frac{Ca(t_{n+1})-Ca(t_n)}{t_{n+1}-t_n} = \frac{f}{T_n}\left[-\alpha\int_{t_n}^{t_{n+1}} I_{Ca}dt - k_{Ca}\int_{t_n}^{t_{n+1}} Ca\ dt\right] \qquad (2.4)$$

Now we identify c(t) as the short-term time-average of Ca where "short-term" means on the time scale of a single spike. Thus we approximate (2.4) by the averaged equation:

$$dc/dt = f(-\alpha\bar{I}_{Ca} - k_{Ca}\,c) \qquad (2.5)$$

where $\bar{I}_{Ca}$ is the average Ca-influx over a spike trajectory at a specific mean Ca level, i.e. $\bar{I}_{Ca}=\bar{I}_{Ca}(c)$. To implement this averaging procedure we must determine $\bar{I}_{Ca}$. We have done this as a subsidiary step in our numerical calculation (using AUTO [7]) of the bifurcation structure; that is,

$$\bar{I}_{Ca}(c) = \frac{1}{T(c)}\int_0^{T(c)} \bar{g}_{Ca}m_\infty^3(V(t))h_\infty(V(t))(V(t)-V_{Ca})\ dt \qquad (2.6)$$

where V(t) and T(c) are the time course and period of the fast oscillation at the Ca-level c. The description (2.5) also applies trivially to steady state solutions of the fast subsystem.

The dependence of $\bar{I}_{Ca}$ upon c for the steady state and oscillatory states of the fast subsystem is shown in Fig. 4A. Notice that the periodic branch terminates at each end by joining a steady state branch. This is expected since at c_{HB} $(=Ca_{HB}$ in Fig. 3A) the oscillation emerges with small amplitude and with mean values of V and n

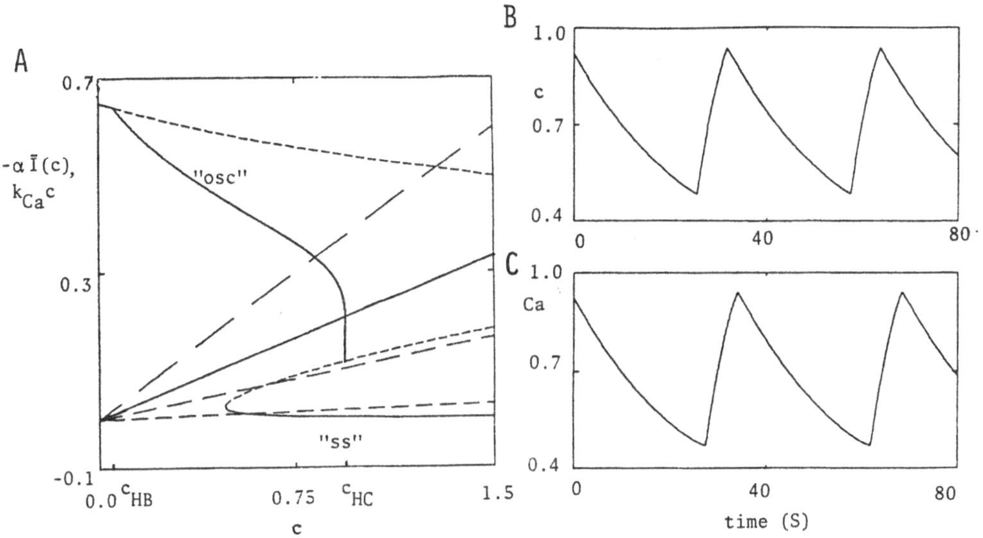

Fig. 4 Representation of "averaged" description, (2.5)-(2.6), for Chay-Keizer model. (A) Dependence upon c of two terms in (2.5). Time averaged calcium current evaluated (with AUTO [7]) for steady state and periodic solution branches of Fig. 3. Straight lines through origin, k_{Ca} c versus c, for k_{Ca}(x100)=0.233, 0.513, 1.03, 1.4 (lower to upper). Intersections of two sets of curves represent steady state solutions of (2.5); these correspond to a steady state or an approximation of beating (continuous spiking) response of full model (2.1)-(2.3). For k_{Ca}=0.00513, the steady state of (2.5) is unstable. Averaged equation, in this case, has a periodic solution which alternates between $\bar{I}(c)$ from "osc" and "ss" branches; time course, c(t), is shown in (B). The Ca(t) time course for corresponding burst solution (Fig. 1A) of full model is shown in (C).

corresponding to the upper steady state. Then, as c tends to c_{HC}, the oscillation trajectory spends essentially all of its time near the saddle point so that the mean influx is that associated with the middle steady state.

To understand the dynamics of (2.5) we represent the uptake term graphically by the straight line k_{Ca} c in Fig. 4A for several different values of k_{Ca}. An intersection of this line with a branch of the $\bar{I}_{Ca}$ curve corresponds to a steady state of (2.5). For low k_{Ca} this steady state (on the lower branch) is stable and corresponds to a stable steady state of (2.1)-(2.3); this represents the resting state of the β-cell in low glucose. For high k_{Ca}, (2.5) may have two steady states. The intersection on the upper dashed branch corresponds to an unstable state. The intersection with "osc", although a stable steady state of (2.5), corresponds to a periodic solution (with period=T(c)) of (2.1)-(2.3); it represents the continuous beating mode of the β-cell in the presence of high glucose. For intermediate k_{Ca}, there is an unstable steady state of (2.5) on the middle dashed branch. In this case the description yields a periodic response which corresponds to bursting.

To compute the time course of increasing c (active phase), the periodic branch of $\bar{I}_{Ca}$ is utilized in (2.5) and for c decreasing (silent phase) we use, in (2.5), $\bar{I}_{Ca}$ from the lower steady state branch. Figure 4B compares the calcium time course predicted by averaging with that calculated from (2.1)-(2.3) for the periodic burst shown in Fig. 1A.

We have found that straightforward application of "averaging" provides an incomplete description in some cases. It predicts a stable beating response for k_{Ca} greater than about 0.13 when (2.5) has a steady state on the "osc" branch in Fig. 4A. This prediction is accurate for $k_{Ca} > 0.22$ (i.e. when the slope of $k_{Ca}c$ vs c exceeds that of the solid line). But as k_{Ca} decreases below 0.22, the beating solution of (2.1)-(2.3) loses stability through a period-doubling bifurcation (independent AUTO calculation). For k_{Ca} just below 0.22 there are further period-doublings and some chaotic behavior (not accounted for by (2.5)). For $0.21 > k_{Ca} > 0.13$, Gear numerical integration of (2.1)-(2.3) predicts bursting rather than beating. Various factors may contribute to such discrepancies. For orbits which linger near the homoclinic, the fast oscillator's time scale may compare to the slow time scale and then averaging may break down. Also, the step-like (but small) increments in Ca may cause the trajectory to leave the periodic branch by overshooting Ca_{HC} (note steepness of branch in Fig. 4A).

3. Bursting Driven by a Slow Wave.

The R-15 neuron of the Aplysia abdominal ganglion has been widely studied as an experimental model system for neuronal bursting activity [1]. In this system, application of certain drugs can eliminate the spikes during the active phase and thereby expose what seems to be an underlying "slow wave" of membrane potential. Thus it appears as though the slow wave drives the potential above threshold for the spike generating mechanism. Plant [15] formulated a model to account for bursting and for the the slow wave; his model takes the following form:

$$C_m\dot{V} = -\bar{g}_{Na}m_\infty^3(V)h(V-V_{Na})-\bar{g}_{Ca}x(V-V_{Ca}) \qquad (3.1)$$
$$-[\bar{g}_K n^4 + \bar{g}_{K-Ca}Ca/(0.5+Ca)](V-V_K)-\bar{g}_L(V-V_L)$$

$$\dot{h} = \lambda[h_\infty(V)-h]/\tau_h(V) \qquad (3.2)$$

$$\dot{n} = \lambda[n_\infty(V)-n]/\tau_n(V) \qquad (3.3)$$

$$\dot{x} = [x_\infty(V) - x]/\tau_x \qquad (3.4)$$

$$\dot{Ca} = \rho[K_c\ x(V_{Ca}-V) - Ca] \qquad (3.5)$$

This HH-like model incorporates V-dependent Na-channels (with activation and inactivation) and K-channels (non-inactivating). It has a slowly activating conductance (x-variable) for inward current, identified as mixed sodium and calcium and it has a

Ca-activated K-channel as in the Chay-Keizer model. This model thus has two variables, x and Ca, which are much slower than the others: τ_x = 235 msec, and $1/\rho = O(10^3)$ msec. The time scales of V, h, n are on the order of 1 to 10 msec.

Note, the equations are displayed here with minor differences from the form given in [15]. We use the classical HH notation and also use x instead of x_T. Plant's factor of 12.5 in the time constant functions is identified as $1/\lambda$ in (3.2)-(3.3). For simplicity, we attribute the slow inward current entirely to calcium; consequently the reversal potential V_{Ca} appears in both (3.1) and (3.5). The conductance $\bar{g}_{Ca}$ has been reduced slightly to compensate for the increased driving potential in (3.1). With values for the other parameters as in [15], the burst solution of Fig. 1B is nearly indistinguishable from that of Plant [15, Fig. 1].

By numerical computation Plant demonstrated that the model exhibits bursting behavior. He also considered a reduced two-variable version of it to study slow wave activity (with $\bar{g}_{Na}$=0 to model the effect of TTX). He showed that a V-Ca subsystem, with the other variables at "rapid equilibrium", can support an oscillation. We have considered alternatively the more obvious slow subsystem formed by x amd Ca (with the "fast" variables V, n, and h at rapid equilibrium). By extending the conceptual approach outlined in Section 2 we provide a systematic dissection of the model which shows how the slow and fast dynamics interact to produce the slow wave and bursting. Here, we out-

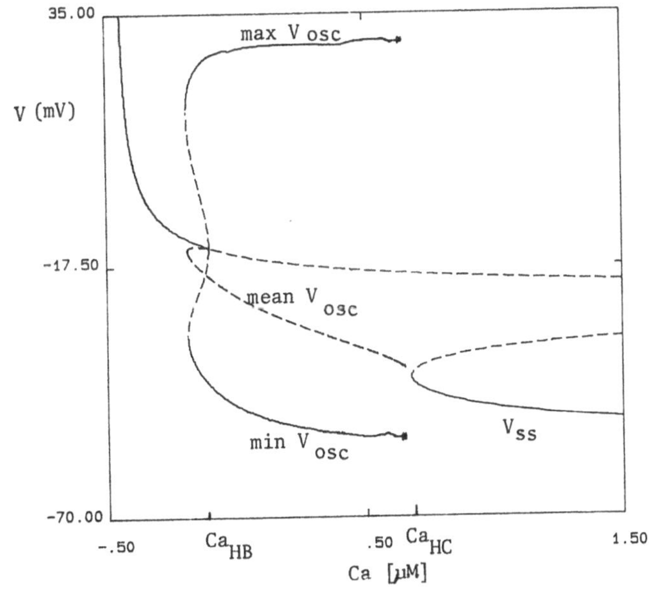

Fig. 5 Bifurcation diagram for fast subsystem, (3.1)-(3.3), of Plant's model with x fixed, x = 0.7. Steady state and periodic solution amplitudes plotted as function of Ca. Note, Hopf bifurcation at Ca_{HB} is subcritical and homoclinic connection at Ca_{HC} is degenerate (i.e., occurs at a saddle-node coalescence of two steady states).

line some of our results; a more complete treatment will be published elsewhere.

We proceed as in Section 2. First, if x is held fixed and Ca is considered as a parameter then the fast subsystem has solutions with a bifurcation structure (Fig. 5) which appears not unlike that shown in Fig. 2. A noticeable quantitative difference is that here the homoclinic orbit occurs (to the accuracy of our numerical calculations) at the coalescence of two steady states. (Such a homoclinic is sometimes called degenerate since it has one zero eigenvalue.) This means that if Ca were swept very slowly back and forth across Ca_{HC} the fast system would switch between the repetitive firng mode and the steady state mode but without exhibiting any hystersis behavior.

Given this slow manifold structure, it is not obvious how one could formulate, with Ca as the only slow variable, an autonomous mechanism to generate bursts with many spikes during the active phase. However, Plant's model has two slow variables and we show next how their dynamics can generate a slow wave. For this, we consider simultaneously x and Ca as parameters. Then the curves in Fig. 5 represent a section of a two-parameter bifurcation surface of the fast subsystem. We have computed this surface and its dependence upon certain physiological parameters; we summarize critical features (points of bifurcation) of the surface in Fig. 6A. The curve labeled

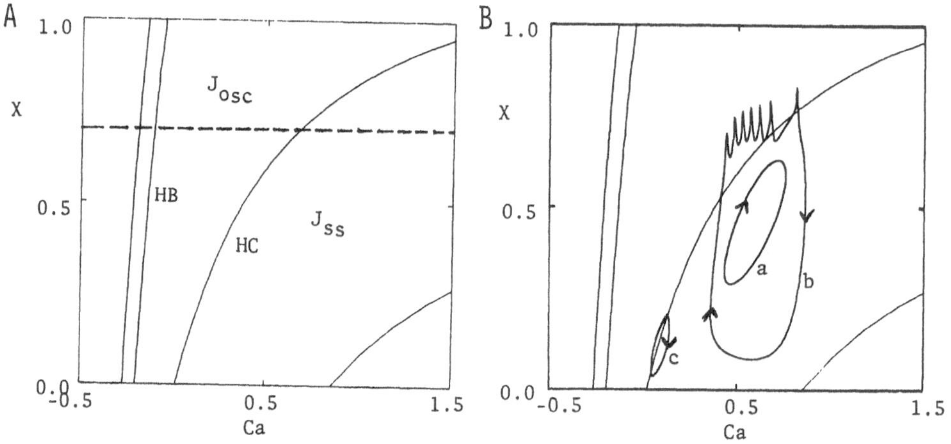

Fig. 6 Projection of bifurcation surface, V versus Ca and x, of fast subsystem, (3.1)-(3.3), for Plant's model. In A, lower right curve (unlabeled) corresponds to fold (i.e., limit points) where upper and middle branches of steady state surface coalesce. Lower fold is labeled HC because it coincides with (degenerate) homoclinic connection where branch of periodic solutions terminates. Leftmost curve (unlabeled) is for limit points of periodic solution branches which arise via subcritical Hopf bifurcation along curve HB. Curves in Fig. 5 represent a section through bifurcation surface for x = 0.7 (dashed line). Regions J_{ss} and J_{osc} of slow variables (described generally in Introduction) are shown explicitly here; J_{ss} is below HC and J_{osc} is between HC and leftmost curve. In B, solution trajectories of full model, (3.1)-(3.5), for three parameter sets, are projected onto Ca-x plane. Case (b) is for burst response of Fig. 1B. For case (a), ρ is replaced by 5ρ and K_c by $0.8\ K_c$ to generate a slow wave pattern without spikes. Case (c) corresponds to the parabolic burst pattern of Fig. 7A.

"HC" indicates the homoclinic connection which is "degenerate" over this entire range of Ca and x. This curve represents the boundary between the disjoint parameter regions (denoted as J_{osc} and J_{ss} in the Introduction) for oscillatory and steady state (low V)/excitable behavior of the fast subsystem. It makes intuitive sense that the boundary curve is monotonic increasing in Ca. As Ca increases so does the hyperpolarizing/inhibiting effect of the K-Ca conductance and therefore a greater activation (x) of calcium current (depolarizing effect) is required to bring the membrane to the threshold for repetitive firing.

If the fast subsystem is in the steady state mode (i.e., right hand sides of (3.1)-(3.3) set to zero) then the slow dynamics live on a triple-branched surface in V-Ca-x space. By adjusting the parameters in (3.4)-(3.5) we can ensure that these dynamics create an oscillation which remains on the lower (stable) branch, i.e. its projection lies inside J_{ss} (Fig. 6B, case a). Now by readjusting the parameters back to their values of Fig. 1B, we see that the slow wave intrudes into the region J_{osc} (Fig. 6B, case b) and thus produces the burst of spikes which rides the depolarizing phase of the slow wave.

By tuning the slow system parameters to yet other values we can control the period of the burst pattern and the number of spikes in a burst. (In some cases the projected response may remain entirely in J_{osc} so that no quiet phase is identifiable.) We point out that, during a burst pattern, the interspike interval (ISI) varies

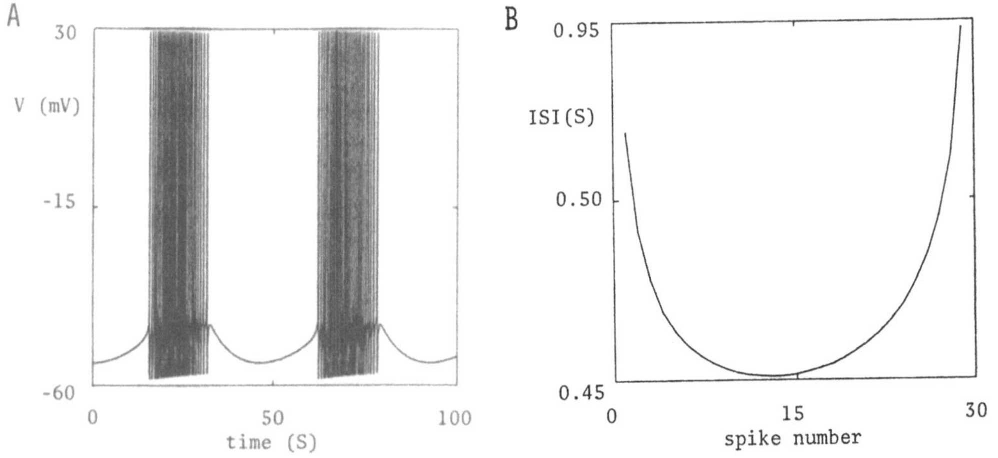

Fig. 7 (A) Parabolic burst pattern for Plant model, (3.1)-(3.5), obtained with parameter adjustments: ρ replaced by $\rho/4$, τ_x replaced by 40 τ_x, and, in $x_\infty(V)$ function, A = 0.30, B = -40mV. (B) Interspike interval (time between successive spike upstrokes) versus spike number for burst response in (A).

from one spike to the next. These ISI's should be predictable from data of our bifurcation calculation (AUTO) and from the approximate location of the projected trajectory on the Ca-x plane. Consistent with this observation, we see (Fig. 6B, case b) that as the trajectory crosses the HC curve at the end of the burst the ISI indeed lengthens dramatically (see also Fig. 1B). Similarly, one might expect long ISI's also at the beginning of the burst. It appears however, for these parameters, that the onset of the burst is rapid; x increases so abruptly that the long ISI's just near the HC curve are not really represented in the time course. By slowing the time scale of x, and by moving the slow wave to a region where x_∞ is less steeply V-dependent (case c, Fig. 6B), we can generate a "parabolic" burst pattern (Fig. 7), i.e., one for which the ISI's first decrease and then increase during the burst. Such patterns have been seen in many recordings of Aplysia bursting activity (e.g., see Fig. 1 of [1]) but not previously in computations with Plant's model.

4. Discussion.

We have exploited the fast and slow time scales to identify and to analyze qualitatively the underlying mathematical structure in two different models for bursting activity of excitable membrane systems. By making this identification and by using numerical bifurcation/branch-tracking methods we can illuminate various qualitative features of the burst pattern. For example, because of the homoclinic orbit in the fast subsystem, the spike frequency drops dramatically as the active phase ends. In Plant's model, the spike frequency may be low also at the beginning of a burst in which case the burst pattern appears parabolic (Fig. 7). Such qualitative features are consistent with experimental observations (see Fig. 1 of [1,3]). By our approach we also gain insight into the effects of changing separately parameters in the fast and slow subsystems. For example, if only slow-subsystem parameters are varied then the response may be predicted without recomputing the fast-subsystem bifurcation structure. As an application of this observation suppose we seek the theoretical dependence of the burst pattern upon glucose in the Chay-Keizer model, say, by treating (as in [5]) k_{Ca} as the glucose sensitive parameter. (For an alternative interpretation and treatment of glucose sensitivity see Chay [this volume].) From Fig. 3A we deduce the following qualitative scenario (also see [17]). For low glucose (low k_{Ca}) there is a stable steady state of low V, the β-cell's rest state. At high glucose (high k_{Ca}), the Ca-nullcline intersects the oscillation branch of the fast subsystem and the response is continuous spiking or beating. Intermediate glucose leads to bursting. Moreover, we conclude that the voltage range of activity during bursting does not depend significantly upon glucose; this agrees with experimental results [3]. This analysis further shows that the range of values traversed by Ca is also independent of glucose; this prediction is unconfirmed experimentally because it is not yet possible to obtain measurements of the Ca time course.

Our approach can also simplify the determination of quantitative information. One measure of burst activity, obtained from experiments [3], is the ratio of active phase duration to total burst period (active plus silent phase durations). By using the averaged description (2.5) (where applicable), one can approximate easily the dependence of the percent active phase upon k_{Ca}. This method avoids computing, for each k_{Ca}, the full burst pattern (and therefore, the time consuming numerical integration for the many many spikes in a burst).

We have contrasted in mathematical terms some of the differences between the two models. The fast subsystem of the Chay-Keizer model exhibits hysteresis (coexistence of a stable steady state and a stable oxcillation) while the Plant model does not. Without this hysteresis the Chay-Keizer model would not exhibit robust bursting for any parameter values of the slow subsystem; there would be no mechanism to sweep the single slow variable Ca into and out of the spike-oscillation mode. Therefore, by adjusting slow-system parameters alone it is not possible to obtain a non-spiking slow wave with this model. Because the Plant model has two slow variables, a slow oscillation may exist and may be manipulated, by adjusting only slow-system parameters, so that the wave (although affected by the fast subsystem) remains below the threshold for spiking or so that it drives the fast subsystem in and out of its spiking mode.

Another difference between the two models has to do with resetting behavior. Consider the response to a very brief pulse of current (δ-function) which instantaneously displaces the membrane potential by an amount dV. It is easy to see, for the Chay-Keizer model, that switching between the active and silent phases can be induced. However for Plant's model, because there is no hysteresis, such switching and resetting will not occur. For longer duration stimuli, say on the time scale of one or a few spikes, then both models would show nontrivial resetting behavior.

While we have distinguished the different structures in these two models one should not view this as a rigorous classification; such a categorization would require more complete constraints and assumptions on the parameters. We have focused on certain parameter ranges but since these models have a number of adjustable parameters and since they are synthesized from some basically similar components (fast V-dependent channel kinetics, slow calcium regulation, a K-Ca channel), they might behave similarly in other parameter regimes. For example, we expect that Plant's model may exhibit hystersis. In this case, the homoclinic connection would be made away from the knee and on the middle branch for appropriate parameter values in the fast system. Then for different slow-system parameters, this subsystem may or may not exhibit an oscillation. In the latter case, it might then operate more like the Chay-Keizer model.

There have been a few other qualitative treatments attempting to expose the mathematical structure of bursting models. Three variable models have been shown to exhibit two different types of hysteresis in their fast subsystems. In addition to the mechanism described here for the Chay-Keizer model (and for the model in [10], which we have also analyzed), hysteresis based upon a subcritical Hopf bifurcation in the fast subsystem is sufficient to yield bursting. This mechanism occurs in a model of the Belousov-Zhabotinskii chemical system [16]. We had also exploited it much earlier in developing a FitzHugh-Nagumo qualitative model for bursting (FitzHugh and Rinzel, unpublished results). In that case, the usual two-variable FitzHugh-Nagumo [9,13] model acts as the fast subsystem and the bifurcation parameter, applied current, is considered as a very slow (autonomous) dynamic variable. Bursting may occur when the fast subsystem has a single steady state (not three as in the models studied here) and no homoclinic orbit. Thus, the spike frequency was seen to be fairly constant during a burst. Recently, Honerkamp, Mutschler and Seitz [12] compared such a model to Plant's model; they saw qualitative similarities between the models rather than the differences which we have described. Ermentrout and Kopell [8] have treated rigorously a class of models for which the fast subsystem resembles that of Plant's model, in that there is a degenerate homoclinic. The slow subsystem (with at least two variables) is hypothesized to have an oscillation which exists when the fast variables are fixed at their values for the saddle-node associated with the homoclinic. This class of models can generate parabolic bursting. Although Plant's model can exhibit parabolic bursting it does not satisfy the above hypothesis; if V is fixed in (3.4)-(3.5) then this slow subsystem has a stable steady state but no oscillation. We suspect however that the treatment in [8] could be extended to include a case like Plant's, for example, in parameter regimes where the slow wave is of small amplitude (i.e., suppose the slow system exhibits a Hopf bifurcation from a steady state in J_{ss} but near the homoclinic boundary with J_{osc}).

Acknowledgement
We thank Steven Baer for a careful reading of the manuscript.

1. Adams, W. B., and J. A. Benson. 1985. The generation and modulation of endogenous rhythmicity in the Aplysia bursting pacemaker neurone R15. Prog. Biophys. Molec. Biol. 46:1-49.
2. Atwater, I., C. M. Dawson, A. Scott, G. Eddlestone, E. Rojas. 1980. The nature of the oscillatory behavior in electrical activity for pancreatic β-cell. J. of Hormone and Metabolic Res., Suppl. 10:100-107.
3. Beigelman, P. M., B. Ribalet, and I. Atwater. 1977. Electrical activity of mouse pancreatic beta-cells II. Effects of glucose and arginine. J. Physiol., Paris 73:201-217.
4. Both, R., W. Finger, and R. A. Chaplain. 1976. Model predictions of the ionic mechanisms underlying the beating and bursting pacemaker characteristics of mulloscan neurons. Biiol. Cybernetics 23:1-11.
5. Chay, T. R., and J. Keizer. 1983. Minimal model for membrane oscillations in the pancreatic β-cell. Biophys. J. 42:181-190.
6. Chay, T. R., and J. Keizer. Theory of the effect of extracellular potassium on oscillations in the pancreatic β-cell. Biophys. J. (in press).
7. Doedel, E. J. 1981. AUTO: A program for the automatic bifurcation and analysis of autonomous systems, (Proc. 10th Manitoba Conf. on Num. Math. and Comput., Winnipeg, Canada), Cong. Num. 30:265-284.
8. Ermentrout, G. B., and N. Kopell. Parabolic bursting in an excitable system coupled with a slow oscillation. SIAM J. Applied Math. (in press).

9. FitzHugh, R. 1961. Impulses and physiological states in models of nerve membrane. Biophys. J. 1:445-466.
10. Hindmarsh, J. L., and R. M. Rose. 1984. A model of neuronal bursting using three coupled first order differential equations. Proc. R. Soc. Lond. B 221:87-102.
11. Hodgkin, A. L., and A. F. Huxley. 1952. A quantitative description of membrane current and its application to conduction and excitation in nerve. J. Physiol. (Lond) 117:500-544.
12. Honerkamp, J., G. Mutschler, and R. Seitz. 1985. Coupling of a slow and a fast oscillator can generate bursting. Bull. Math. Biol. 47:1-21.
13. Nagumo, J. S., S. Arimoto, and S. Yoshizawa. 1962. An active pulse transmission line simulating nerve axon. Proc. IRE. 50:2061-2070.
14. Plant, R. E., and M. Kim. 1976. Mathematical description of a bursting pacemaker neuron by a modification of the Hodgkin-Huxley equations. Biophys. J. 16:227-244.
15. Plant, R. E. 1981. Bifurcation and resonance in a model for bursting nerve cells. J. Math. Biology 11:15-32.
16. Rinzel, J., and W. C. Troy. 1982. Bursting phenomena in a simplified Oregonator flow system model. J. Chem. Phys. 76:1775-1789.
17. Rinzel, J. Bursting oscillations in an excitable membrane model. In Proc. 8th Dundee Conf. on the Theory of Ordinary and Partial Differential Equations (eds., B. D. Sleeman, R. J. Jarvis, and D. S. Jones). Springer-Verlag (in press).
18. Scott, A. M., I. Atwater, and E. Rojas. 1981. A method for the simultaneous measurement of insulin release and β-cell membrane potential in single mouse islet of Langerhans. Diabetologia 21:470-475.

A MITOTIC OSCILLATOR WITH A STRANGE ATTRACTOR AND DISTRIBUTIONS OF CELL CYCLE TIMES

Michael C. Mackey, Martin Santavy, and Pavla Selepova
Department of Physiology
McGill University
3655 Drummond
Montreal, Quebec
Canada H3G 1Y6

I. Introduction.

The concept of the existence of the cell cycle appeared shortly after the advent of light microscopy in the last century when natural scientists first described the intranuclear events involved in mitosis and cytokinesis. It was not long before significant intercellular variation in the cell generation time (the elapsed time between cell birth and the production of two daughter cells, also called the intermitotic time) was described in a variety of cell types.

The next major advance in cataloging the events of the cell cycle came when Howard and Pelc [1] utilized radioactively labelled compounds to demonstrate that DNA synthesis occupied a discrete portion of the cell cycle, and used this observation to divide the cell cycle into four discrete phases. The period between cell birth and the initiation of DNA synthesis was denoted G1, the DNA replication and mitotic phases were called S and M respectively, and the period between the completion of S phase and the initiation of mitosis was named G2. Later work [2] indicated that the G1 phase might consist of two functionally different phases in series, with the phase preceeding G1 denoted by G0. Detailed studies of many different cell lines using a variety of techniques have revealed that there is significant variation in the duration of all phases of the cell cycle [3].

Given the nature of the experimental data, it is not unnatural that most interpretations of this variability in cell phase duration were based on probabilistic considerations [4-9]. However an alternate assumption, invoking the existence of an intracellular oscillator timing the cell cycle, has also been proposed [10-14] and criticized [15]. The nature of the oscillators considered ranges from 'limit cycle' to 'relaxation' types, though the distinction is more one of degree than of type. Others, to mimic the variability of cell cycle events, have assumed that this intracellular oscillator has superimposed 'noise' [16-20]. However, this assumption begs the question of the origin of cell cycle variability and is unable to account for the observed differences in the coefficient of variation of the densities of the distributions of the various cell cycle phase durations [3].

In [21], the hypothesis that there exists a mitotic oscillator with a strange attractor timing the cell cycle was shown to lead to results identical with those derived from the purely probabilistic model presented in [9]. Interestingly, the existence of such an oscillator seems not to have been previously considered even though it was all but explicitly postulated in [22], and all of the dynamical ingredients are present in the cell for its occurrence.

Here, a mitotic oscillator model consistent with the general formulation in [21] is used to analyze data for the duration of the cell cycle in a variety of cellular populations. Specific assumptions concerning the nature of the intracellular production of a substance or property called mitogen, and the dependence of this hypothetical oscillator on mitogen levels, allow a unique and simple deterministic model to replace a probabilistic approach.

II. The Model

In this section, the model is specified. The model rests on three hypotheses previously presented in a more general context [21].

Hypothesis 1. There exists some substance(s) (mitogen) necessary, but not sufficient, for mitosis to take place. (There is ample experimental support for this concept, as reviewed in [23]).

Consider a cell in a large population, born at time $t = 0$ with mitogen content r. Assume that mitogen levels are normalized and bounded on the closed interval $[0,2]$, and that the evolution of mitogen following birth is governed by

$$B \frac{dm}{dt} = m(2-m), \qquad m(0) = r. \tag{1}$$

where $B > 0$ is a parameter to be determined from the data. The solution of (1), denoted by $m(r,t)$, is

$$m(r,t) = 1 + \tanh \{[t - \bar{t}(r)]/B\} \tag{2}$$

where $\bar{t}(r) = (B/2) \ln [(2 - r)/r]$ and $m(r,\bar{t}) = 1$.

Hypothesis 2. There exists an oscillating intracellular variable sufficient to trigger mitosis once it exceeds a threshold value.

More precisely, let $x(t)$ denote the value of this variable at time t, and t_n, $n = 0,1,\ldots$, denote the times at which $x(t)$ attains a relative maximum. If $x_n = x(t_n)$ are the values of these relative maxima, then it is assumed that

$$x_{n+1} = m(t_n)S(x_n) \tag{3}$$

where $S:[0,1] \to [0,1]$ is a Rényi transformation, and that mitosis takes place whenever $x_n > 1$. (A mapping $S:[0,1] \to [0,1]$ which satisfies:

i) There exists a partition $0 = a_0 < a_1 < \ldots < a_p = 1$ of $[0,1]$ such that for each integer i, $i = 1,\ldots,p$, the restriction S_i of S to the open interval (a_{i-1}, a_i) can be extended as a C^2 function to $[a_{i-1}, a_i]$;

ii) $S_i([a_{i-1}, a_i]) = [0,1]$, $i = 1,\ldots, p$; and

iii) $\inf_{(a_{i-1},a_i)} |S'(x)| > 1 \qquad i = 1,\ldots,p$

is called a Rényi transformation, see [24]).

As shown in [21], the consequence of this hypothesis in conjunction with a theorem of Lasota and Yorke [25] is that once the mitogen level exceeds the threshold value of 1, the mitotic rate is S(m-1), where $S > 0$ is constant. Therefore, if $\alpha(r,t)$ denotes the fraction of cells born with mitogen level r that have not divided by time t, then

$$\alpha(r,t) = \begin{cases} 1 & 0 < t < \bar{t}(r) \\ \exp\left[-S\int_{\bar{t}}^{t} [m(r,y)-1]\, dy\right] & \bar{t}(r) \leq t. \end{cases} \tag{4}$$

Note that $-\alpha_t(r,t) = -\partial\alpha(r,t)/\partial t = S[m(r,t)-1]\alpha(r,t)$ is the density function for the distribution of intermitotic (generation) times in this population of cells having initial mitogen levels r.

Using the solution for m(r,t) given in equation 2, $\alpha(r,t)$ assumes the explicit form

$$\alpha(r,t) = \begin{cases} 1 & 0 < t < \bar{t}(r) \\ \{\cosh[t-\bar{t}(r)]/B\}^{-SB} & \bar{t}(r) \leq t. \end{cases} \tag{5}$$

Generally, the initial mitogen level r in a population of cells will be distributed on [0,1) with a density f(r), so $\alpha(t)$ for the entire population is given by

$$\alpha(t) = \int_0^1 \alpha(r,t)f(r)dr. \tag{6}$$

Finally, the density function for the distribution of generation times for the entire population is given by

$$\psi(t) = -\int_0^1 \alpha_t(r,t)f(r)dr. \tag{7}$$

To complete the specification of this model, the distribution f(r) of initial mitogen levels in the population of cells is required. This results in the third and final hypothesis of the model:

Hypothesis 3. Each sister cell receives exactly one-half of the mitogen present in the mother cell at mitosis.

As shown in [9], from this and the previous hypotheses there is a globally asymptotically stable distribution of mitogen f in the population of cells given by

$$f(r) = \begin{cases} 0 & 0 < r < 1/2 \\ 2q(2r)\exp\left[\int_1^{2r} q(z)dz\right] & 1/2 \leq r < 1. \end{cases} \tag{8}$$

where q(y) = SB(y-1)/y(y-2). Carrying out the integration in equation 8 gives

$$f(r) = \begin{cases} 0 & 0 < r < 1/2 \\ 2SB\,\frac{2r-1}{4r(1-r)}\,[4r(1-r)]^{\frac{1}{2}SB} & 1/2 \leq r < 1. \end{cases} \tag{9}$$

Another statistic, widely used by cell kineticists in characterizing populations of renewing cells, is the fraction of sibling cell pairs whose intermitotic times differ by at least a time t. This fraction is denoted by β(t). The derivation of β(t) given in [9] for this hypothetical cellular population leads directly to

$$\beta(t) = -2\int_0^\infty \int_0^1 \alpha_t(r,x)\alpha(r,x+t)f(r)dxdr,$$

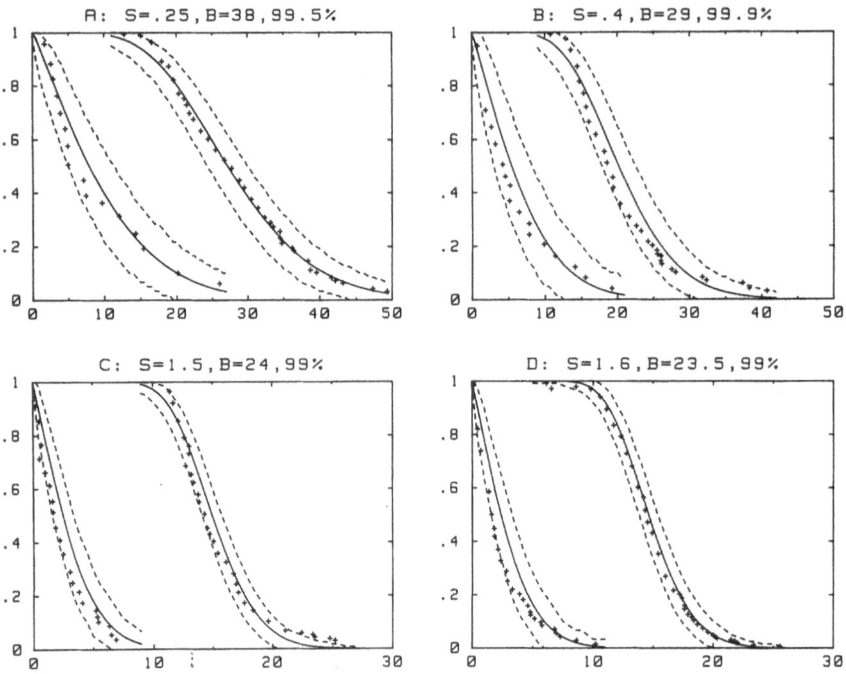

Figure 1. Correspondence between model predictions and data from SV3T3 cells grown at four different concentrations of fetal calf serum. A: Serum = 0.05%; B: 0.25%; C: 1.0%, and D: 10.0%. Data from [27]. In each panel of this and succeeding figures, the vertical axis is α or β and the horizontal axis is cell age in hours, the data points are indicated by (+), and the best fitting α(t) and β(t) curves predicted by the model, based on the indicated parameter values, are represented by solid lines. The dashed (---) lines bounding the predicted α(t) and β(t) curves give the confidence intervals corresponding to the percentages indicated at the top of each panel. [A confidence interval value of 99% means that if the data can be described by α(t), then the probability that, at a given t_j, the observed value $\alpha(t_j)$ will be either above or below the confidence interval is 1% or less.]

which, under the assumptions of this paper, takes the form

$$\beta(t) = 2\int_0^\infty x \tanh(x/B) \{\cosh(x/B) \cosh [(x+t)/B]\}^{-SB} dx. \qquad (10)$$

III. Analysis of the Data.

Equations 5, 6, and 9 and 10, respectively specify the functions α(t) and β(t) and involve only two parameters, S and B. We examined the ability of this model of the cell cycle to account for 18 existing published data sets $\{t_j, \alpha(t_j)\}$,

$\{t_j, \beta(t_j)\}$, though here we present only some of the results of our analysis.

We digitized and stored the data sets $\{t_j,\alpha(t_j)\}$, $\{t_j,\beta(t_j)\}$, and then determined the parameter values that gave the best fit to the data using nonlinear regression techniques [Santavy, in preparation]. We first determined the values of S and B that gave the best fit to the $\{t_j, \alpha(t_j)\}$ data, and then compared the resulting predicted $\beta(t)$ with the actual $\{t_j,\beta(t_j)\}$ data set.

The results of this procedure are illustrated in Figure 1 for SV3T3 cells grown at four different concentrations of fetal calf serum. Note that with increasing serum levels, there is a progressive elevation in the parameter S and a concomitant fall in the parameter B.

The same procedure was followed in an examination of the ability of this oscillator model of the cell cycle to account for data from eight other cell lines as shown in Figure 2.

For the data analyzed in Figures 1 and 2, as well as other cases that we have not presented, we are able to achieve a more than satisfactory correspondence between the model predictions and the $\{t_j,\alpha(t_j)\}$ data. However, the predicted $\beta(t)$ often deviates significantly above the data, i.e. sister cell pairs have differences in intermitotic times significantly less than predicted based on the population response.

Noting this, we followed two different data fitting procedures, illustrated for the BHK 21 cells of Figure 2 (reproduced in Figure 3A). First, the parameters S and B were determined to give the best fit to the $\{t_j,\beta(t_j)\}$, and then the $\alpha(t)$ curve was predicted (Figure 3B). Second, S and B were determined to give the best fit to both the $\{t_j,\alpha(t_j)\}$ and the $\{t_j,\beta(t_j)\}$ data (Figure 3C). It is clear that neither procedure gives a satisfactory fit to the data.

One might question whether the failure of this model to account for the existing sets of data is due to the improper choice of the assumed dynamics of mitogen production (Equation 1). However, it is easy to show that this is not the case. Consider the following. Assume that a population of N cells is composed of m different subtypes, with each different subtype characterized by a different mitotic rate. Let $h_i(t)dt$ be the fraction of the ith cell type with intermitotic time T_i such that $t < T_i < t+dt$ so $h_i(t)$ is the mitotic rate for the ith type of cell. Further take $h_i(t) = 0$, $0 < t \leq t_1$, $h_i(t) \leq H_i$ for $t_1 < t \leq t_2$, and $h_i(t) = H_i$ for $t_2 < t$. Then the mitotic rate for the entire cellular population satisfies $h(t) \leq H = \sum w_i H_i$ for $t_1 < t \leq t_2$ and $h(t) = H$ for $t_2 < t$, where w_i is the fraction of the total population comprised of the ith cell type. Thus, the logarithmic slope of the $\alpha(t)$ curve for the entire population will be constant and equal to -H for sufficiently large t. Now consider sister cell pairs drawn from this same population. Then, the mitotic rate for a cell of type i at time t+X whose sister divided at time X is simply $h_i(t+X)$ which, for $t_2-t_1 < t$, is simply H_i. A straightforward argument then shows that for sufficiently large times t the

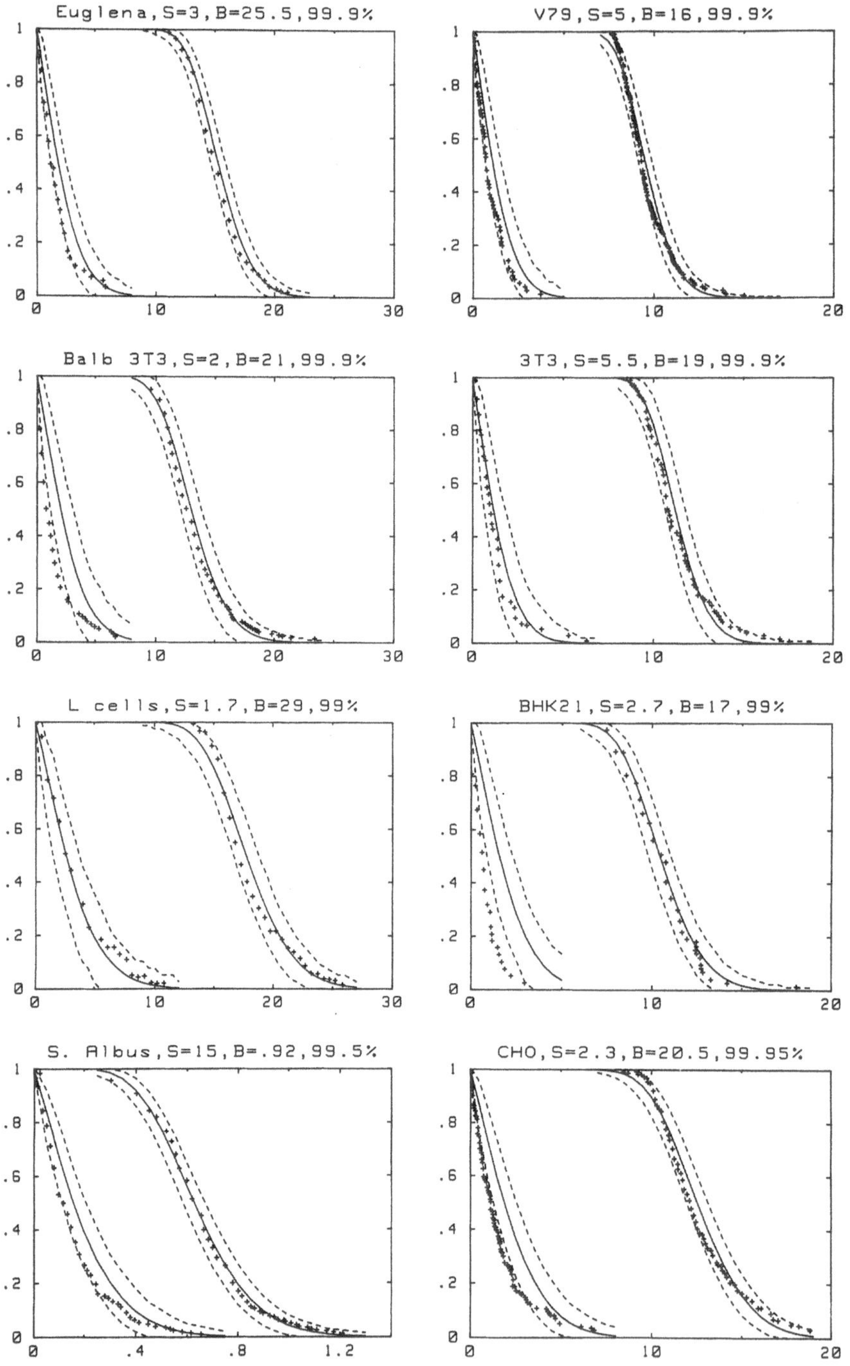

Figure 2. The results of determining the best model fit to eight sets of data. Data sources: Euglena gracilis [26], Balb 3T3 [27], L cells [28], S. Albus [29], V79 [30], 3T3 [31], BHK21 [32], and CHO [31].

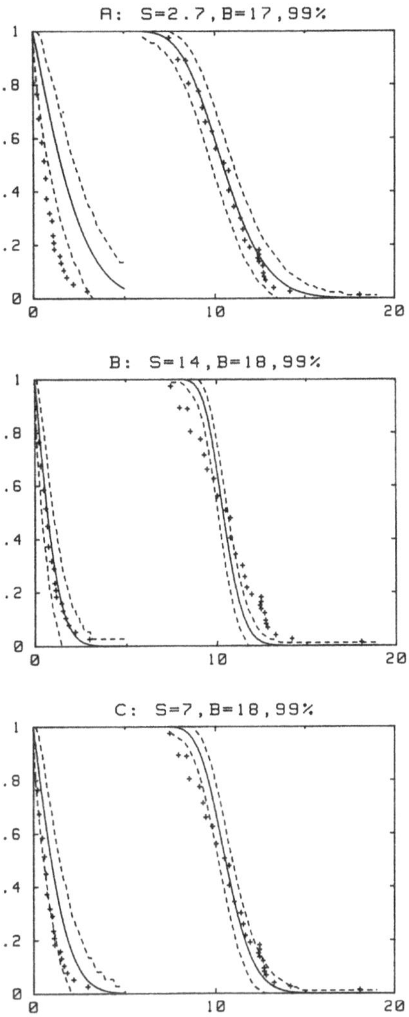

Figure 3. The results of three different procedures to fit data from BHK 21 cells [27]. A: The result of fitting the α data and predicting β, reproduced from Figure 2; B: The fit to β and the predicted α; and C: The consequence of fitting the α and β data together.

logarithmic slope of the $\beta(t)$ function must be identical to -H. Thus, an improper choice of the mitogen dynamics in equation 1 cannot be the source of the discrepancy between the observed and predicted behaviour of sister cell pair intermitotic times.

To have more control over t_1 and t_2 than was available in our formulation of the model, we introduced a third parameter D such that the mitotic rate is given by $S[(m(r,t)-1)/D]$ for $1 \leq m(r,t) < 1+D$ and is just S for $m(r,t) \geq 1+D$. With this modification it was possible to achieve equal predicted logarithmic slopes for both $\alpha(t)$ and $\beta(t)$ at smaller values of t and not just in the limiting ($t\to\infty$) case. However, this modification was insufficient to give a satisfactory agreement between the model predictions and the existing data, for the $\beta(t)$ data once again displayed a significantly greater (in absolute value) slope than predicted from the model.

IV. Discussion.

Of the three hypotheses used in the development of this model, the major mathematical and biological questions revolve around Hypothesis 3.

First, note that probabilistic formulations of cell cycle models are mathematically identical with a formulation based on the existence of a mitotic oscillator with the characteristics detailed in Hypothesis 3 (compare [9] and [21]). However, there are significant and profound differences in the interpretation of the underlying biology within the context of the two formulations.

Second, is there any oscillating dynamical or semi-dynamical system with successive maxima determined by a Rényi transformation? Any continuous time system with this property must, by necessity, be of dimension 3 or greater. Though there is no analytic proof of the existence of such systems at this time, there is good numerical evidence for their existence as shown in [33] and [34] from an analysis of the Lorenz equations.

Third, do intracellular oscillations exist? Again, the answer is yes. For example, Rapp [35] has cataloged literally hundreds of reported oscillations of cellular events occurring on relatively short time scales. More recently, a variety of investigators (see Lloyd and Edwards [36]) have noted oscillations in cellular respiration, a variety of enzymes, the adenine nucleotides, and ATPase activity to name but a few. Finally, Klevecz (see [13,37]) has collected data from a number of cell lines which he interprets as evidence for the existence of a quantal oscillator clocking the cell cycle with a period of the order of 4 hours.

Fourth, do there exist known intracellular biochemical control pathways capable of generating irregular oscillations? Biochemical control loops within the cycling cell are numerous, richly interconnected, and nonlinear. Further, many display mixed positive/negative feedback with or without significant time delays in their feedback pathways [38-40]. As has been repeatedly shown numerically [41,42] these are exactly the conditions under which strange attractors and the attendent properties required by the theory developed in [21] may be encountered. Two examples immediately come to mind.

The general scheme for the production of repressible enzumes put forward in Tyson and Othmer [38] contains multiple nonlinearities as well as time delays required for transcription and translation. A subsequent application of this model to the tryptophan operon in E. coli [40] experimentally demonstrated the instability of this gene control network and the existence of oscillations.

A second example involves the biochemical control networks for purine and pyrimidine synthesis which supply necessary precursors for DNA replication and the initiation and maintenance of transcription. These networks are highly interconnected with multiple positive, negative, and mixed positive-negative feedback loops. Preliminary simulations of the dynamics of this system [Mackey, unpublished] with known or estimated parameter values and feedback functions indicate that there are large regions of parameter space in which steady states are unstable and highly irregular oscillations may occur.

All of the failures, noted in the previous section, to obtain better agreement between the existing data and the one dimensional oscillator model formulated here may be traced to the fact that the oft claimed equality of the limiting logarithmic slopes of the $\alpha(t)$ and $\beta(t)$ data is only approximate. This experimental observation thus disqualifies all one dimensional models (either probabilistic or deterministic) for the statistics of the cell cycle, and suggests that a multi-dimensional model would be more appropriate. However, with the introduction of a multi-dimensional mitotic model we would also be in the position of not being able to check the validity of such a model based only on existing $\alpha(t)$ and $\beta(t)$ data. For example, a two-dimensional model will be able to easily fit both the $\alpha(t)$ and $\beta(t)$ data (even if they are independent as assumed in the Brooks two random transition model [31]), but the validation of such an extended model would require the availability of a third statistic in addition to $\alpha(t)$ and $\beta(t)$.

Of the known events associated with the cell cycle, the two most recognizable are mitosis and DNA synthesis. We feel that the observations cited above concerning the potential behaviour of the biochemical control networks involved in the synthesis of DNA precursors, along with the failure of a single threshold oscillator model as presented here to account for the sister-sister intermitotic time data, offer a strong motivation for expanding the model to include two oscillators -- one for the initiation of DNA synthesis, and the second for the initiation of mitosis.

There are a number of attractive aspects to the hypothesis that variability within the cell cycle may arise from the operation of an oscillator with a strange attractor.

First, if the present formulation were extended to a higher dimensional threshold crossing oscillator model it offers the possibility to understand the deterministic origin of the variability found within specific phases of the cell cycle [3]. Second, if amplitude characteristics of the oscillator and/or threshold levels are responsive to a variety of extracellular factors (eg., the serum effects shown in Figure 1) then the model offers a different interpretation of the current view of the G0 state. Third, if intracellular oscillators are assumed able to cross threshold for DNA synthesis and to occasionally miss a mitotic threshold, then the hypothesis offers a qualitative explanation for the existence of polyploid cells in normal tissue. An alteration of oscillator and/or threshold characteristics in transformed cells could explain the increased frequency of polyploid cells in abnormal tissues. A similar explanation would suffice to explain the existence of reduction divisions. Fourth, irregular oscillations in the pools of DNA precursors in conjunction with a threshold for the initiation of DNA synthesis could serve to explain data that have often been interpreted to indicate that DNA synthesis may be intermittent. Finally, the model would have sufficient flexibility to explain the fact that DNA synthesis and mitosis may not always occur in a sequential fashion as noted, for example, in rapidly growing bacteria.

Acknowledgements. This research was supported by the Natural Sciences and Engineering Research Council of Canada through grant A-0091. We would like to thank Robert F. Brooks for generously supplying original data sets published in [30] and [31] for analysis.

References

1. A. Howard, S.R. Pelc: Heredity (Suppl.) 6, 261 (1953)
2. L.G. Lajtha, R. Oliver, C.W. Gurney: Brit. J. Haemat. 8, 442 (1962)
3. M. Guiguet, J.J. Kupoiec, A.J. Valleron, in Cell Cycle Clocks (ed. L.N. Edmunds Jr.), Marcel Dekker Inc., New York, 1984, p. 97.
4. O. Rahn: J. Gen. Physiol. 15, 257 (1932)
5. D.G. Kendall: Biometrika 35, 316 (1948)
6. A.L. Koch, M. Schaechter: J. Gen. Microbiol. 29, 435 (1962)
7. F.J. Burns, I.F. Tannock: Cell Tissue Kinet. 3, 321 (1970)
8. J.A. Smith, L. Martin: Proc. Natl. Acad. Sci. USA 70, 1263 (1973)
9. A. Lasota, M.C. Mackey: J. Math. Biology 19, 43 (1984)
10. S. Kauffman: Bull. Math. Biol. 36, 171 (1974)
11. S. Kauffman, J.J. Willie: J. theor. Biol. 55, 47, (1976)
12. J. Tyson, S. Kauffman: J. Math. Biology 1, 289 (1975)
13. R.R. Klevecz: Proc. Natl. Acad. Sci. USA 73, 4012 (1976)
14. L.N. Edmunds, K.J. Adams: Science 211, 1002 (1981)
15. J. Tyson, W. Sachsenmaier: J. theor. Biol. 73, 723 (1978)
16. B.C. Goodwin: Temporal organization in Cells, Academic Press, London-New York, 1963
17. B.C. Goodwin: J. theor. Biol. 28, 375 (1970)
18. W.H. Woolley, A.G. De Rocco: J. theor. Biol. 39, 73 (1973)
19. A.G. De Rocco, W.H. Woolley: Math. Biosci. 18, 77 (1973)
20. R.M. Shymko, R.R. Klevecz: in Biomathematics and Cell Kinetics (ed. M. Rotenberg), Elsevier/North Holland, Amsterdam-New York, 1981, p. 329.
21. M.C. Mackey: in Temporal Order (eds. L. Rensing, N.I. Jaeger), Springer-Verlag, Berlin-New York, 1985, p. 315
22. J. Engelberg: J. theor. Biol. 20, 249 (1968)
23. J.M. Mitchinson: The Biology of the Cell Cycle, Cambridge University Press, London-New York, 1973
24. A. Lasota, M.C. Mackey: Probabilistic Behavior in Deterministic Systems, Cambridge University Press, London-New York, in press 1985
25. A. Lasota, J.A. Yorke: Rend. Sem. Mat. Univ. Padova 64, 141 (1981)
26. J.R. Cook, B. Cook: Exp. cell Res. 28, 524 (1962)
27. R. Shields, J.A. Smith: J. cell Physiol. 91, 345 (1977)
28. H. Miyamoto, E. Zeuthen, L. Rasmussen: J. cell Sci. 13, 879 (1973)
29. R. Shields: Nature 273, 755 (1978)
30. R.F. Brooks, P.N. Riddle, F.N. Richmond, J. Marsden: Exp. Cell Res. 148, 127 (1983)

31. R.F. Brooks, D.C. Bennett, J.A. Smith: Cell 19, 493 (1980)
32. P.D. Minor, J.A. Smith: Nature 248, 241 (1974)
33. J.A. Yorke, E.D. Yorke: J. Stat. Phys. 21, 263 (1979)
34. E.N. Lorenz: J. Atmos. Sci. 20, 130 (1963)
35. P.E. Rapp: J. Exp. Biol. 81, 281 (1979)
36. D. Lloyd, S.W. Edwards: in Cell Cycle Clocks (ed. L.N. Edmunds Jr.), Marcel Dekker Inc., New York, 1984, p. 27
37. R.R. Klevecz: in Cell Cycle Clocks (ed. L.N. Edmunds Jr.), Marcel Dekker Inc., New York, 1984, p. 47
38. J.J. Tyson, H.G. Othmer: in Progress in Theoretical Biology, volume 5 (eds. R. Rosen, F.M. Snell), Academic Press, New York-London, 1978, p. 1
39. J.M. Mahaffy: J. Math. Anal. Appl. 74, 72 (1980)
40. R.D. Bliss: Anal. Biochem. 93, 390 (1979)
41. U. an der Heiden: J. Math. Biology 8, 345 (1979)
42. U. an der Heiden, M.C. Mackey: J. Math. Biology 16, 75 (1982)

AN ANALYSIS OF ONE- AND TWO-DIMENSIONAL PATTERNS IN A MECHANICAL MODEL FOR MORPHOGENESIS

by

P. K. Maini and J. D. Murray
Centre for Mathematical Biology
Mathematical Institute
University of Oxford
Oxford OX1 3LB

G. F. Oster
Department of Biophysics
University of California
Berkeley, CA 94720

ABSTRACT

In early embryonic development, fibroblast cells move through an extracellular matrix (ECM) exerting large traction forces which deform the ECM. We model these mechanical interactions mathematically and show that the various effects involved can combine to produce pattern in cell density. A linear analysis exhibits a wide selection of dispersion relations, suggesting a richness in pattern forming capability of the model. A nonlinear bifurcation analysis is presented for a simple version of the governing field equations. The one-dimensional analysis requires a non-standard element. The two-dimensional analysis shows the possibility of roll and hexagon pattern formation. A realistic biological application to the formation of feather germ primordia is briefly discussed.

1. Introduction.

Several models have been proposed to describe the mechanisms involved in morphogenesis - the development of biological pattern and form. The majority of these view morphogenesis as a two stage process. In the first, a pre-pattern is set up in some chemical (morphogen) concentration. In the second, cells respond to the local concentration of morphogen and differentiate according to their 'positional information' (Wolpert (1969), (1981)).

The pre-pattern may be set up by simple diffusion or in a typical Turing reaction-diffusion way (Turing (1952), Murray (1977),(1981 a,b), Meinhardt (1982), Tickle et al (1975), Wolpert et al (1971)).

An alternative approach has been made by Oster et al (1983, 1985) and Murray and Oster (1984 a, b), based on the following experimental observations (Harris et al (1981)): 1) cells spread and migrate within a substratum consisting of a fibrous extracellular matrix and 2) they generate large contractile forces which deform the ECM.

For completeness, section 2 contains a brief derivation of the model equations (see Oster et al (1983) for fuller details). In section 3 the results of a linear analysis are given which show the wide selection of dispersion relations possible. In section 4, we carry out a nonlinear bifurcation analysis on a simple but practical version of the model. In section 5, we investigate two dimensional patterns in a caricature of the model analyzed in section 4, and show the possibility of roll and hexagonal structure in cell density populations. Finally, we briefly discuss the biological application to the problem of feather germ formation.

2. Model Equations.

The model is based on the three field variables:

$n(\underline{x},t)$ = density of mesenchymal cells at position $\underline{x}$ and time t.

$\rho(\underline{x},t)$ = density of ECM at position $\underline{x}$ and time t.

$\underline{u}(\underline{x},t)$ = displacement at time t of a material point in the matrix initially at $\underline{x}$.

The model equations are:

Cell conservation: $\frac{\partial n}{\partial t} = \nabla\cdot[D_1\nabla n - D_2\nabla\nabla^2 n - \alpha n\nabla\rho - n\frac{\partial \underline{u}}{\partial t}] + rn(N-n)$ (2.1a)

Mechanical balance: $\nabla\cdot[\mu_1\frac{\partial \varepsilon}{\partial t} + \mu_2\frac{\partial \theta}{\partial t}\mathbf{I} + \frac{E}{1+\nu}\{\varepsilon + \hat{\nu}\theta\mathbf{I}\}$

$$+ \frac{\tau n}{(1+\lambda n^2)}\{\rho + \beta\nabla^2\rho\}\mathbf{I}] = s\underline{u}\rho \qquad (2.1b)$$

Matrix conservation: $\frac{\partial \rho}{\partial t} + \nabla\cdot\{\rho\,\frac{\partial \underline{u}}{\partial t}\} = 0$ (2.1c)

where $\varepsilon = \frac{1}{2}\{\nabla\underline{u} + \nabla\underline{u}^T\}$ is the linear strain tensor and D_1, D_2, α, r, N, μ_1, μ_2, E, ν, $\hat{\nu}$, λ, τ and β are parameters which we describe below.

We briefly motivate the various contributions to equations (2.1a)-(2.1c).

Cell conservation. The equation for cell conservation is of the form

$$\frac{\partial n}{\partial t} = -\nabla\cdot\underline{J} + rn(N-n) \qquad (2.2)$$

where $\underline{J}$ is the cell flux through a volume element of matrix and, to be specific, we have taken a logistic growth term where r, the mitotic rate, and N are positive constants. The cell flux, $\underline{J}$, is made up of a number of terms:

Random dispersal: In populations of low cell density, random dispersal of cells may be modelled with a Fickian flux $-D_1\nabla n$ where D_1 is the diffusion coefficient. However, with high mesenchymal cell densities, we should take non-local effects into account since mesenchymal cells have long finger-like extensions (filopodia) which can detect non-local cell densities. Thus cell 'diffusion' depends also on the average cell density in the immediate surrounding. We model this non-local (long range) diffusion by $D_2\nabla(\nabla^2 n)$ where D_2 is the coefficient of long range diffusion.

Haptotaxis. Cells actively move by attaching their filopodia to adhesive sites in the ECM. The tendency of cells to move up a gradient in adhesive sites (Harris (1973)) is called haptotaxis. Assuming, reasonably, that the density of adhesive sites is proportional to the matrix density, we model the haptotactic flux as $\alpha n\nabla\rho$ where α is the coefficient of haptotaxis. We could, of course, include a long range haptotactic effect but for simplicity do not do so here.

Convection. Cells are passively carried on the substratum and this convective motion is given by $n\frac{\partial \underline{u}}{\partial t}$. Here $\frac{\partial \underline{u}}{\partial t}$ is the velocity. This is probably the most important dispersal component.

Substituting these contributions into (2.2) gives (2.1a)

Mechanical balance. We are dealing with systems with very low Reynolds number (Purcell (1977), Odell et al (1981)) so the viscous and elastic forces dominate the inertial terms and cell motion instantly ceases when the applied forces are turned off. Therefore there is a balance between the cell contractile forces deforming the ECM and the viscous and elastic restoring forces in the ECM. There is experimental evidence for assuming small strains. We therefore model the cell-matrix composite as a linear, isotropic, viscoelastic material with stress tensor $\boldsymbol{\sigma} = \boldsymbol{\sigma}_{matrix} + \boldsymbol{\sigma}_{cell\text{-}matrix}$, where

$$\boldsymbol{\sigma}_{matrix} = \underbrace{\mu_1 \frac{\partial \boldsymbol{\varepsilon}}{\partial t} + \mu_2 \frac{\partial \theta}{\partial t}\mathbf{I}}_{\text{viscous}} + \underbrace{\frac{E}{1+\nu}\{\varepsilon + \hat{\nu}\theta\mathbf{I}\}}_{\text{elastic}},$$

where $\theta = \text{div}\,\underline{u}$, the dilatation, E is Young's modulus, μ_1 and μ_2 the shear and bulk viscosities respectively, ν the Poisson ratio, $\hat{\nu} = \frac{\nu}{1-2\nu}$, and I the unit tensor.

The stress due to the contractile forces exerted by the cells is taken as $\sigma_{cell\text{-}matrix} = \tau(n)n[\rho+\beta\nabla^2\rho]\mathbf{I}$ where $\tau(n) = \frac{\tau}{1+\lambda n^2}$ is the traction/unit length/unit cell, τ, λ and β are positive constants. The motivation for this term is that mesenchymal cells exert contractile forces by attaching their filopodia to the adhesive sites and compressing the matrix. Therefore, the contractile force is proportional to the density of adhesive sites which in turn, we take as proportion to ρ. Long range effects should again be taken into account: this gives the term $\beta\nabla^2\rho$. As cell density increases, the traction/unit cell decreases due to cell-cell inhibition (Trinkaus (1984)), hence the qualitative form of $\tau(n)$ taken. The equation for mechanical equilibrium is

$$\nabla\cdot\sigma + \rho\underline{F} = \underline{0}$$

where $\underline{F}$ is a body force. Typically the ECM is attached elastically to a subdermal layer which we model as a linear spring and so $\underline{F} = -s\underline{u}$, where s is a positive constant.

Matrix conservation. The matrix density ρ satisfies the usual conservation equation

$$\frac{\partial \rho}{\partial t} + \nabla\cdot\{\rho \frac{\partial \underline{u}}{\partial t}\} = 0$$

With the processes and time scales we are concerned with here no matrix is being secreted.

3. Linear Dispersion Relations.

It is convenient to non-dimensionalise the system to highlight the dimensionless parameter groupings in a biologically significant way: they indicate which biological processes have equivalent effects. The system reduces to

$$\frac{\partial n}{\partial t} = D_1\nabla^2 n - D_2\nabla^4 n - \alpha\,\nabla\cdot\{n\nabla\rho\} - \nabla\cdot\{n\frac{\partial \underline{u}}{\partial t}\} + rn(1-n) \tag{3.1a}$$

$$\nabla\cdot[\mu_1 \frac{\partial \varepsilon}{\partial t} + \mu_2 \frac{\partial \theta}{\partial t}\mathbf{I} + (\varepsilon + \hat{\nu}\theta\mathbf{I}) + \frac{\tau n}{(1+\lambda n^2)}\{\rho + \beta\nabla^2\rho\}\mathbf{I}] = s\underline{u}\rho \tag{3.1b}$$

$$\frac{\partial \rho}{\partial t} + \nabla\cdot\{\rho \frac{\partial \underline{u}}{\partial t}\} = 0 \tag{3.1c}$$

where D_1, D_2, α, r, μ_1, μ_2, λ, β, τ and s are non-dimensional parameter groupings (see Appendix (A.1)).

Linearising about the biologically relevant steady state $n = 1 = \rho$, $\underline{u} = \underline{0}$ in the usual way we obtain the dispersion relation for the growth rate σ as a function of the wave number k. (Appendix (A.2)).

Setting various parameters equal to zero gives rise to varied behaviour for $\sigma(k^2)$: Table 1 illustrates a few of these. Spatial patterns evolve when the parameters are such that $R\ell\sigma(k^2) > 0$, $k^2 \neq 0$.

Table 1: (a) Examples of possible types of dispersion relation $\sigma(k^2)$ (Appendix (A.2)) with <u>finite</u> ranges of unstable k (wave number) in the case $\lambda = \beta = 0$.

-+++- denotes unstable wave number

● denotes non-zero parameter

○ denotes parameter set to zero

D_1	D_2	α	r	μ	s	Conditions on τ	$\sigma(k^2)$
●	○	○	○	●	●	$1/2 + \mu D_1 s < \tau < 1$	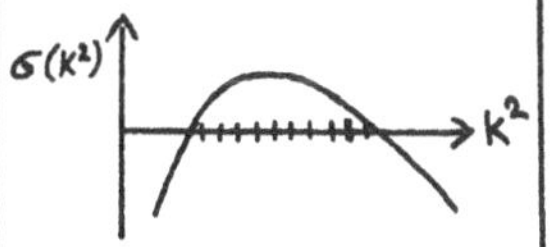
●	○	○	○	●	○	$1/2 < \tau < 1$	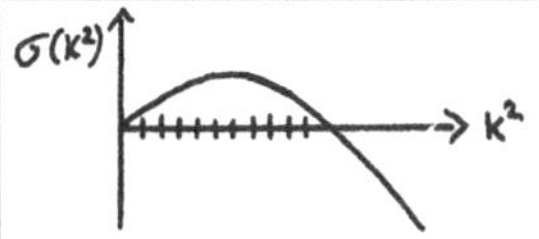
●	●	●	○	○	●	$(D_1 + sD_2)/(D_1 + \alpha) < \tau < 1/2$ $[(D_1 + \alpha)\tau - (D_1 + sD_2)]^2 > 4sD_1D_2(1 - \tau)$	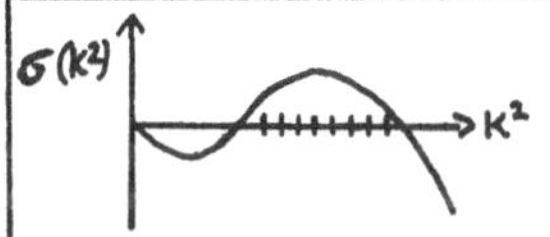
○	●	●	○	○	●	$\tau > \max[1, sD_2/\alpha]$	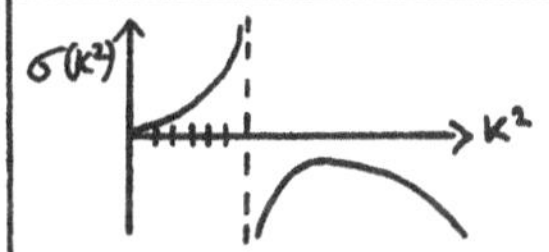
●	●	●	●	○	●	$1/2 < \tau < 1$ + quadratic condition on τ	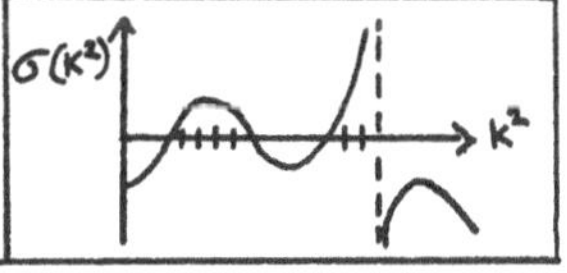

(b) Examples of the dispersion relation $\sigma(k^2)$ (Appendix (A.2)) with <u>infinite</u> ranges of unstable k (wave number) in the case $\lambda = \beta = 0$.

D_1	D_2	α	r	μ	s	Conditions on τ	$\sigma(k^2)$
●	●	●	○	○	●	$1/2 < \tau < 1$ $\tau > (D_1 + sD_2)/(D_1 + \alpha)$ + two quadratic conditions on τ	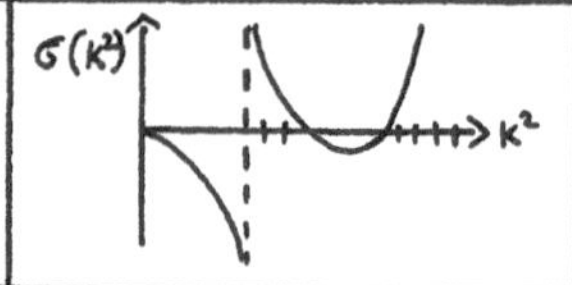
●	●	●	○	○	●	$1/2 < \tau < 1$ $\tau > (D_1 + sD_2)/(D_1 + \alpha)$ + two quadratic conditions on τ	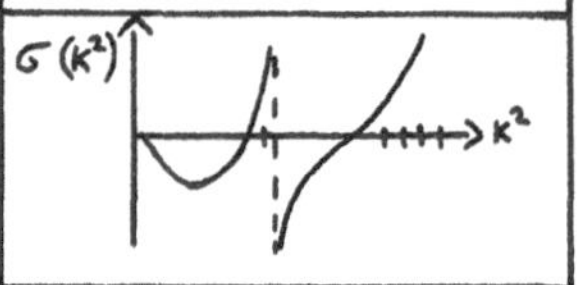
○	●	●	○	○	●	$1/2 + sD_2/2\alpha < \tau < 1$ $\tau > sD_2/\alpha$	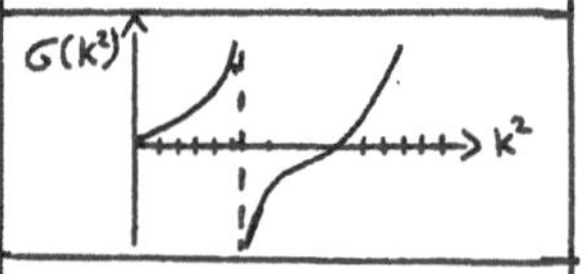
●	●	●	●	○	●	$\tau < 1/2$ + condition that $c(k^2)=0$ has 3 real roots	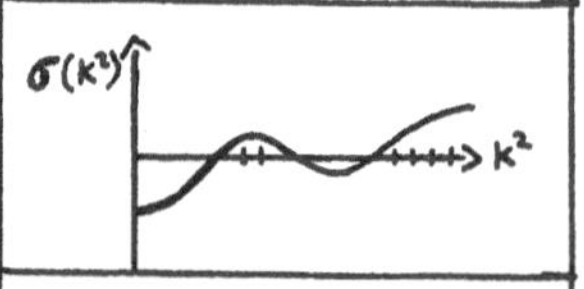
○	●	●	○	○	●	$1/2 < \tau < 1/2 + sD_2/2\alpha$	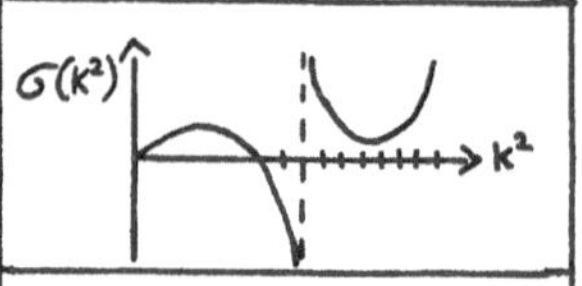
○	○	○	●	●	●	$1 < \tau < (\mu r + 1)/2$	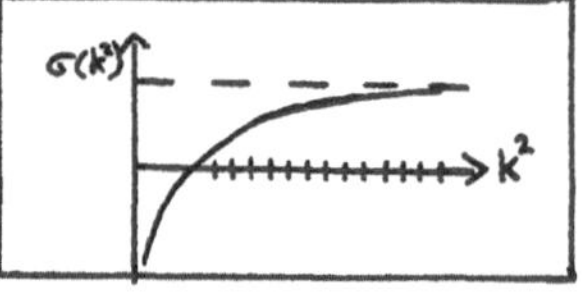

[The form of some of these dispersion relations indicates the lack of validity of a linear theory.]

4. One-Dimensional Nonlinear Analysis.

In this section we briefly summarize some results from Maini et al (1984) for a simple version of the model of biological relevance in which τ, μ, β, and s are the only non-zero parameters. We take τ to be the bifurcation parameter. The one-dimensional system we consider is

$$\frac{\partial n}{\partial t} + \frac{\partial}{\partial x}\left\{n\,\frac{\partial u}{\partial t}\right\} = 0 \qquad (4.1a)$$

$$\frac{\partial^3 u}{\partial x^2 \partial t} + \frac{\partial^2 u}{\partial x^2} + \tau\,\frac{\partial}{\partial x}\left[n\left\{\rho+\beta\,\frac{\partial^2 \rho}{\partial x^2}\right\}\right] - su\rho = 0 \qquad (4.1b)$$

$$\frac{\partial \rho}{\partial t} + \frac{\partial}{\partial x}\left\{\rho\,\frac{\partial u}{\partial t}\right\} = 0 \qquad (4.1c)$$

where we have rescaled time by dividing through by μ .

The dispersion relation (from (A.2)) is

$$\sigma = 0,\ \sigma(k^2) = -\,\frac{\beta\tau k^4 + (1-2\tau)k^2 + s}{k^2} \qquad (4.2)$$

The trivial solution corresponds to neutral stability of the steady state and is not considered in the following analysis. Figure 1 illustrates the behaviour of the dispersion relation as τ increases.

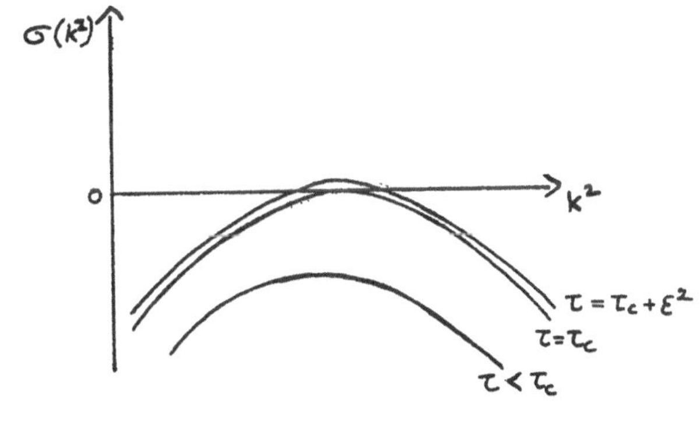

Figure 1. Behaviour of the non-zero solution for $\sigma(k^2)$ (see(4.2)) as τ increases. There is a critical value for τ (τ_c) such that if $\tau < \tau_c$, the uniform steady state ($n = \rho = 1$, $u = 0$) is stable. For $\tau = \tau_c + \varepsilon^2$ ($0 < \varepsilon \ll 1$) the uniform steady state is unstable and spatial disturbances of wave number k_c grow fastest where $k_c^2 = \sqrt{\dfrac{s}{\beta\tau_c}}$ and $\tau_c = \frac{1}{2}\left[1+s\beta+\sqrt{(1+s\beta)^2-1}\right]$.

Linear analysis therefore, predicts that if $\tau = \tau_c + \varepsilon^2\delta$, where $0 < \varepsilon < 1$ and $\delta = +1$, then the uniform steady state goes unstable and the fastest growing unstable wave number is k_c, where

$$\tau_c = \frac{1}{2}\left[1+s\beta+\sqrt{(1+s\beta)^2-1}\,\right] \quad \text{and} \quad k_c^2 = \sqrt{\frac{s}{\beta\tau_c}} \tag{4.3}$$

Expanding σ about (k_c^2,τ) in a Taylor series we have

$$\sigma(k_c^2,\tau_c+\varepsilon^2\delta) = \underbrace{\sigma(k_c^2,\tau_c)}_{0} + \varepsilon^2\delta\left.\frac{\partial\sigma}{\partial\tau}\right|_{k_c^2,\tau_c} + O(\varepsilon^4) \tag{4.4}$$

Thus the exponential growth term is $\exp(O(\varepsilon^2)t)$ and this suggests the usual long time scale $T = \varepsilon^2 t$ (Lara Ochoa and Murray (1983)). As k_c is the fastest growing wave number, we assume that on the long time scale T, the nonlinear solution will have wave number k_c (Matkowsky (1970)). To investigate the nonlinear behaviour of (4.1a)-(4.1c), we use the method of balancing harmonics and substitute

$$n(x,T,\varepsilon) = 1 + \sum_{j=1} \varepsilon^j\{A_j(\varepsilon,T)\cos jk_cx + D_j(\varepsilon,T)\sin jk_cx\}$$

$$u(x,T,\varepsilon) = \sum_{j=1} \varepsilon^j\{B_j(\varepsilon,T)\sin jk_cx + E_j(\varepsilon,T)\cos jk_cx\}$$

$$\rho(x,T,\varepsilon) = 1 + \sum_{j=1} \varepsilon^j\{C_j(\varepsilon,T)\cos jk_cx + F_j(\varepsilon,T)\sin jk_cx\} \tag{4.5}$$

where $T = \varepsilon^2 t$ and $A_j = \sum_{i=0} A_j^i(T)\varepsilon^i$, etc.

into (4.1a)-(4.1c) and equate coefficients of ε. This leads to a hierachy of <u>linear</u> equations for the coefficients $A_j^i(T)$, $B_j^i(T)$, $C_j^i(T)$, $D_j^i(T)$, $E_j^i(T)$ and $F_j^i(T)$ which we can solve. To lowest order in ε, we have

$$\begin{aligned} &\frac{d}{dT}\left(A_1^0(T) + k_cB_1^0(T)\right) = 0\\ &k_c\tau_cA_1^0(T) + (k_c^2+s)B_1^0(T) + k_c\tau_c(1-k_c^2\beta)C_1^0(T) = 0\\ &\frac{d}{dT}\left(C_1^0(T) + k_cB_1^0(T)\right) = 0 \end{aligned} \tag{4.6}$$

and a similar set of equations for $\{D_1^0(T), E_1^0(T), F_1^0(T)\}$. For the remaining calculation we shall only consider $\{A_j^i(T), B_j^i(T), C_j^i(T)\}$ to simplify the analysis. The analysis may be repeated exactly for $\{D_j^i(T), E_j^i(T), F_j^i(T)\}$.

Order ε^2 terms give

$$\frac{d}{dT}\{\overset{0}{A}_2(T) + 2k_c\overset{0}{B}_2(T)\} + k_c\overset{0}{A}_1(T)\frac{d\overset{0}{B}_1(T)}{dT} = 0$$

$$2k_c\tau_c\overset{0}{A}_2(T) + (4k_c^2+s)\overset{0}{B}_2(T) + 2k_c\tau_c(1-4\beta k_c^2)\overset{0}{C}_2(T) =$$

$$-k_c\tau_c(1-\beta k_c^2)\overset{0}{A}_1(T)\overset{0}{C}_1(T) - \frac{s}{2}\overset{0}{B}_1(T)\overset{0}{C}_1(T) \qquad (4.7)$$

$$\frac{d}{dT}\{\overset{0}{C}_2(T) + 2k_c\overset{0}{B}_2(T)\} + k_c\overset{0}{C}_1(T)\frac{d\overset{0}{B}_1(T)}{dT} = 0$$

and at order ε^3, (4.1b) gives

$$-k_c\frac{d\overset{0}{B}_1(T)}{dT} + k_c\delta\{\overset{0}{A}_1(T) + \overset{0}{C}_1(T)\} + \beta\delta k_c^3\overset{0}{C}_1(T) +$$

$$k_c\tau_c(2\beta k_c^2 - \tfrac{1}{2})\overset{0}{A}_1(T)\overset{0}{C}_2(T) + \frac{k_c\tau_c}{2}\{\beta k_c^2 - 1\}\overset{0}{A}_2(T)\overset{0}{C}_1(T) \qquad (4.8)$$

$$-\frac{s}{2}\{\overset{0}{C}_1(T)\overset{0}{B}_2(T) - \overset{0}{B}_1(T)\overset{0}{C}_2(T)\} = 0$$

Standard nonlinear analysis simply requires successive suppression of secular terms. With the structure of our equations this is not sufficient to determine the amplitude equations. We have to use an integrated form of the conservation equations. Integrating the first and third of (4.6) we have three simultaneous equations for $\overset{0}{A}_1(T)$, $\overset{0}{B}_1(T)$ and $\overset{0}{C}_1(T)$, namely

$$\begin{aligned}&\overset{0}{A}_1(T) + k_c\overset{0}{B}_1(T) = \overset{0}{\gamma}_1\\&k_c\tau_c\overset{0}{A}_1(T) + (k_c^2+s)\overset{0}{B}_1(T) + k_c\tau_c(1-\beta k_c^2)\overset{0}{C}_1(T) = 0\\&k_c\overset{0}{B}_1(T) + \overset{0}{C}_1(T) = \overset{0}{\gamma}_3\end{aligned} \qquad (4.9)$$

where $\overset{0}{\gamma}_1$ and $\overset{0}{\gamma}_3$ are constants. The system (4.9) is degenerate and has nontrivial solution if and only if

$$[k_c^2(1-\tau_c) + s]\overset{0}{\gamma}_3 + k_c^2\tau_c\overset{0}{\gamma}_1 = 0 \qquad (4.10)$$

that is, there is a constraint on the initial conditions. This is what we would expect because of the intimate relationship between $u(x,t)$ and $\rho(x,t)$.

Integrating the first and third of (4.7) gives

$$\begin{aligned} \overset{0}{A}_2(T) + 2k_c\overset{0}{B}_2(T) &= \frac{\{\overset{0}{A}_1(T)\}^2}{2} + \overset{1}{\gamma}_1 \\ \overset{0}{C}_2(T) + 2k_c\overset{0}{B}_2(T) &= \frac{\{\overset{0}{C}_1(T)\}^2}{2} + \overset{1}{\gamma}_3 \end{aligned} \tag{4.11}$$

where $\overset{1}{\gamma}_1$ and $\overset{1}{\gamma}_3$ are constants.

Note that $\overset{0}{\gamma}_1 = \overset{0}{A}_1(0) + k_c\overset{0}{B}_1(0)$. If we assume initial perturbations to be $O(\varepsilon^2)$, then $\overset{0}{\gamma}_1 = \overset{0}{\gamma}_3 = 0$. Moreover, assuming initial perturbations to be $O(\varepsilon^3)$ implies $\overset{1}{\gamma}_1 = \overset{1}{\gamma}_3 = 0$. Making these assumptions is not necessary but they simplify the analysis.

We solve system (4.9) and the integrated form of (4.7) for $\overset{0}{B}_1(T)$, $\overset{0}{C}_1(T)$, $\overset{0}{A}_2(T)$, $\overset{0}{B}_2(T)$ and $\overset{0}{C}_2(T)$ in terms of $\overset{0}{A}_1(T)$ and, substituting into (4.8), we have the usual Landau equation

$$\frac{d\overset{0}{A}_1(T)}{dT} = \delta X\overset{0}{A}_1(T) + Y\{\overset{0}{A}_1(T)\}^3 \tag{4.12}$$

where $X = \dfrac{2\tau_c+1}{2\tau_c}$ and $Y = \dfrac{\{14\beta s\tau_c+24\tau_c-63\beta s-12\}}{72\beta s}$

The behaviour of (4.12) is summarised in Table 2.

Table 2. Behaviour of $A_1^0(T)$ from equation (4.12).

	$Y < 0$	$Y > 0$
$\delta > 0$	A_1^0 evolves to $\{X/\lvert Y\rvert\}^{1/2}$	$A_1^0 \to \infty$
$\delta < 0$	$A_1^0 \to 0$	Threshold in $A_1^0(0)$: $A_1^0(0) < \{X/Y\}^{1/2} \Rightarrow A_1^0 \to 0$ $A_1^0(0) > \{X/Y\}^{1/2} \Rightarrow A_1^0 \to \infty$

If we are in the parameter space P (Figure 2) the cell density evolves to the bounded steady state

$$n = 1 + \varepsilon\sqrt{\frac{X}{|Y|}}\cos k_c x \tag{4.13}$$

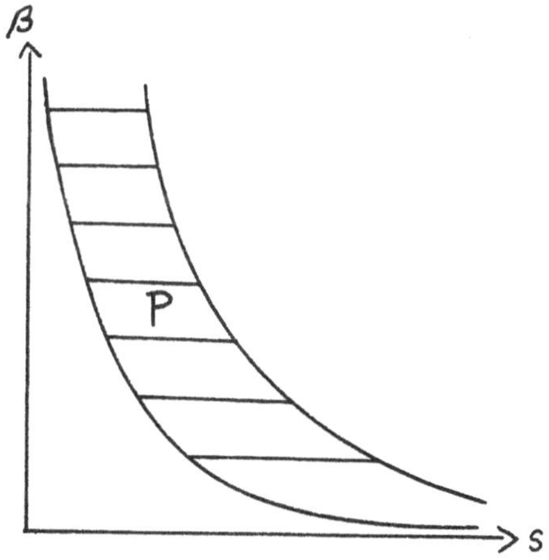

Figure 2. Parameter space P, (shaded), where $Y > 0$ ((4.12)) and the homogeneous steady state $n(x,t) = 1$ evolves to the heterogeneous solution (4.13).

If we include all of the initial constants, we would have finished up with the perturbed version of the above Landau equation, namely

$$\frac{d\overset{0}{A_1}(T)}{dT} = C_0 + \delta(X+X_0)\overset{0}{A_1}(T) + Z_0\{\overset{0}{A_1}(T)\}^2 + Y\{\overset{0}{A_1}(T)\}^3 \qquad (4.14)$$

where C_0, X_0 and Z_0 are functions of γ_1^i and γ_3^i , $i = 0,1$.

In this case the homogeneous steady state would evolve to a heterogeneous steady state dependent on initial perturbations. As we are dealing with small perturbations, these variations are small.

5. Two-Dimensional Nonlinear Analysis.

In this section we analyze a caricature of the above model (3.1) in which (3.1a) involves only convection and consider it in two dimensions to investigate the possibility of a regular tesselation pattern. We linearise the conservation equations $\frac{\partial n}{\partial t} + \nabla\cdot(n\frac{\partial \underline{u}}{\partial t}) = 0 = \frac{\partial \rho}{\partial t} + \nabla\cdot(\rho\frac{\partial \underline{u}}{\partial t})$ about the steady state $n = \rho = 1$, $\underline{u} = \underline{0}$ which on integration give

$$n = 1 - \theta, \ \rho = 1 - \theta \qquad (5.1)$$

where $\theta = \text{div } \underline{u}$, and in which we assume $\theta < 1$ as n and ρ are necessarily non-negative.

We replace the $s\underline{u}\rho$ term in the mechanical balance equation by the linearised term $s\underline{u}$ and take the divergence of the resulting equation. Using the identity $\nabla\cdot\varepsilon = \text{grad div } \underline{u} - \frac{1}{2} \text{ curl curl } \underline{u}$ for the divergence of the linear strain tensor, the equation becomes

$$\nabla^2 \frac{\partial\theta}{\partial t} + \nabla^2\theta + \tau\nabla^2\{[1-\theta]^2 - \beta[1-\theta]\nabla^2\theta\} - s\theta = 0 \qquad (5.2)$$

Clearly the dispersion relation for this equation is (4.2) and for $\tau > \tau_c$, where τ_c is given by (4.3), the uniform steady state is linearly unstable. To study the full nonlinear system, we substitute

$$\tau = \tau_c + \sum_{i=1} \varepsilon^i\tau_i, \ \theta(\underline{x},t) = \sum_{i=1} \varepsilon^i\theta_{i-1}(\underline{x},t) \quad \text{where} \quad 0 < \varepsilon \ll 1 \qquad (5.3)$$

into equation (5.2). As in section 4, this gives rise to a hierachy of linear equations which we can solve. We follow in part the process used by Busse (1983).

The only regular patterns in the plane are rolls, triangles (hexagons), squares, and tesselations of these. Therefore we look for $\theta_0(\underline{x},t)$ of the form

$$\theta_0 = \frac{1}{2} a_1(t)\{\cos(kx + ly) + \cos(ly - kx)\} + a_2(t)\cos 2ly \qquad (5.4)$$

where the fastest growing unstable mode has wave number κ $(= (s/\beta\tau_c)^{1/4})$ and $k^2 + \ell^2 = \kappa^2$, $4\ell^2 = \kappa^2$. Putting $a_1(t) = 0$, $a_2(t) = 0$, $a_1(t) = 2a_2(t)$ into (5.4) gives roll, rhombic and hexagonal structures respectively (Christopherson (1940)).

With $\theta_0(\underline{x},t)$ as above, $\theta_1(\underline{x},t)$ must have the form

$$\theta_1 = b_1(t)\cos 4ly + b_2(t)\cos(3ly+kx) + b_3(t)\cos(3ly-kx)$$
$$+ b_4(t)\cos 2(ly+kx) + b_5(t)\cos 2(ly-kx) + b_6(t)\cos 2kx$$

where, on equating coefficients of ε^2, we can find $b_i(t)$, $i = 1, 2...6$ in terms of $a_1(t)$ and $a_2(t)$: it is a simple but tedious calculation.

Taking powers of ε up to ε^3 into account, to suppress secular terms $a_1(t)$ and $a_2(t)$ must satisfy the coupled system of ordinary differential equations

$$\kappa^2 \frac{da_1(t)}{dt} = Xa_1(t) - Ya_1(t)a_2(t) - 2Za_1(t)a_2^2(t) - \frac{1}{4}(2Z+R)a_1^3(t)$$
$$\kappa^2 \frac{da_2(t)}{dt} = Xa_2(t) - \frac{Y}{4}a_1^2(t) - Za_1^2(t)a_2(t) - Ra_2^3(t) \qquad (5.5)$$

where $X = \frac{(2\tau_c+1)}{2\tau_c}\{\varepsilon\tau_1+\varepsilon^2\tau_2\}\kappa^2$, $Y = \frac{1}{2\tau_c}\{\varepsilon\tau_c+\varepsilon^2\tau_1\}\kappa^2$,

$Z = \frac{3\varepsilon^2}{16\beta\tau_c}\{\tau_c-1\}$ and $R = \frac{\varepsilon^2}{36\beta\tau_c}\{6\tau_c-5\}$.

We can take $\tau_1 > 0$, thus X and Y are positive and the sign of R and Z depend on the value of τ_c.

The system (5.5) has the following steady states

I. $a_1 = a_2 = 0$;

II. a,b $a_1 = 0,\ a_2 = \pm\sqrt{X/R}$

$$\text{III. a,b}\quad a_1 = 2a_2,\ a_2 = \frac{-Y\pm\sqrt{Y^2+4X(4Z+R)}}{2(4Z+R)}; \tag{5.6}$$

$$\text{IV. a,b}\quad a_1 = -2a_2,\ a_2 = \frac{-Y\pm\sqrt{Y^2+4X(4Z+R)}}{2(4Z+R)};$$

$$\text{V. a,b}\quad a_1 = \frac{\pm 2\sqrt{X-\frac{RY}{R-2Z}}}{\sqrt{R+2Z}},\ a_2 = \frac{Y}{R-2Z}$$

where, for example, IIa,b exists if and only if $\tau_c > \frac{5}{6}$. Note that to lowest order in ε, IIIa is $a_1 = 2a_2$, $a_2 = \frac{-Y_0+\sqrt{Y_0^2}}{2(4Z+R)} = 0$, where $Y_0 = \frac{\varepsilon}{2}\kappa^2$, and this is the same as I. A similar argument can be applied to IVa and, in the following analysis, we will not distinguish between states I, IIIa and IVa.

We can analyze the stability of states I-IV in (5.6) by calculating the eigenvalues (λ_s) of the appropriate matrix. This gives a quadratic for λ_s, with coefficients dependent on ε. To simplify the solution of the quadratic we approximate the coefficients to lowest order in ε. Table 3 summarises the results of the stability analysis.

Table 3. Summary of stability analysis of the steady states I-IV (5.6) of the coupled system of ordinary differential equations (5.5).

$\tau_c < 5/6$	$5/6 < \tau_c < 47/51$	$47/51 < \tau_c < 32/33$	$\tau_c > 32/33$
I unstable star			
IIa,b do not exist	IIa stable node IIb saddle point		
IIIb and IVb unstable node		IIIb and IVb saddle point	IIIb and IVb stable node

Note that for $\frac{5}{6} < \tau_c < \frac{32}{33}$ the only stable regular pattern is a roll, while for $\tau_c > \frac{32}{33}$ rolls and hexagons are stable. In the latter case, the evolved pattern depends on initial conditions.

6. Biological Application to the Formation of Skin Organ Primordia.

In the early stages of skin organ development (hair, teeth, feathers, scales) dermal cells aggregate to form a regular spatial pattern. These aggregations (papillae), in association with overlapping arrays of columnar epidermal cells (placodes), lead to the formation of skin organ primordia (e.g. Rawles (1963), Wessels (1965)).

Rows of feather primordia develop sequentially to form a hexagonal pattern within well-defined regions of chicken skin (pterylae). The pattern is initiated by a single row of feather primordia forming along the dorsal midline in the posterior part of the spinal pteryla (Stuart and Moscona (1967), Davidson (1983)) and successive rows form on either side of this initial row.

We apply our model to this with the following scenario: initially there is a uniform density of dermal cells along the dorsal midline. As the cell traction increases (or other parameters involved in the dimensionless traction parameter change appropriately) this homogeneous steady state bifurcates into a heterogeneous pattern of isolated clumps (cf. section 4). This parameter evolution may be due, for example, to cell maturity: cells "age" into the unstable regime in parameter space. The tractions produced by these aggregates strain the matrix and a secondary row of papillae are encouraged to form at loci midway between the primary papillae, where the strain is a local minimum. This recruits other cells and thereby forms a hexagonal pattern. The scenario is illustrated in Figure 3.

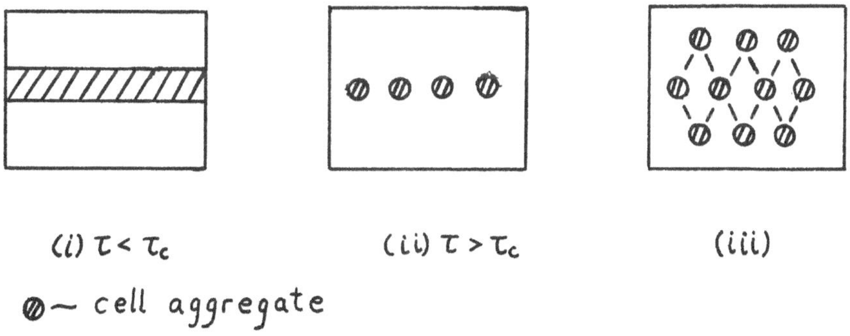

Figure 3. Scenario for hexagonal pattern formation in an idealized section of chick pteryla. As traction increases, the uniform cell density (i) becomes unstable and evolves into a row of isolated aggregations (ii); this sets up a strain field causing condensations along a neighbouring row at intermediate points; (iii) illustrates how hexagonal patterns can arise.

7. Conclusions.

We have presented a model for cell aggregation based on well documented mechanical properties of cells and extracellular matrix. We have illustrated how cell traction on a viscoelastic substratum can produce aggregations in one- and two-dimensions. Two dimensional patterns may be produced synchronously (section 5) or asynchronously (section 6). No directed cell migration is necessary. Numerical simulations (in one dimension) of the full equations (to be presented elsewhere) show the aggregation patterns. Numerical studies are underway to verify the two-dimensional patterns.

In this model, the extracellular matrix behaves as a passive viscoelastic material. However, in cartilage formation in chick limbs, the osmotic component of the extracellular matrix plays an active role in aggregation formation (Oster, et al (1985)). Some of the predictions made by these mechanical models can be (and are being) investigated experimentally.

Acknowledgements: P. K. Maini and J. D. Murray would like to thank the Mathematics Department, University of Utah for supporting a visit during which some of this research was carried out. P. K. Maini wishes to thank the Department of Education of N. Ireland for a postgraduate studentship. G. F. Oster was supported by NSF grant #MCS-8110557. G. F. Oster would also like to acknowledge support from the Science and Engineering Research Council of Great Britain (grant GR/c/63595) for a visit to the Centre for Mathematical Biology in Oxford.

APPENDIX

(A.1) To non-dimensionalise the model (2.1a)-(2.1c), let L and T_0 by typical length and time scales respectively and let ρ_0 be a typical matrix density. The dimensionless quantities are

$$\tilde{\underline{x}} = \frac{\underline{x}}{L}, \quad \tilde{t} = \frac{t}{T_0}, \quad \tilde{n}(\tilde{\underline{x}},\tilde{t}) = \frac{n(\underline{x},t)}{N}, \quad \tilde{\rho}(\tilde{\underline{x}},\tilde{t}) = \frac{\rho(\underline{x},t)}{\rho_0}, \quad \tilde{\underline{u}}(\tilde{\underline{x}},\tilde{t}) = \frac{\underline{u}(\underline{x},t)}{L},$$

$$\tilde{D}_1 = \frac{D_1T_0}{L^2}, \quad \tilde{D}_2 = \frac{D_2T_0}{L^4}, \quad \tilde{\lambda} = \lambda N^2, \quad \tilde{\beta} = \frac{\beta}{L^2}, \quad \tilde{s} = \frac{s\rho_0L^2(1+\nu)}{E}, \quad \tilde{\alpha} = \frac{\alpha\rho_0T_0}{L^2},$$

$$\tilde{r} = rNT_0, \quad \tilde{\mu}_1 = \frac{\mu_1(1+\nu)}{T_0E}, \quad \tilde{\mu}_2 = \frac{\mu_2(1+\nu)}{T_0E}, \quad \tilde{\tau} = \frac{\tau N\rho_0(1+\nu)}{E}$$

Dropping the tildes, the model system reduces to (3.1a)-(3.1c).

(A.2) The dispersion relation satisfied by the linear growth rate $\sigma(k^2)$ is

$$\sigma(k^2) = 0; \quad \text{or} \quad \sigma(k^2) = \frac{-b(k^2) \pm \sqrt{b^2(k^2) - 4\mu k^2c(k^2)}}{2\mu k^2}$$

where $b(k^2) = \mu D_2k^6 + \{\frac{\tau\beta}{(1+\lambda)} + \mu D_1\}k^4 + \{1 + \mu r - \frac{2\tau}{(1+\lambda)^2}\}k^2 + s$ and

$$c(k^2) = \frac{\tau\beta}{(1+\lambda)} D_2k^8 + \{(\frac{\tau}{1+\lambda})[\beta D_1 - D_2] + D_2\}k^6$$

$$+ \{(\frac{\tau}{1+\lambda})[r\beta - D_1 - \alpha(1- \frac{2\lambda}{1-\lambda})] + sD_2 + D_1\}k^4$$

$$+ \{sD_1 + r - \frac{r\tau}{(1+\lambda)}\}k^2 + rs$$

where $\mu = \mu_1 + \mu_2$ and we have divided μ, τ and s by $(1+\hat{\nu})$.

REFERENCES

[1] Busse, F. H. (1983). Patterns of bifurcations in plane layers and spherical shells, in Modelling of Patterns in Space and Time (ed. W. Jager & J. D. Murray, (1984)) Proceedings, Springer-Verlag, Heidelberg. pp 51-60.

[2] Christopherson, D. G. (1940). Note on the vibration of membranes. Quart. J. of Math., 11, 63-65.

[3] Davidson, D. (1983). The mechanism of feather pattern development in the chick. 1. The time determination of feather position. J. Embryol. Exp. Morph., 74, 245-259.

[4] Harris, A. K. (1973). Behaviour of cultured cells on substrata of variable adhesiveness. Exp. Cell Res., 77, 285-297.

[5] Harris, A. K., Stopak, D. & Wild, P. (1981). Fibroblast traction as a mechanism for collagen morphogenesis. Nature, 290, 249-251.

[6] Lara Ochoa, F. & Murray, J. D. (1983). A non-linear analysis for spatial structure in a reaction diffusion model. Bull. Math. Biol. 45, 917-930.

[7] Maini, P. K., Murray, J. D. & Oster, G. F. (1985). A mechanical model for biological pattern formation: A nonlinear bifurcation analysis. Dundee Conference (1984) on "Partial and Ordinary Differential Equations." (to appear in the proceedings.) Springer-Verlag, Heidelberg.

[8] Matkowsky, B. J. (1970). A simple non-linear dynamic stability problem. Bull. Amer. Math. Soc., 620-625.

[9] Meinhardt, H. (1982). Models of biological pattern formation. Academic Press, London.

[10] Murray, J. D. (1977). Lectures on non-linear differential equation models in biology. Oxford University Press: Oxford.

[11] Murray, J. D. (1981 a). On pattern formation mechanisms for Lepidopteran wing patterns and mammalian coat markings. Phil. Trans. Roy. Soc. Lond. B295, 473-496.

[12] Murray, J. D. (1981 b). A pre-pattern formation mechanism for animal coat markings. J. Theor. Biol., 88, 161-199.

[13] Murray, J. D. & Oster, G. F. (1984 a). Generation of biological pattern and form. I. M. A. J. Maths. Appl. to Biol. & Med., 1, 51-75.

[14] Murray, J. D. & Oster, G. F. (1984 b). Cell traction models for generating pattern and form in morphogenesis. J. Math. Biol., 19, 265-279.

[15] Odell, G., Oster G. F., Burnside, B. & Alberch, P. (1981). The mechanical basis of morphogenesis I: Epithelial folding and invagination. Devl. Biol., 85, 446-462.

[16] Oster, G. F., Murray J. D. & Harris, A. K. (1983). Mechanical aspects of mesenchymal morphogenesis. J. Embryol. Exp. Morph. 78, 83-125.

[17] Oster, G. F., Murray, J. D. & Maini, P. K. (1985). A model for chondrogenic condensations in the developing limb: The role of extracellular matrix and cell tractions. J. Embryol. Exp. Morph. (to appear).

[18] Purcell, E. M. (1977). Life at low Reynolds number. Amer. J. of Phys., 45, 3-11.

[19] Rawles, M. (1963). Tissue interactions in scale and feather development as studied in dermal-epidermal recombinations. J. Embryol. Exp. Morph., 11, 765-789.

[10] Stuart, E. S. & Moscona, A. A. (1967). Embryonic morphogenesis: Role of fibrous lattice in the development of feathers and feather patterns. Science, 157, 947-948.

[21] Tickel, C., Summerbell, D. & Wolpert, L. (1975). Positional signalling and specification of digits in chick limb morphogenesis. Nature Lond., 254, 199-202.

[22] Trinkaus, J. P. (1984). Cells into organs: The forces that shape the embryo. (Prentice-Hall).

[23] Turing, A. M. (1952). The chemical basis of morphogenesis. Phil. Trans. Roy. Soc. Lond., B237, 37-73.

[24] Wessels, N. K. (1965). Morphology and proliferation during early feather development. Dev. Biol., 12, 131-153.

[25] Wolpert, L. (1969). Positional information and the spatial pattern of cellular differentiation. J. Theor. Biol., 25, 1-47.

[26] Wolpert, L. (1981). Positional information and pattern formation. Phil. Trans. Roy. Soc. Lond. B295. 441-450.

[27] Wolpert, L., Hicklin, J. S. Hornbruch, A. (1971). Positional information and pattern regulation in regeneration of hydra. Symp. Soc. exp. Biol., 25, 391-415.

CHEMICAL SYSTEMS

ELECTRICALLY COUPLED BELOUSOV-ZHABOTINSKII OSCILLATORS: EXPERIMENTAL OBSERVATION OF CHAOS IN A CHEMICAL SYSTEM AND IDENTIFICATION OF ITS SOURCE IN THE FIELD-NOYES EQUATIONS[1]

Michael F. Crowley
and
Richard J. Field
Department of Chemistry
University of Montana
Missoula, MT 59812/USA

The Belousov-Zhabotinskii (BZ) reaction is by far the chemical oscillator that is best characterized experimentally and best understood mechanistically (Field et al. 1972; Field 1985; Field and Boyd, 1985). It is the metal-ion-catalyzed oxidation by bromate ion (BrO_3^-) of any of a large class of organic materials in a strongly acidic, aqueous medium. In the experiments reported here, the Ce(IV)/Ce(III) couple is used as the metal-ion catalyst and acetylacetone (CH_3COCH_2-$COCH_3$), which we will refer to as AA, is used as the organic material to avoid bubble formation. All experiments were carried out in 2.73 M H_2SO_4 and in continuous-flow, stirred tank reactors (CSTR) driven by peristaltic pumps. The general behavior of oscillating chemical reactions in CSTR experiments has been reviewed by DeKepper and Boissonade (1985), and the detailed CSTR behavior of the BZ reaction with malonic acid has been described by DeKepper and Bar-Eli (1983).

The major experimentally observable, oscillatory variables in the BZ reaction are the ratio [Ce(IV)]/[Ce(III)] and [Br^-]. The value of [Ce(IV)]/[Ce(III)] may be monitored either potentiometrically with a shiny platinum electrode, as is done here, or spectrophotometrically. A Ag/AgBr electrode or a bromide-ion-selective electrode may be used to monitor [Br^-]. Figure (1) shows oscillations in redox potential obtained experimentally at various CSTR pumping rates. The oscillations are clean, well-behaved, and stable.

---

[1]This work partially supported by the National Science Foundation under Grant CHE 80-23755.

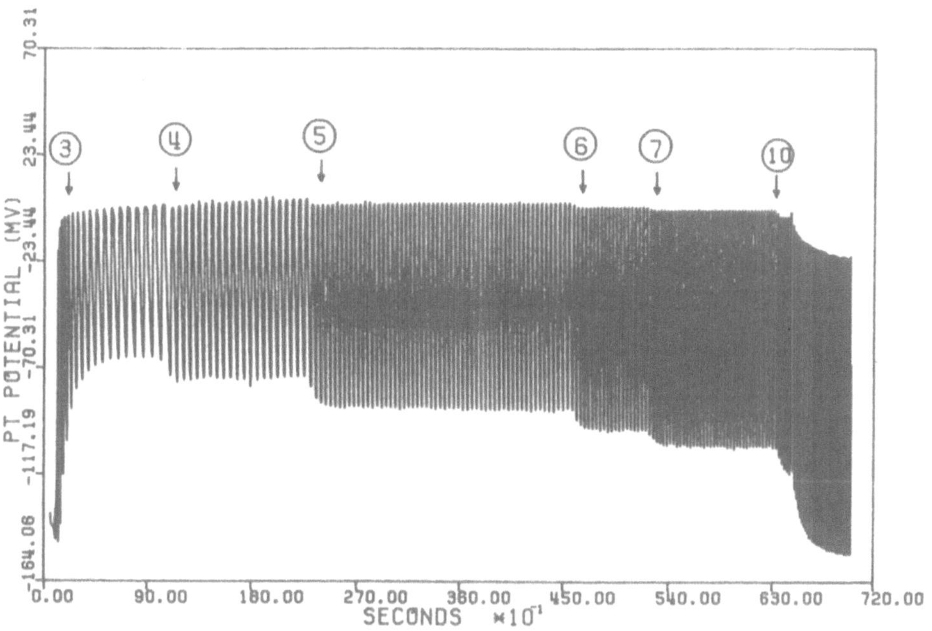

Figure 1. Experimental oscillations in redox potential exhibited by the Belousov-Zhabotinskii Reaction in a CSTR and monitored by a shiny platinum electrode. The experimental conditions were: Solution A (See Figure 4), $[CH_3COCH_2CO\ CH_3]$ = 0.03 M, [Ce(III)] = 0.0016 M, and $[H^+]$ = 2.73 M; Solution B, $[BrO_3^-]$ = 0.096 M and $[H^+]$ = 2.73. The residence times are in the range of 15 to 3 minutes.

I. Coupling of Chemical Oscillators

As the contents of this volume indicate, the coupling of oscillators is of considerable general theoretical and practical interest in a number of contexts, especially biological. Coupled chemical oscillators offer examples of the rich dynamic behaviors possible and are often well enough understood mechanistically for the dynamic structure of the coupling to be analyzed. The general area of periodically perturbed and coupled chemical oscillators has been reviewed and systematized by Rehmus and Ross (1985).

A number of experiments and calculations have been carried out with coupled BZ oscillators, but all previous work has been done using coupling by mass transfer (exchange of reaction mixture) between two or more reactors. Figure (2) shows a schematic diagram of such an apparatus.

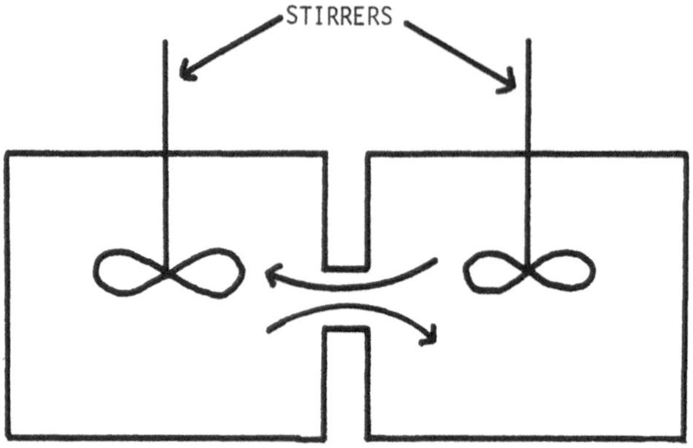

Figure 2. Schematic diagram of an apparatus for the coupling of chemical oscillators by mass transfer.

Because the Belousov-Zhabotinskii reaction is well understood mechanistically (Field, 1985) and can be modeled in terms of a relatively tractable set of differential equations (Field and Noyes, 1974; Troy, 1985; Tyson, 1985) it is extensively used as a model of nonlinear dynamic behavior (DeKepper and Boissonade, 1985; Becker and Field, 1985). We report here experiments on electrically coupled BZ oscillators. Chaos appears in our experiments, and we are able to understand its source in the dynamic structure of the Field-Noyes (1974) model of the BZ chemistry by analogy to an analysis due to Levi (1981) that identifies a Smale (1965, 1967) horseshoe in a forced Van der Pol oscillator.

Mass-transfer-coupled BZ oscillator experiments have been carried out by Marek and Stuchl (1975), Fujii and Sawada (1978), Nakajima and Sawada (1980), Stuchl and Marek (1982), and Bar-Eli and Reuveni (1985). Simulations of mass-transfer-coupled BZ oscillators on the basis of the Field, Körös and Noyes (1972) Chemistry (see also: Bar-Eli and Haddad, 1979, and Bar-Eli and Ronkin, 1984) have been carried out by Bar-Eli (1984a, 1984b, 1985). The effect of periodic perturbation of the BZ reaction in a CSTR experiment by an oscillatory pumping rate has been explored both experimentally (Buchholtz and Schneider, 1983) and by simulation methods (Bar-Eli, 1985).

However, mass-transfer-coupled BZ oscillator experiments suffer from a number of severe disadvantages. The coupling is gross; all reactant, catalyst, intermediate, and product species are exchanged. It is thus difficult to model the effect of coupling in terms of skeleton chemical mechanisms such as the Field-Noyes model (1974). The actual mechanism of coupling is convective reaction-mixture exchange between the two reactors, but this is most difficult to model in any way other than as a linear diffusion-like coupling. If the two oscillators are quite different in composition, as is most desirable to study the interaction of oscillators of dissimilar frequency and amplitude, the situation becomes more complicated as principal reactants (e.g. BrO_3^- and organic material) are exchanged so that the gross features of the oscillators are changed. In this regard, Bar-Eli (1985) has shown in simulations that violation of the rapid, uniform-mixing assumption leads to considerable complication that is essentially impossible to model quantitatively.

What is clearly necessary for useful coupling experiments to be carried out with the BZ reaction is a method by which physically isolated oscillators may be coupled by exchange of a single chemical species whose rate of transfer can be instantaneously and quantitatively related to the condition of each oscillator. Several years ago we proposed electrical coupling of BZ oscillators as a way to do this (Crowley and Field, 1981). We now report implementation of this method.

II. Electrical Coupling of Belousov-Zhabotinskii Oscillators

The electrical coupling of BZ oscillators running in separate CSTR's is based on their oscillatory redox potentials (Fig. 1). Unless two oscillators are identical and in phase, there must be an instantaneous redox potential difference between them. If large-area platinum working electrodes are placed in both oscillators, and if the circuit is completed by a wire between the electrodes and an ion bridge between the CSTR's, then this potential difference will cause a current to flow. The resulting electrochemistry at the working electrodes causes the concentration changes that are the actual chemical mechanism of coupling. A schematic diagram of the oscillatory redox potentials of two identical oscillators running 180° out-of-phase and their instantaneous relative potential is shown in Fig. (3).

The electrochemistry of electrical coupling is simple in the BZ reaction. It can be shown by comparison of the oscillatory redox potential with simultaneous spectrophotometric measurements of [Ce(IV)] that the redox potential in a single reactor is proportional to [Ce(IV)]/[Ce(III)] so long as reactant concentrations are adjusted

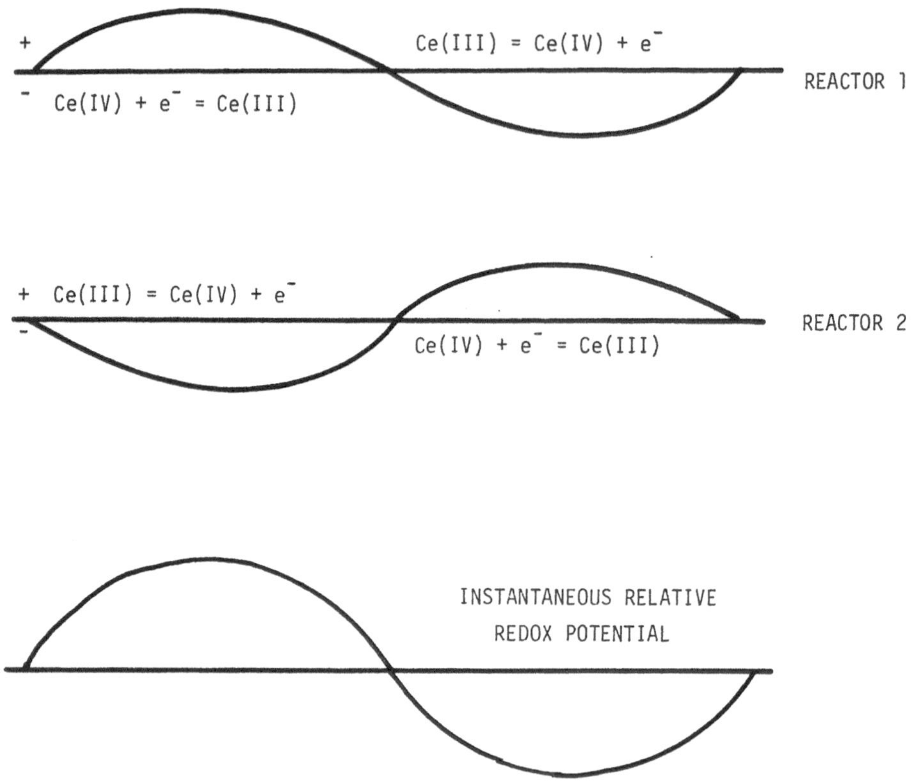

Figure 3. Schematic redox potential oscillations in two identical oscillators running 180° out-of-phase and (lowest curve) their instantaneous relative potential. The electrochemistry occurring in each reactor is shown for each sign of the relative potential. The major effect of electrical coupling is to drive [Ce(IV)]/[Ce(III)] toward the same value in both reactors.

such that only small concentrations of elemental bromine develop (Field and Boyd, 1985). Furthermore the concentrations of Ce(III) and Ce(IV) are much higher ($\geq$ 1000 times) than the concentrations of other oscillatory intermediates, and the electron-transfer reactions of these ions at electrode surfaces are rapid. Thus it is the redox reactions of Ce(III) and Ce(IV) at the working electrodes in each reactor that carry the current and are the chemical source of coupling. That is, for one sign of the relative potential, Ce (III) is oxidized to Ce(IV) in one reactor while Ce(IV) is reduced to Ce(III) in the other, and vice-versa when the relative potential changes its sign. Thus, since [Ce(III)] + [Ce(IV)] is constant in each reactor, coupling affects only one dynamic variable ([Ce(IV)]/[Ce(III)]) and is easily modeled.

The coupling current is given by $E/\mathcal{R}$ where E is the relative potential between the two oscillators and $\mathcal{R}$ is the resistance of the coupling circuit. The relative potential in our system with low currents and large-area electrodes is related to the concentrations of Ce(III) and Ce(IV) in each reactor by the Nernst equation (Bard and Faulkner, 1980). The instantaneous rate of change of [Ce(IV)] and [Ce(III)] in each reactor (in mol/liter sec and assuming that V is the volume of both reactors) is given by Equation (1),

$$-\frac{d[Ce(IV)]_1}{dt} = \frac{d[Ce(IV)]_2}{dt} = \frac{d[Ce(III)]_1}{dt} = -\frac{d[Ce(III)]_2}{dt} = \frac{RT}{\mathcal{R}VF} \ln \left\{\frac{[Ce(IV)]_2 \; [Ce(III)]_1}{[Ce(IV)]_1 \; [Ce(III)]_2}\right\} \tag{1}$$

where R is the ideal gas constant, T is the temperature and F is the Faraday constant.

Details of the construction of the electrically coupled BZ oscillator experiment are given elsewhere (Crowley and Field, 1985). A schematic diagram of the apparatus is shown in Fig. (4). The reactors are very small ($V \sim 2.5$ ml) so that the relatively small coupling currents developed can significantly affect cerium ion concentrations in them and so that the resistance of the coupling circuit (which mainly results from ion-transport processes between the electrodes) is as small as possible. The ion bridge between the two CSTR's is through a relatively large port connecting them. Convective transport between the reactors is eliminated by a 25 μ millipore filter covering the entrance to the port in each reactor. There is a space between the two millipore filters that is continually flushed by a H_2SO_4 solution of the same concentration as in the reactors. Current flow between the reactors is carried by H_3O^+ ion, whose high concentration in the reactors is not significantly changed. Any reactant solution that leaks from the reactors through the millipore filter is washed away by the flowing H_2SO_4, thus assuring that the reactors are physically isolated from each other. We are unable to detect any differences between the behavior of two uncoupled oscillators when they are isolated by flowing H_2SO_4 and when the H_2SO_4 is replaced by a solid piece of Plexiglas.

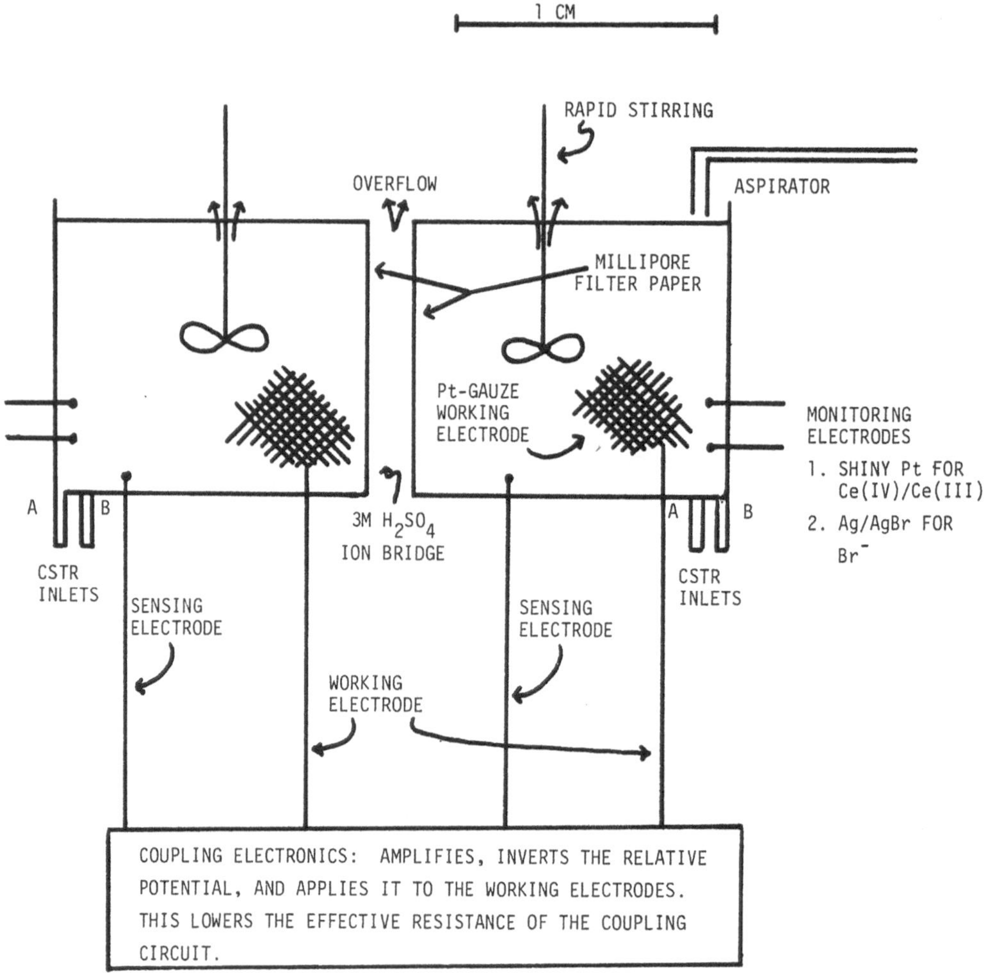

Figure 4. Schematic diagram of electrically coupled BZ oscillator apparatus. Note how small the reactors are.

Even with this arrangement the resistance of the coupling circuit remains high enough that it is difficult to obtain large and easily controlled coupling currents. Thus the virtual coupling resistance is lowered by amplifying the relative potential. Sensing electrodes are placed in each CSTR and their relative potential amplified, inverted and applied to the working electrodes. The amplification is done through an opto-coupler which guarantees that there is no direct electrical connection between the sensing electrodes and the working electrodes. The input and output sides of the coupling circuit (as well as each of the analog-digital converters used for data collection) have separate and isolated power supplies. We are able to find no artifacts introduced into the experiment by the coupling electronics, whose net effect is only a virtual reduction of the resistance of the coupling circuit. This set-up also allows considerable flexibility because the actual strength of the coupling can be adjusted by simply inserting an appropriate resistor (usually in the range 2-20 KΩ) into the coupling circuit. Coupling strengths are reported here as such a resistance. It should be remembered that the degree of coupling is inversely proportional to the resistance of the coupling circuit.

We point out that this same experimental apparatus can be used to perturb a single BZ oscillator with an applied potential of any desired amplitude and form. This experiment is somewhat simpler to interpret than the coupled oscillator experiment in which both oscillators are affected by the coupling.

III. Survey of Experimental Results

We are mainly interested here in experiments showing chaos, but a number of other types of coupled behavior are observed. We shall consider the experiments showing chaos in some detail and only give a survey of the other phenomena observed. A more comprehensive description of our experiments and results is given elsewhere (Crowley and Field, 1985).

The major features of the two oscillators when not coupled which determine their coupled behavior are: (1) their natural (uncoupled) frequency ratio, f_o, (2) their relative amplitudes, and (3) their mean redox potentials. All of these are affected strongly by the feed-stream [AA], and this is the principal variable used to adjust the characteristics of the individual oscillators. An increase in [AA] causes an increase in amplitude and a decrease in period of the oscillations. Each reactor is fed from its own reservoir so that the oscillations can be quite different in them even at the same flow-rate. The lower frequency oscillator is usually more strongly affected by coupling than the higher frequency one.

Extensive use is made of power spectra to characterize our experimental and numerical results. An IMSL Fast Fourier Transform (Jenkins, 1969) program was used to calculate power spectra from both experimental data and simulations. Examples of such power spectra are shown in Figs. (5c)-(5h). The quality of our experimental data is reflected by the fact that the fundamental peaks in power spectra of unperturbed oscillators (Fig. 5c) are higher than the baseline (experimental) noise by factors often greater than 10^4.

All experiments were started with a very large ($\sim$ 1 MΩ) coupling resistance, i.e. in an essentially uncoupled state. The CSTR's were filled and the flowrate adjusted until oscillations appeared in both reactors. The oscillations were then allowed to stabilize for about an hour before coupling was introduced. A sequence of behaviors is then observed as coupling is increased in steps. A typical sequence is the following. (1) Drifting in which the two oscillators remain close to their limit cycles but are slightly perturbed by weak coupling. The effect of such weak coupling can usually be seen only as an increase in noise level in the corresponding power spectra. (2) Quasiperiodic behavior in which the oscillations appear aperiodic but are still composed of a finite number of frequencies without a common divisor. The power spectrum of a quasiperiodic oscillator is noisy but still dominated by a number of sharp peaks. (3) Chaos which often can be seen in the waveform of the oscillations but is reflected in power spectra as a very high baseline noise level and essentially no major peaks or harmonics. (4) Entrainment. Depending on the characteristics of the oscillators and their relative phase when the coupling is turned on, entrainment can be: 1:1 in-phase (indicated by a flat relative potential curve), 1:1 out-of-phase (indicated by an oscillatory relative potential curve), 2:1, 3:2, etc. Entrained oscillators are characterized by power spectra as clean as those of uncoupled oscillators. (5) Annihilation of oscillations in one or both oscillators. (6) A bursting phenomena in which oscillations come and go. This phenomenon presumably results (Janz <u>et al</u>. 1980; Rinzel and Troy, 1982) from the accumulation of a reaction product that inhibits the oscillations (Heilweil and Epstein, 1979), and is not modeled here. A state diagram can be constructed in a coupling strength-f_o space (Crowley and Field, 1985) that shows which of these phenomena appear for particular values of these variables.

Data from a typical experiment involving the coupling of dissimilar oscillators (f_o = 2.52:1) are shown in Fig. (5). In Figs. (5a) and (5b) the lower two curves are the measured redox potentials of each oscillator and the top curve is their relative potential. The absolute values of these curves are shifted for ease of display, but the relative magnitudes are correct. Figures (5c)-(5h) are power spectra calculated for both oscillators from data at the various coupling resistances shown in Figs. (5a) and (5b). The absolute magnitudes of the power spectra are also shifted for ease of display. The lower spectra correspond to the bottom trace in Figs. (5a) and (5b).

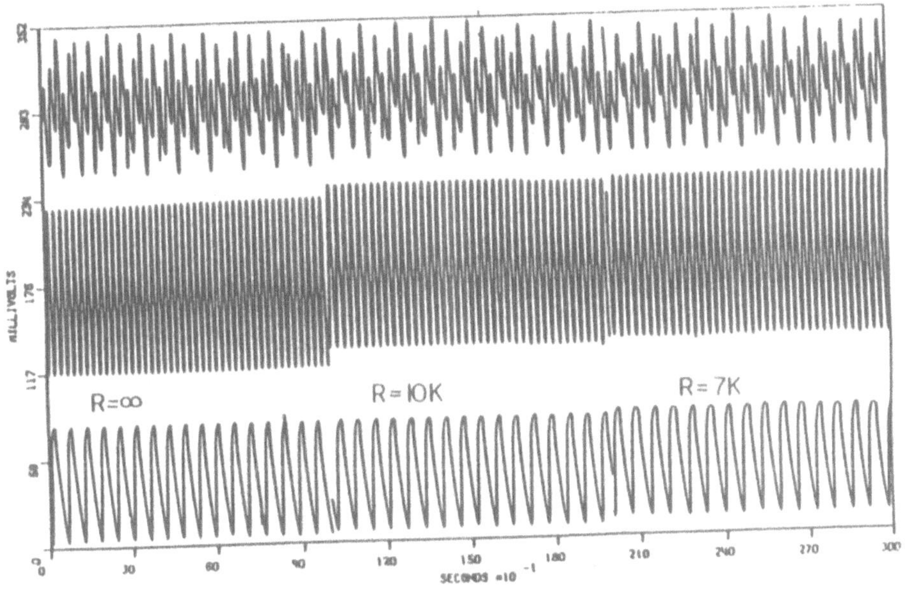
MILLIVOLTS
SECONDS ×10^-1
R=∞
R=10K
R=7K

(5a)

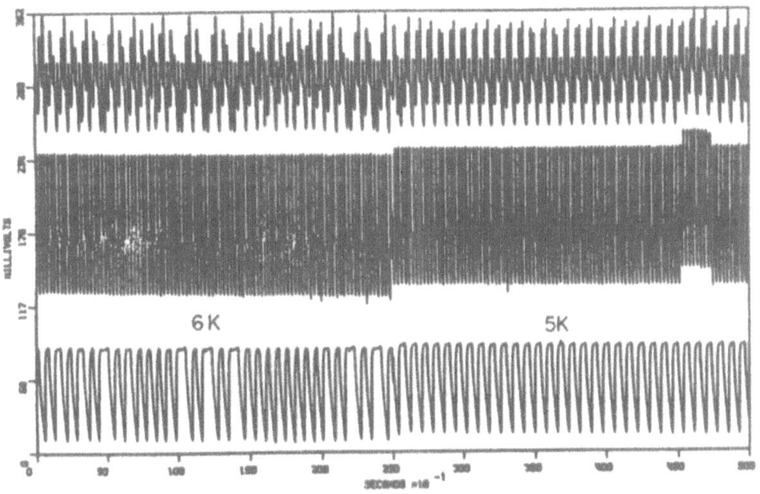
MILLIVOLTS
SECONDS ×10^-1
6K
5K

(5b)

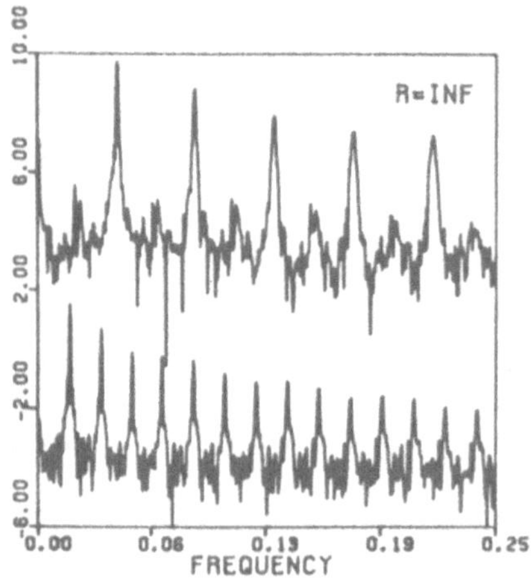
R=INF
10.00
6.00
2.00
-2.00
-6.00
0.00
0.06
0.13
0.19
0.25
FREQUENCY

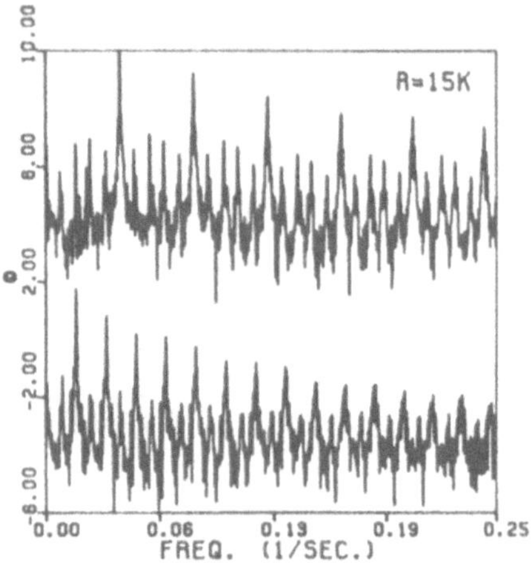
R=15K
10.00
6.00
2.00
-2.00
-6.00
0.00
0.06
0.13
0.19
0.25
FREQ. (1/SEC.)

(c) (d)

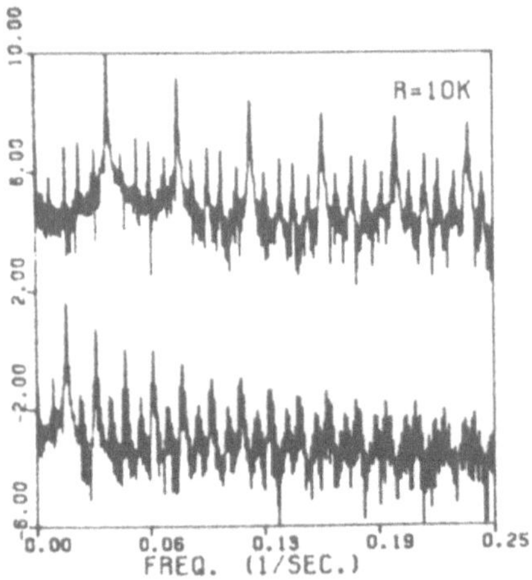
R=10K
10.00
6.00
2.00
-2.00
-6.00
0.00
0.06
0.13
0.19
0.25
FREQ. (1/SEC.)

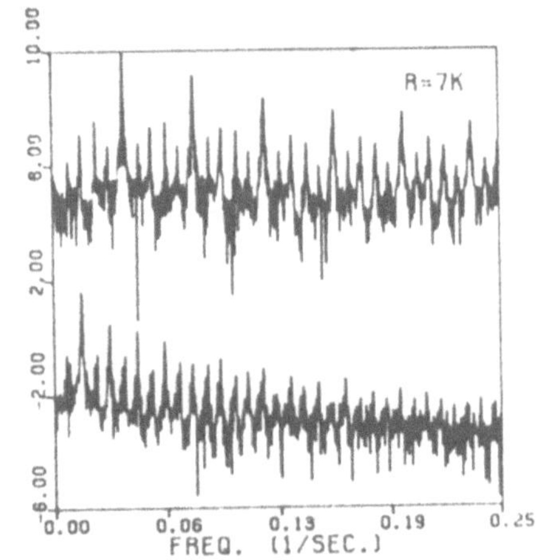
R=7K
10.00
6.00
2.00
-2.00
-6.00
0.00
0.06
0.13
0.19
0.25
FREQ. (1/SEC.)

(e) (f)

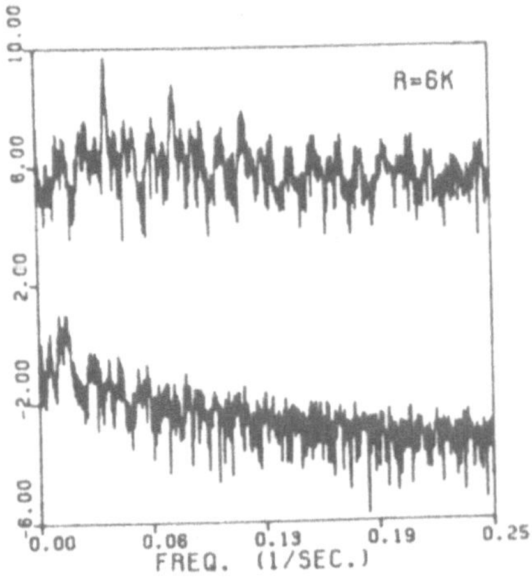
R=6K
10.00
6.00
2.00
-2.00
-6.00
0.00
0.06
0.13
0.19
0.25
FREQ. (1/SEC.)

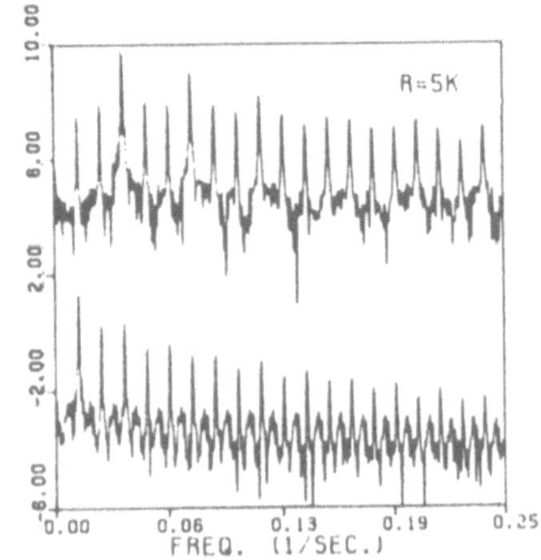
R=5K
10.00
6.00
2.00
-2.00
-6.00
0.00
0.06
0.13
0.19
0.25
FREQ. (1/SEC.)

(g) (h)

Figure 5. Electrical coupling of dissimilar oscillators with f_o = 2.52 through the series of coupling resistance (5a) ∞, 15 KΩ, 10 KΩ, and 7 KΩ, (5b) 6 KΩ and 5 KΩ. The lower trace is the measured redox potential of the lower oscillator and the center trace is the measured redox potential of the faster oscillator. The upper trace is the relative potential of the two oscillators. The corresponding power spectra are shown in (5c)-(5h). Feedstream concentrations were: Reactor 1 (slower oscillator), Solution A, [AA] = 0.030 M, $[Ce_2(SO_4)_3]$ = 0.00084 M and $[H^+]$ = 2.73 M. Solution B, $[KBrO_3]$ = 0.072 M and $[H^+]$ = 2.73 M; Reactor 2 (faster oscillator), Solution A, [AA] = 0.070 M, $[Ce_2(SO_4)_3]$ = 0.00084 M and $[H^+]$ = 2.73 M; Solution B, $[KBrO_3]$ = 0.084 M and $[H^+]$ = 2.73 M. Residence times were 10 and 11 minutes, respectively.

The experiment is started with $\mathcal{R}$=∞. The power spectra of both oscillators (Fig. 5c) show fundamental peaks and harmonics at least 10^4 times higher than the baseline noise. At $\mathcal{R}$=15 KΩ no effect of coupling is apparent in the oscillatory waveforms (not shown) as the oscillators drift, but the power spectrum (Fig. 5d) of the slower oscillator shows peaks at the fundamental and harmonic frequencies of the faster oscillator. At $\mathcal{R}$=10 KΩ, and more so at 7 KΩ, the effect of coupling can be seen in the waveform of the slower oscillator. The power spectra (Figs. 5e and 5f) show quasiperiodicity at these coupling strengths.

Chaos appears at $\mathcal{R}$=6KΩ. The power spectrum of the slower oscillator (bottom, Fig 5g) is essentially flat, there are no fundamental peaks and harmonics, and the amplitude of the baseline noise has increased by several orders of magnitude compared to nonchaotic states. The empirical source of the chaos is apparent in the waveform of the slower oscillator (Fig. 5b, bottom trace). The major effect is at its peaks, which are lengthened by apparently random amounts by the coupling. The exact dynamic source of this lengthening can be seen in the Field-Noyes (1974) model of the BZ reaction and will be discussed in a later section. Two other features of this chaos are worth mentioning. Chaos always follows a quasiperiodic region, and chaos appears only in the waveform of the slower oscillator. The faster oscillator retains its periodicity (Fig. 5g).

At $\mathcal{R}$=5 KΩ the two oscillators entrain 3:1. The noise level in the power spectra (Fig. 5h) drops dramatically, and essentially identical fundamental peaks and harmonics describe both oscillators. At even higher coupling strengths (not shown in Fig. 5), oscillations are annihilated in both reactors with one in a high [Ce(IV)] state and the other in a low [Ce(IV)] state. This general trend from drifting through quasiperiodicity to chaos and finally to entrainment and/or annihilation of oscillations appeared in all of our experiments and is reproduced by the Field-Noyes (1974) model.

IV. Simulation in Terms of the Field-Noyes Equations

Electrically coupled oscillator experiments are useful mainly to the extent that they can be understood in terms of the dynamics of the BZ reaction. We find that the Field-Noyes (1974) model (usually referred to as the Oregonator) of the BZ chemistry (Field *et al.*, 1972; Field, 1985) does an excellent job of interpreting our results.

The principal oscillatory variables in the BZ reaction are [Br^-], [$HBrO_2$], and [Ce(IV)]/[Ce(III)]. Assigning X≡$HBrO_2$, Y≡Br^-, Z≡2Ce(IV), A≡BrO_3^-, and P≡HOBr, the Field-Noyes model is written as reactions (M1-M5).

$$\left.\begin{array}{ll} A + Y \longrightarrow X + P & \quad (M1) \\ X + Y \longrightarrow 2P & \quad (M2) \end{array}\right\} \text{Process A}$$

$$\left.\begin{array}{ll} A + X \longrightarrow 2X + Z & \quad (M3) \\ X + X \longrightarrow A + P & \quad (M4) \end{array}\right\} \text{Process B}$$

$$Z \longrightarrow fY \qquad \text{Process C} \qquad (M5)$$

The f in reaction (M5) is an expendable parameter which corresponds to the number of bromide ions (Y) produced per 2Ce(IV) (Z) consumed. A 5-variable version of the Field-Noyes model that does not contain an expendable parameter is now available (Field and Boyd, 1985).

The model is a classic relaxation oscillator. Bistability is a fundamental feature of it (Boissonade and DeKepper, 1980; DeKepper and Boissonade, 1985) with Process A (reactions M1 and M2) representing one state and Process B (reactions M3 and M4) the other. Transitions between Processes A and B are very rapid. Bromide ion concentration is the control intermediate such that Process A is dominant at high [Br^-] and Process B is dominant at low [Br^-]. The net effect of Process A is the removal of Br^-, leading to the onset of Process B, which is autocatalytic in X because of reaction (M3). Process C is a delayed negative feedback that regenerates Br^- from the products of Process B, leading to the return to dominance of Process A.

There has been some controversy recently concerning the validity of the Field-Noyes model. Several modifications of the BZ reaction (Noszticzius, 1979; Noszticzius and Bódiss, 1979) were not easily fit into the Field-Noyes framework, leading to the proposal of an alternate model (called the Explodator) with a dynamic structure entirely different from the Field-Noyes model and based on Br_2 rather than Br^- control (Noszticzius et al., 1984) of the oscillations. The Explodator was initially criticized by Noyes (1984). Bromine control was disproved experimentally by Nosticzius et al. (1985). All of the difficult experiments have now been understood in terms of the Field-Noyes model (Field and Boyd, 1985; Varga et al. 1985; Ruoff and Schwitters, 1984). Finally, it has been shown (McKinnon and Field, 1985) that the Explodator cannot reproduce the bistability that is a dominant feature of the BZ reaction. There seems to be little doubt remaining that the Field-Noyes model incorporates the fundamental dynamic structure of the BZ reaction (Tyson, 1985; Troy, 1985) and is able to rationalize essentially all of its experimental behavior. Indeed, the major effect of the questions raised by the Explodator has been to emphasize the validity and versatility of the Field-Noyes model.

It is often assumed in the Field-Noyes model that only a relatively small fraction of the Ce(III) is oxidized to Ce(IV) in each cycle. Thus [Ce(IV)] is assumed to be the important dynamic variable and [Ce(III)] is assumed to be constant. However, using this assumption, we are unable to reproduce some important features of our experiments. In particular, the major effect of coupling appears at the troughs of the slower oscillator rather than at the peaks as observed, and chaos does not appear. This result is not surprising. Relatively small concentrations of cerium ion catalyst are used in our experiments in order to increase the effect of coupling. Furthermore, the major effect of electrical coupling is on [Ce(IV)]/[Ce(III)]. Thus, following the lead of Tyson (1982), we use here a version of the Field-Noyes model where metal-ion conservation is enforced by requiring [Ce(III)] + [Ce(IV)] = C, where C is the total cerium-ion concentration.

The differential equations corresponding to the Field-Noyes model for one reactor assuming the conservation of metal ion and that the effect of coupling is given by Eq. (1) are:

$$\frac{dX}{dt} = k_{M1}AY - k_{M2}XY + k_{M3}AX - 2k_{M4}X^2 \tag{2i}$$

$$\frac{dY}{dt} = -k_{M1}AY - k_{M2}XY + fk_{M5}Z \tag{2ii}$$

$$\frac{dZ}{dt} = k_{M3}AX - k_{M5}Z - \frac{RT}{\mathcal{R}VF} \ln \frac{Z'(C-Z)}{Z(C'-Z')} \tag{2iii}$$

The primes indicate concentrations in the other reactor. The CSTR pumping rates used in our experiments were always very slow, and we assume that the dynamic effect of pumping is only to maintain the concentrations of reactants at constant values. Spot checks indicate that this assumption is valid. Thus there are no flow terms in Eqs. (2).

Equations (2) can be scaled according to Tyson (1982) to yield Eqs. (3) that describe both reactors.

Reactor 1

$$p_x\dot{x} = x(1-2z-x) - y(x-q) \tag{3i}$$

$$p_y\dot{y} = bgz - y(x+q) \tag{3ii}$$

$$\dot{z} = 2x(1-2z)-bz + \rho\,(T_o/C)\ \ln\,((1-z)z'/z(1-z')) \tag{3iii}$$

Reactor 2

$$(T_o'/T_o)p_{x'}\dot{x}' = x'(1-2z'-x') - y'(x'-q') \tag{3iv}$$

$$(T_o'/T_o)p_{y'}\dot{y}' = b'g'z' - y'(x'+q') \tag{3v}$$

$$(T_o'/T_o)\dot{z}' = 2x'(1-2z')-b'z'-\rho(T_o'/C)\ln((1-z)z'/z(1-z')) \tag{3vi}$$

In Eqs. (3) the unprimed variables refer to one reactor and the primed variables refer to the other, $(\dot{\ })$ = d/dT, and the scaled variables are related to the real variables by: $\tau = t/T_o$, $x = X/X_o$, $y = Y/Y_o$, $z = Z/Z_o$, and $\rho = RT/\mathcal{R}VF^2$. The quantity (T_o'/T_o) appears because $[BrO_3^-]\equiv A$ enters into the time scaling, and this term assures that the scaled time is the same in both reactors when $[BrO_3^-]$ is not.

The relationship of the scaling factors in Eqs (3) to the Field-Noyes model using the numbering of Tyson (1982) is given in Table (1). The empirical values of the scaling factors given by Tyson (1982) are used here and are also given in Table (1). While there has been some controversy (Bar-Eli and Ronkin, 1985; Tyson 1985; Field and Boyd, 1985) concerning the relationship of these parameters to the detailed chemistry of the Field, Körös and Noyes (1972) mechanism of the BZ reaction, they do give excellent quantitative agreement with experiment (Tyson, 1982, 1985).

Electrical coupling experiments can be simulated by numerical integration of Eqs. (3). The problem is stiff and the Gear (1971) algorithm must be used for its integration. Simulations of uncoupled oscillators give good quantitative agreement with experiment, and reproduce the trends observed when the CSTR flowrate (Fig. 1) and the concentrations of major reactants are varied. The results of a typical simulation of an electrical coupling experiment for a series of increasing values of ρ are shown in Fig. (6). In Fig (6a) ρ=o and the uncoupled oscillators show sharply peaked power spectra. Figure (6b) at $\rho=2x10^{-8}$ shows drifting while Fig. (6c) $\rho=2.4x10^{-8}$ is quasiperiodic. Figure (6d) at $\rho=2.5x10^{-8}$ is chaotic while Fig. (6c) shows 7:5 entrainment. The system then passes through (Fig. 6f, $\rho=2.7x10^{-8}$) more chaos to 3:2 (Fig. 6g), and finally (Fig. 6h) 4:3 entrainment. Oscillations are annihilated in both reactors at even higher values of ρ with one reactor in a high Z state and the other in a low Z state.

Table 1. Scaling Factors and Parameter Values According to Tyson (1982).

Parameter	Definition	Approximate Value
T_o	$2k_4C/(k_{-5}HA)^2$	800 sec
X_o	$k_{5a}HA/2k_4$	$1x10^{-7}$ M
Y_o	$k_{5a}A/k_2$	$5x10^{-9}$ M
Z_o	C	total [Cerium]
p_x	X_o/C	$1x10^{-4}$ M
p_y	Y_o/C	5.10^{-6} M
q	$2k_3k_4/k_2k_{5a}$	$4x10^{-4}$ M
g	$k_{10}/(6k_9 + 4k_{10})$	0.10 M
b	$(6k_9 + 4k_{10})2k_4C/(k_{5a}HA)^2$	10.0 M

These values are calculated for $H\equiv[H^+]$ = 1 M, $A\equiv[BrO_3^-] = 5x10^{-2}$ M, $[CH_3COCH_2COCH_3] = [CH_3COCHBrCOCH_3] = 10^{-2}$ M. The latter two concentrations enter through k_9 and k_{10}. See Tyson (1982).

The important features to note in the simulations shown in Fig (6) are the following. The experimentally observed sequence of behaviors (drifting, quasiperiodicity, chaos, entrainment, annihilation) is

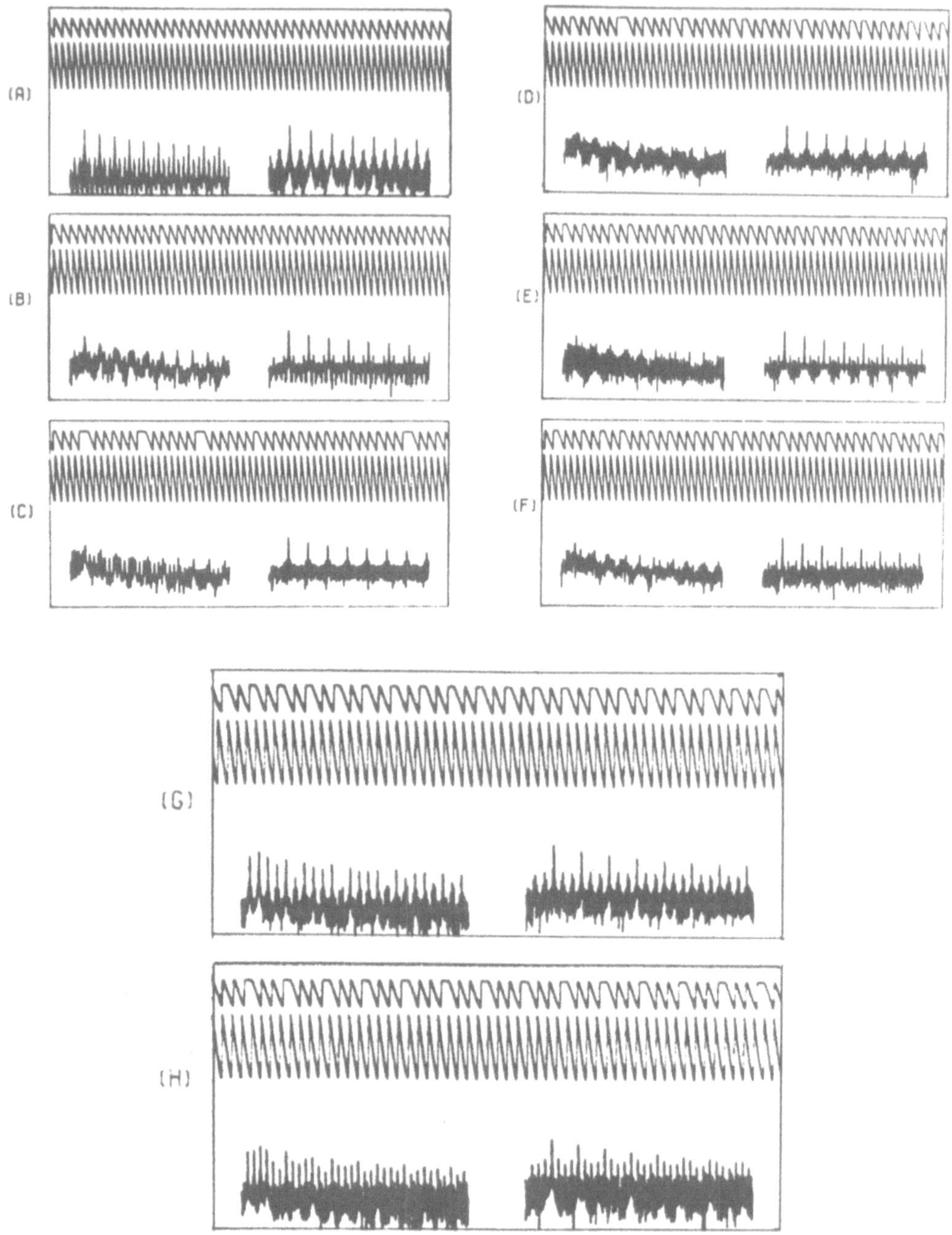

Figure 6. Results of numerical simulations of electrically coupled BZ oscillators using Eqs. (4). For the upper (faster) oscillator ($6\ k_9 + 4\ k_{10}$) = 0.03, A = 0.03, C = 0.00005, g = 0.28 and $[H^+]$ = 1.00 M. For the lower (slower) oscillator ($6\ k_9 + 4\ k_{10}$) = 0.10, A = 0.30, C = 0.0001, g = 0.30, and $[H^+]$ = 1.00 M. The other parameters are as in Table 1.

reproduced over a quite reasonable range of ρ values, although it is difficult to relate absolute values of ρ to the actual coupling resistance, $\mathcal{R}$, because of the amplification circuit. The calculated (f_o vs ρ) state diagram is very similar to the corresponding experimental diagram (Crowley and Field, 1985). Comparison of the experimentally observed chaos (Fig. 5g) with the simulated chaos (Figs. 6d and 6f) reveals considerable similarities. In both cases the effect of the coupling is mainly at the peaks of the slower oscillator, and the aperiodicity results from random-length delays at these peaks. The Field-Noyes model does an excellent job of reproducing all major dynamic features of our experiments, and with the Tyson (1982) parameterization quantitative agreement is also quite good.

V. Phase-Plane Analysis

Equations (3) are susceptible to a phase-plane analysis that makes quite clear the dynamic source of the experimentally observed chaos. The parameters p_x and p_y in Eqs (2) are both very small. In Table (1) $p_x = 1x10^{-4}$ and $p_y = 5x10^{-6}$. These numbers suggest that the dimensionality of Eqs. (2) can be reduced from 3 to 2 by setting either p_x or $p_y = 0$. Field and Noyes (1974) originally suggested reduction by setting $p_x = 0$. This assumes that x follows y, in keeping with the Br^--control view of the chemistry. However, doing this leads to a set of equations that are difficult to analyze, i.e.

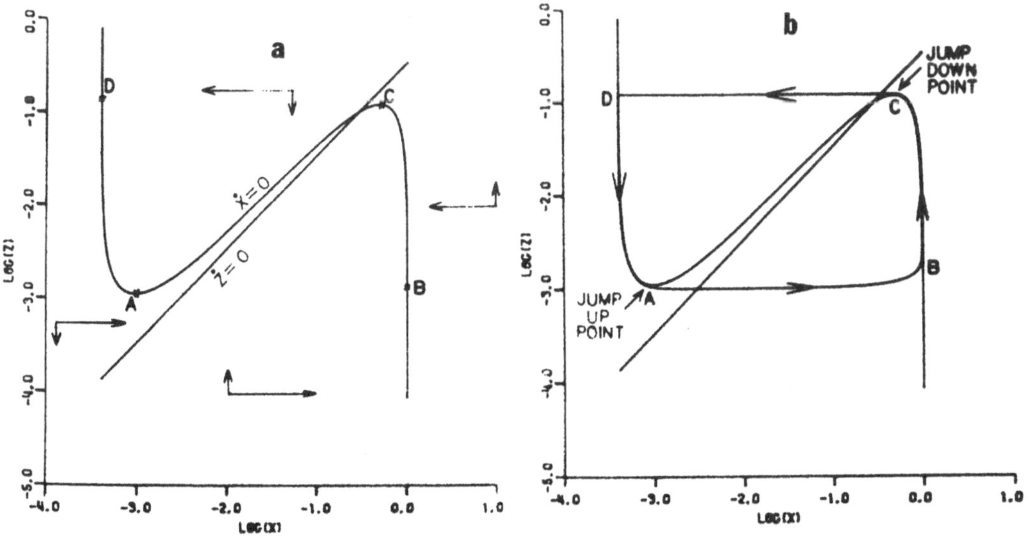

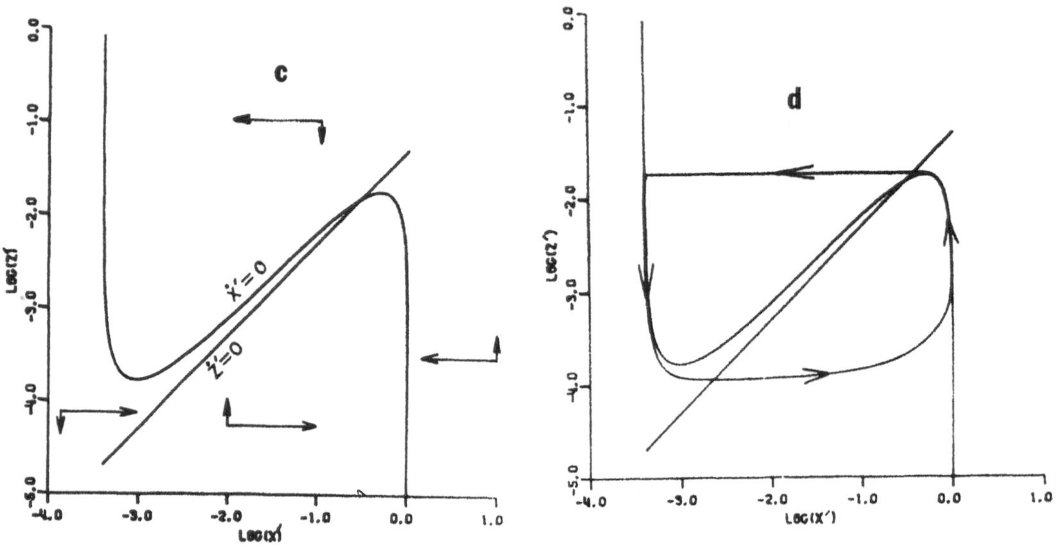

Figure 7. Nullclines in the z-x plane based on equations (4i)-(4iv) but with the effect of coupling ignored. Parts (b) and (d) show the calculated limit cycle superimposed on the nullclines.

Eqs (2i) yields a quadratic in x. Tyson and Fife (1980) and Tyson (1982) noted that p_y is even smaller than p_x and so set $p_y = 0$. This assumes that y follows x, which is certainly true during the transition from Process A to Process B.

An analytical solution to this reduced set of equations has been obtained by Crowley and Field (1984), who also demonstrated that numerical integrations of both the 2 and 3-dimensional systems yield essentially the same limit cycle. Equations (4) for the electrically coupled system result from this approximation.

$$p_x\dot{x} = x(1-2z-x) - bgz((x-q)/(x+q)) \qquad (4i)$$

$$\dot{z} = 2x(1-2z)-bz-(T_o\rho/C)\ln(z(1-z')/z'(1-z)) \qquad (4ii)$$

$$(T_o/T_o')p_x'\dot{x}' = x'(1-2z'-x')-b'g'z'((x-q)/(x+q)) \qquad (4iii)$$

$$(T_o/T_o')\dot{z}' = 2x'(1-2z')-b'z' + (T_o'\rho/C)\ln(z(1-z')/z'(1-z)) \qquad (4iv)$$

A phase-plane analysis of Eqs. (4) can be carried out by drawing the $\dot{x}=0$ and $\dot{z}=0$ nullclines in one plane and the $\dot{x}'=0$ and $\dot{z}'=0$ nullclines in another plane. These nullclines are shown in Figs. (7a) and (7c) for two dissimilar oscillators <u>ignoring the coupling terms</u>. The calculated limit cycles are superimposed on the nullclines for each oscillator in Figs. (7b) and (7d). When the effect of electrical coupling is introduced the $\dot{x}=0$ and $\dot{x}'=0$ nullclines are not affected because they do not depend upon the value of z(z') in the other oscillator, i.e. they do not depend upon the phase of the other oscillator. The $\dot{z}=0$ and $\dot{z}'=0$ nullclines do depend on z(z') in the other oscillator, however, and do change their form with its phase. Figure (8) shows the effect of coupling on the nullclines of the oscillators in Fig. (7) as well as the form of their oscillations when they are entrained 2:1. The $\dot{z}=0$ and $\dot{z}'=0$ nullclines are shown for both the maximum and minimum values of z(z') in the other oscillator.

The major effects of electrical coupling in both our experiments and simulations are at the peaks of the slower oscillator where both z and y are high. We call this the jump-down point. It is seen in Fig. (8) that there is a crossing of the $\dot{x}=0$ and $\dot{z}=0$ nullclines in this region. Thus we focus our attention on the slower oscillator at the jump-down point.

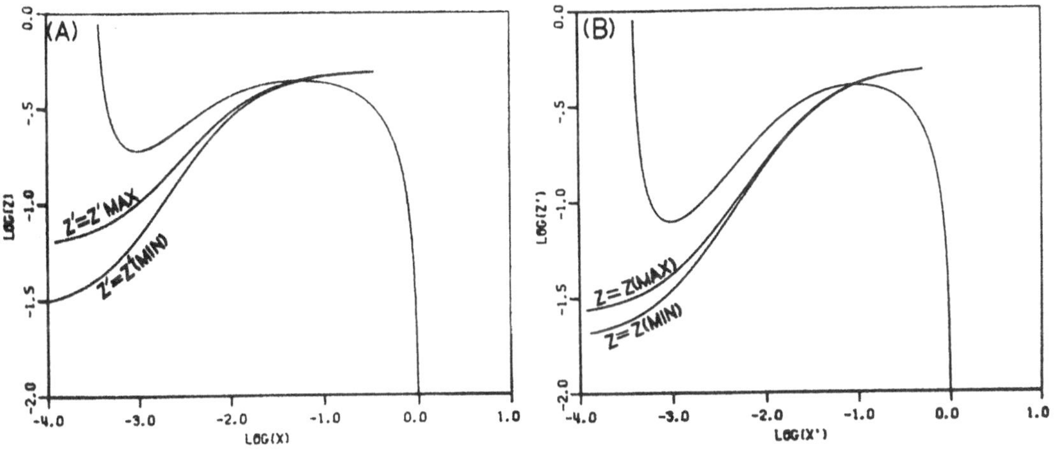

Figure 8. Nullclines in the z-x plane based on Eqs. (4i)-(4vi) including the effect of coupling. The $\dot{z}=0$ nullclines are given (a and b) for both the maximum and minimum values of Z in the other reactor.

Comparing Figs. (7a) and (8a) it can be seen that the principal effect of electrical coupling on the $\dot{z}=0$ nullcline is to move it toward higher z. This effect is particularly noticeable at low x, which corresponds to the troughs of the oscillations. Indeed, when the conservation of cerium ion is not enforced, the $\dot{z}=0$ nullcline actually crosses the $\dot{x}=0$ nullcline twice (one crossing is stable and the other is unstable) in this region, and contrary to experiment the principal effects of coupling in simulations based on this model appear at the troughs of the oscillations. There is no crossing in this region in Fig. (8a), though.

Looking at the jump-down point in Fig. (8a), two things are apparent. The first is that the crossing of the nullclines is very close to a maximum of the $\dot{x}=0$ nullcline. The steady state at this crossing will be stable if it is on the high x side of the $\dot{x}=0$ nullcline maximum and unstable if it is on the low x side of the crossing. The second is that the $\dot{x}=0$ nullcline is relatively flat near its maximum. This means that we expect trajectories to slow down in this region, yielding quite rounded peaks. The effect appears numerically and is an important feature of our analysis.

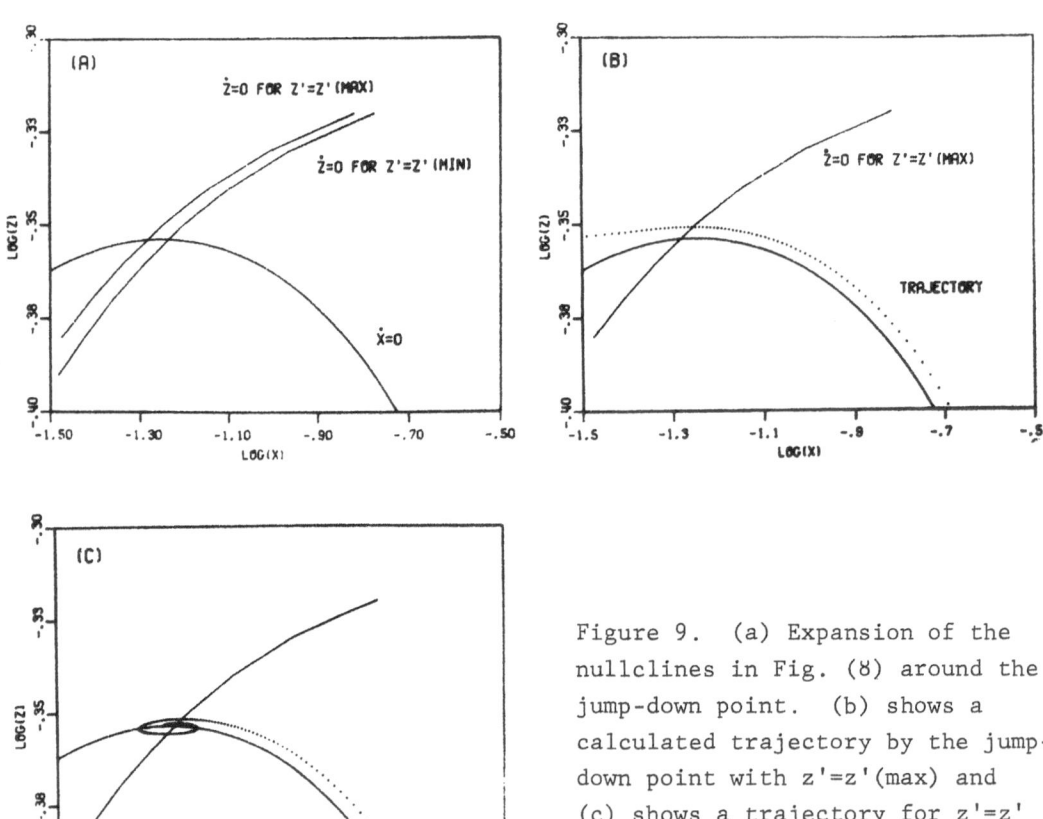

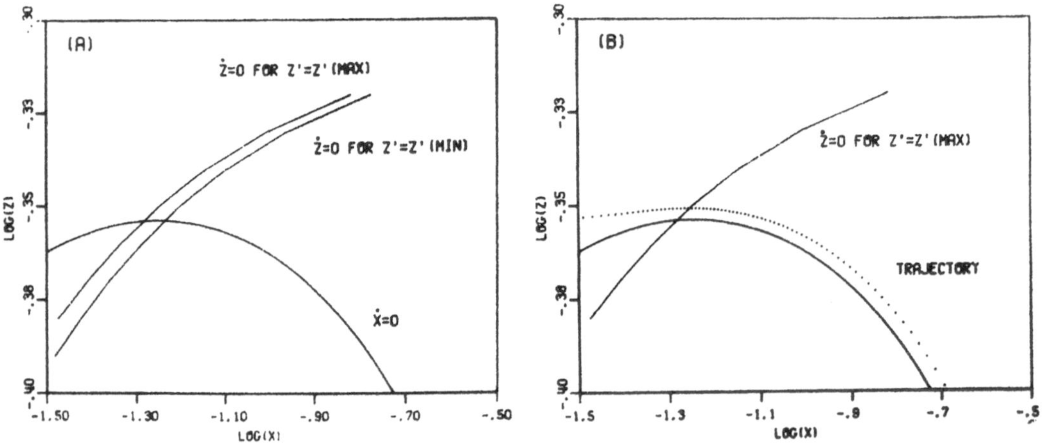

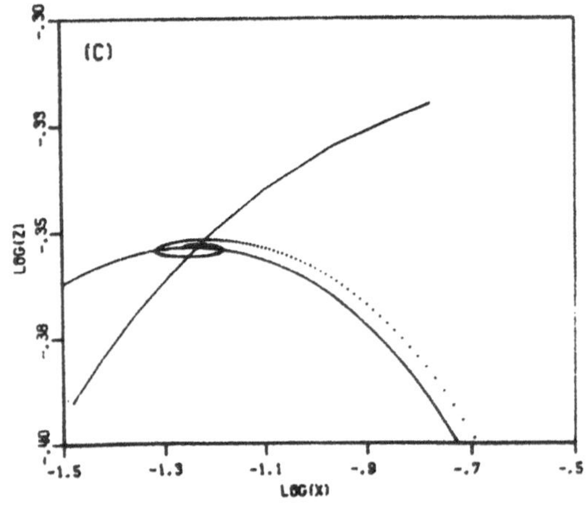

Figure 9. (a) Expansion of the nullclines in Fig. (8) around the jump-down point. (b) shows a calculated trajectory by the jump-down point with z'=z'(max) and (c) shows a trajectory for z'=z'(min).

Figure (9) shows a more detailed view of the region around the $\dot{x}=0$ nullcline maximum. It turns out that the crossing of the two nullclines is on the unstable side of the $\dot{x}=0$ nullcline for $z'=z'(max)$ and on the stable side for $z'=z'(min)$. Figure (9b) shows a calculated trajectory for the slower oscillator with its partner at $z'=z'(max)$. The points in this trajectory are separated by 0.01 s, and it is seen that the trajectory slows down dramatically near the crossing before heading off toward the left branch of the $\dot{x}=0$ nullcline. This slowing down results from the flatness of the $\dot{x}=0$ nullcline. The trajectory with $z'=z'(max)$ is essentially identical to that of the oscillator in an uncoupled state.

Figure (9c) shows a calculated trajectory of the slower oscillator when $z'=z'(min)$. The crossing of the nullclines in this case is on the high x side of the jump-down point and is stable. Thus the trajectory slows down enormously (the points become very close together) and spirals into the steady state, where it stays until z' in the other oscillator increases to the point where the steady state becomes unstable. The trajectory is then allowed to continue on its cycle. It is not apparent in a particular case how long it will take for a trajectory to get out of this spiral (which can be seen in our simulations and experiments as ripples on the peaks of the slower oscillator) and exactly what route it will take when it is released. Thus we see a very complex and interesting behavior in a small region of phase space that has considerable significance to the existence of a strange attractor and the presence of chaos in our experiments and simulations.

VI. Forced Oscillators And A Smale Horseshoe Analogy

A feature of the chaos observed in both our experiments and simulations is that when the slower oscillator is chaotic the faster oscillator is relatively unperturbed. This can be seen in the power spectra displayed in Fig. (5g) (experimental) and Fig. (6d) (simulation). In general, in the case of significantly different oscillators, we note that the faster oscillator is not significantly affected until coupling becomes strong enough for entrainment to occur. Thus it is appropriate to look to the literature of forced oscillators to find an analogy to our results.

Levi (1980) proved the existence of a strange attractor leading to chaos in a forced Van der Pol oscillator. This system is a two-variable relaxation oscillator (Minorsky, 1962) with nullclines that are of the same basic shape as the reduced Field-Noyes model (Eqs. 4i and 4ii), i.e. one s-shaped and the other more-or-less linear and crossing the s-shaped nullcline near a maximum. In Levi's work the forcing affected only the linear nullcline and not the s-shaped nullcline, just as in the electrically coupled BZ oscillator case. The proof of the existence of a strange attractor used by Levi and several other authors in similar systems (Rössler, 1979; Holmes, 1980a and 1980b) is the existence of a so-called "Smale horseshoe" map (Smale, 1965; 1967). A good description of the Smale horseshoe, its dynamics, and its application to systems containing a homoclinic connection is given by Guckenheimer (1979).

A horseshoe map is one that maps a region back into itself in a certain way. In the present case the mapping function, f, is Eqs (4i)-(4ii). Smale proved that the existence of a horseshoe map implies the existence of an infinite number of periodic (of arbitrary period) and aperiodic trajectories. He also proved the existence of so-called dense trajectories which come arbitrarily close to each other in the set of all possible trajectories in its motion through iterates of f.

Levi used this property of horseshoe maps to prove the existence of chaos and a strange attractor in the forced Van der Pol oscillator. He showed that if a series of trajectories that are consecutive with respect to the unforced limit cycle folds back on itself (i.e., some trajectories that were behind others end up ahead of them) sometime during one cycle of the perturbed limit cycle, then a horseshoe map can be constructed.

The map here will be defined in the same way as in Levi's work. It is a point going around the complete cycle of the slower oscillator once. We will choose a line of points near the unperturbed limit cycle and determine how that line deforms as it passes around the perturbed limit cycle to its original position. Of course in the unperturbed limit cycle the line will not be deformed, and there will be no horseshoe.

The result of sending a line of ten points through the perturbed slower oscillator limit cycle is shown in Fig. (10). The mapping is done by starting independent numerical integrations at each of ten consecutive points very close to the limit cycle. The horseshoe comes from passage of the line through the region of the jump-down point, where the spiral occurs in Fig. (9c). We begin the calculation with the line of trajectories near and approaching the jump-down point with the forcing oscillator heading toward its trough (Fig. 10a), at which point the jump-down point is stable. It is expected that if the phase of the faster oscillator is chosen properly, some points will get caught the spiral at the jump-down point while other points will encounter the jump-down point after the forcing oscillator has passed through its minimum and will not get caught.

The initial set of points making up the line is shown in Fig. (10a). In Figs. (10b) and (10c), z' has gotten very small, and the $\dot{z}=0$ nullcline has moved so that points 1 and 2 are forced below the $\dot{x}=0$ nullcline and must reverse their direction. We see the beginning of the horseshoe in Fig. (10c). Remember that points move rather slowly near the jump-down point, allowing its stability properties to strongly affect trajectories. By the time point 5 reaches the jump-down point, z' has passed through its minimum and has risen past the point at which the jump-down point becomes unstable. Thus point 5 passes the jump-down point without interference, and gets ahead of points 1-4, as shown in Figs. (10d) and (10e). If this order persists to the left branch of the $\dot{x}=0$ nullcline (Fig. 10f), then it will remain all the way back to the starting point. The dashed lines in Fig. (10f) shows the expected positions of points before and after the sequence just considered. Fig. (10g) shows the values of z on the left branch of $\dot{x}=0$ nullcline at a time t after the beginning of the calculation. In the absence of the horseshoe point, point 10 should have the highest z, and the values of z should decline monotonically through points 9-1. The presence of the horseshoe causes this order to be broken with point 6 rather than point 1 having the lowest z. We regard this as strong empirical evidence that, just as does the forced Van der Pol oscillator (Levi, 1981), the Field-Noyes model of electrically coupled BZ oscillators contains a horseshoe. This argues strongly that the chaos observed in our experiments and simulations is genuine and not a result of random experimental errors. However, a good deal of mathematical analysis clearly must be carried out before our argument can be regarded as proved.

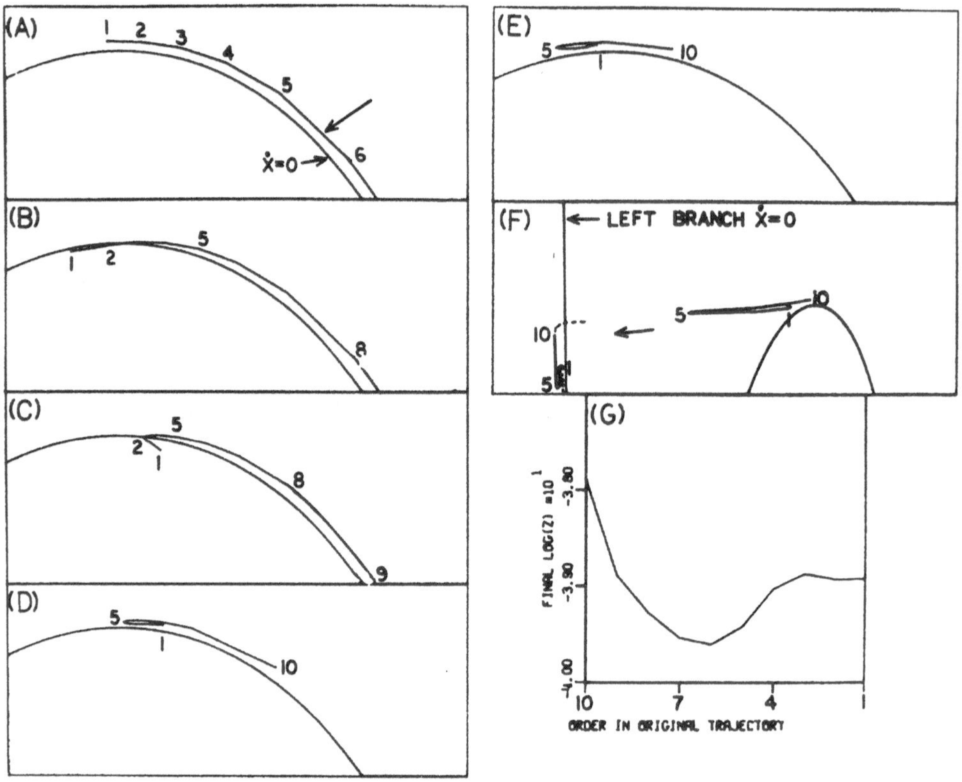

Figure 10. The behavior of a series of trajectories in the vicinity of the jump-down point.

VII. Conclusion

We have demonstrated that electrically coupled Belousov-Zhabotinskii oscillator experiments are feasible and that in fact excellent experimental results can be obtained. A number of interesting coupled behaviors are observed including: drifting, quasiperiodicity, chaos, synchronization, entrainment at various frequency ratios, and annihilation of oscillations in one or both oscillators. We have been particularly interested here in the chaotic results.

It further has been demonstrated that the Field-Noyes model of the Belousov-Zhabotinskii chemistry does an excellent job of interpreting our experimental results. This model also can be reduced to a two-variable system and subjected to a phase-plane analysis that shows the source of the aperiodic behavior in the dynamics of the BZ reaction and the coupling.

We have noted that the nullclines of the reduced Field-Noyes model bear a striking resemblance to the nullclines of the Van der Pol oscillator. By analogy to work of Levi on the forced Van der Pol oscillator, we have presented numerical evidence that the Field-Noyes model contains a "Smale horseshoe" that is the source of the observed chaos. We believe that this is the first case in which experimentally observed chaos can be directly linked to the presence of a strange attractor in the dynamics of a realistic model of the chemical system under investigation. While there are elegant and important experiments done with uncoupled BZ oscillators in CSTR mode that exhibit aperiodic behavior reflecting in detail that expected from the presence of a strange attractor (Roux, 1983; Roux et al. 1983; Swinney et al., 1983), the connection to the dynamic structure of the chemistry involved in this case has not yet been discovered (Ganapathisubramanian and Noyes, 1981), even in a fairly detailed version of the Field-Noyes equations.

We note further that these preliminary experiments show no evidence that the transition to chaos results from either a Feigenbaum cascade or a tangent bifurcation (Eckmann, 1981). It instead is a transition from quasiperiodicity to chaos. The chaos observed here likely results from a torus bifurcation such as that described numerically by Marek and Schrieber (1982) in a set of diffusively coupled Brussulators. Such a transition to chaos is not as well understood as the Feigenbaum cascade or tangent bifurcations, although Guckenheimer (1981) has suggested that it may occur when a tangent bifurcation appears near a steady state.

REFERENCES

1. Bar-Eli, K. and Haddad, S., 1979, The Belousov-Zhabotinskii Reaction. Comparison of Experiments and Calculations, J. Phys. Chem., 83, 2944.

2. Bar-Eli, K., 1984a, Coupling of Chemical Oscillators, J. Phys. Chem., 88, 3616.

3. Bar-Eli, K., 1984b, The Dynamics of Coupled Oscillators, J. Phys. Chem., 88, 6174.

4. Bar-Eli, K. and Ronkin, J., 1984, Oscillations and Steady States in the Bromate-Bromide-Cerous System: Comparison of Experimental and Calculated Data of Different Sets of Rate Constants, J. Phys. Chem., 88, 2844.

5. Bar-Eli, K., 1985, The Peristaltic Effect on Chemical Oscillations, J. Phys. Chem., 89, in press.

6. Bar-Eli, K. and Reuveni, S., 1985, Stable Stationary States of Coupled Chemical Oscillators. Experimental Evidence, J. Phys. Chem., 89, 1329.

7. Bard, A.J. and Faulkner, J.R., 1980, Electrochemical Methods, Wiley, New York.

8. Becker, P.K. and Field, R.J., 1985, Stationary Concentration Patterns in the Oregonator Model of the Belousov-Zhabotinskii Reaction, J. Phys. Chem., 89, 118.

9. Boissonade, J. and DeKepper, P., 1980, Transitions from Bistability to Limit Cycle Oscillations. Theoretical Analysis and Experimental Evidence in An Open Chemical System, J. Phys. Chem., 84, 501.

10. Buchholtz, F. and Schneider, F., 1983, First Experimental Demonstration of Chemical Resonance in An Open System, J. Am. Chem. Soc., 105, 7450.

11. Crowley, M.F. and Field, R.J., 1981, Electrically Coupled Belousov-Zhabotinskii Oscillators: A Potential Chaos Generator, in: Nonlinear Phenomena in Chemical Dynamics (C. Vidal and A. Pacault, Eds) Springer-Verlag, Berlin, Heidelberg, and New York.

12. Crowley, M.F. and Field, R.J., 1984, Asymptotic Solutions of A Reduced Oregonator Model of the Belousov-Zhabotinskii Reaction, J. Phys. Chem., 88, 762.

13. Crowley, M.F. and Field, R.J., 1985, Electrically Coupled Belousov-Zhabotinskii Oscillators. I. Experiments and Simulations, J. Phys. Chem., 89, in press.

14. DeKepper, P. and Bar-Eli, K., 1983, Dynamical Properties of the Belousov-Zhabotinskii Reaction in a Flow System. Theoretical and Experimental Analysis, J. Phys. Chem., 87, 480.

15. DeKepper, P. and Boissonade, J., 1985, From Bistability to Sustained Oscillations in Homogeneous Chemical Systems in Flow Reactor Mode, in: Oscillations and Traveling Waves in Chemical Systems (R.J. Field and M. Burger, Editors), Wiley-Interscience, New York.

16. Field, R.J., Körös, E., and Noyes, R.M., 1972, Oscillations In Chemical Systems. I. Thorough Analysis of Temporal Oscillation in the Ce-BrO_3^--Malonic Acid-System, J. Am. Chem. Soc., 94, 8649.

17. Field, R.J. and Noyes, R.M., 1974, Oscillations in Chemical Systems IV. Limit Cycle Behavior in A Model of A Real Chemical Reaction, J. Chem. Phys., 60, 1877.

18. Field, R.J., 1985, Experimental and Mechanistic Characterization of Bromate-Ion-Driven Chemical Oscillations and Traveling Waves in Closed Systems, in: Oscillations and Traveling Waves in Chemical Systems (R.J. Field and M. Burger, Editors), Wiley-Interscience, New York.

19. Field, R.J. and Boyd, P.M., 1985, Bromine-Hydrolysis Control in the Cerium Ion-Bromate Ion- Oxalic Acid-Acetone Belousov-Zhabotinskii Oscillator, J. Phys. Chem., 89, in press.

20. Fujii, H. and Sawada, Y., 1978, Phase-difference Locking of Coupled Oscillating Chemical Systems, J. Chem. Phys., 69, 3830.

21. Ganapathisubramanian, N. and Noyes, R.M., 1981, A Discrepancy Between Experimental and Computational Evidence for Chemical Chaos, J. Chem. Phys., 76, 1770.

22. Gear, C.W., 1971, The Automatic Integration of Ordinary Differential Equations, Comm. ACM, 14, 176.

23. Guckenheimer, J., 1979, A Brief Introduction to Dynamical Systems, in: Nonlinear Oscillations in Biology, Vol. 17 of Lectures in Applied Mathematics, American Mathematical Society, Providence, Rhode Island.

24. Guckenheimer, J., 1981, On Codimension Two Bifurcation, Dynamical Systems and Turbulence, Warwick, 1980, Lecture Notes in Mathematics, Vol. 898, (D.R. and L.S. Young, Editors) Springer-Verlag, Berlin.

25. Heilweil, E.J. and Epstein, I.R., 1979, Chemical Oscillations and Chaos in A Single System, J. Phys. Chem., 83, 1359.

26. Holmes, P., 1980a, Averaging and Chaotic Motions in Forced Oscillations, SIAM J. Appl. Math., 38, 65.

27. Holmes, P., 1980b, Phase Locking and Chaos in Coupled Limit Cycle Oscillations, Proc. Symp. on Recent Advances on Structural Dynamics, Southampton, Great Britain.

28. Janz, R.D., Vanecek, D.J., and Field, R.J., 1980, Composite Double Oscillation in a Modified Version of the Oregonator Model of the Belousov-Zhabotinskii Oscillator, J. Chem. Phys., 73, 3132.

29. Jenkins, M. and Watts, D.G., 1969, Spectral Analysis and its Applications, Holden-Day, San Francisco.

30. Levi, M., 1981, Qualitative Analysis of Forced Relaxation Oscillators, Mem. Am. Math. Soc., No. 244.

31. Marek, M. and Stuchl, I., 1975, Synchronization in Two Interacting Oscillatory Systems, Biophys. Chem., 3, 241.

32. McKinnon, C.K. and Field, R.J., 1985, CSTR Bistability in the Belousov-Zhabotinskii Reaction: Oregonator and Explodator Models, J. Phys. Chem., 89, in press.

33. Minorsky, N., 1972, Nonlinear Oscillations, Van Nostrand, New York; reprinted by Krieger, Huntington, New York (1974).

34. Nakajima, K. and Sawada, Y., 1980, Experimental Studies on the Weak Coupling of Oscillatory Chemical Reaction Systems, J. Chem. Phys., 72, 2231.

35. Noszticzius, Z., 1979, Non-Br^--Controlled Oscillations in the Belousov-Zhabotinskii Reaction of Malonic Acid, J. Am. Chem. Soc., 101, 3177.

36. Noszticzius, Z. and Bodiss, J., 1979, A Heterogeneous Chemical Oscillator, The Belousov-Zhabotinskii Type Reaction of Oxalic Acid, J. Am. Chem. Soc., 101, 3660.

37. Noszticzius, Z., Noszticzius, E., and Schelly, Z.A., 1984, Explodator: A New Skeleton Mechanism for Halate Driven Chemical Oscillators, J. Chem. Phys., 80, 6062.

38. Noszticzius, Z., Gáspár, V., and Försterling, H.-D., 1985, Experimental Test for the Control Intermediate in the Belousov-Zhabotinskii (BZ) Reaction, J. Am. Chem. Soc., 107, 2314.

39. Noyes, R.M., 1985, An Alternative to the Stoichiometric Factor in the Oregonator Model, J. Chem. Phys., 80, 6071.

40. Rehmus, P. and Ross, J., 1985, Periodically Perturbed Chemical Systems, in: Oscillations and Traveling Waves in Chemical Systems (R.J. Field and M. Burger, Editors) Wiley-Interscience, New York.

41. Rinzel, J. and Troy, W.C., 1982, Bursting Phenomena in a Simplified Oregonator Flow System Model, J. Chem. Phys., 76, 1775.

42. Rössler, O.E., 1979, Chaos, in: Structural Stability in Physics (W. Güttinger and H. Eikemeir, Editors) Springer-Verlag, Berlin, Heildelberg, New York.

43. Roux, J.-C., 1983, Experimental Studies of Bifurcations Leading to Chaos in the Belousov-Zhabotinskii Reaction, Physica D, 7D, 57.

44. Roux, J.-C., Simoyi, R.H., and Swinney, H.L., 1983, Observation of A Strange Attractor, Physica D, 8D, 257.

45. Ruoff, P. and Schwitters, B., 1984, Theoretical Study of Ag^+ - Induced Oscillations and Excitations in the Classical Homogeneous Belousov-Zhabotinsky Reaction Using the Oregonator Model, J. Phys. Chem., 88, 6424.

46. Schreiber, I. and Marek, M, 1982, Strange Attractors in Coupled Reaction Diffusion Cells, Physica D, 5D, 258.

47. Smale, S., 1965, Diffomorphisms with Many Periodic Points, in: Differential and Combinatorial Topology (S.S. Cairns, Editor) Princeton University Press, Princeton, N.J.

48. Smale, S., 1967, Differentiable Dynamical Systems, Bull. Am. Math. Soc., 73, 747.

49. Swinney, H.L., 1983, Observations of Order and Chaos in Nonlinear Systems, Physica D, 7D, 3.

50. Stuchl, I. and Marek, M., 1982, Dissipative Structures in Coupled Cells: Experiments, J. Chem. Phys., 77, 2956.

51. Troy, W.C., 1985, A Quantitative Account of Oscillations, Bistability and Traveling Waves in the Belousov-Zhabotinskii Reaction, in: Oscillations and Traveling Waves in Chemical Systems (R.J. Field and M. Burger, Editors), Wiley-Interscience, New York.

52. Tyson, J.J. and Fife, P., 1980, Target Patterns in a Realistic Model of the Belousov-Zhabotinskii Reaction, J. Chem. Phys., 73, 2224.

53. Tyson, J.J., 1982, Scaling and Reducing the Field-Körös-Noyes Mechanism of the Belousov-Zhabotinskii Reaction, J. Phys. Chem., 86, 3006.

54. Tyson, J.J., 1985, Mathematical Analysis of the Oregonator Model of the Belousov-Zhabotinskii Reaction, in: Oscillations and Traveling Waves in Chemical Systems (R.J. Field and M. Burger, Editors), Wiley-Interscience, New York.

55. Varga, M., Györgyi, L., and Körös, E., 1985, A Thorough Study of Bromide Control in Bromate Oscillators. 2. Simulation by the Oregonator Model of the Behavior of the Reacting Belousov-Zhabotinskii Systems Perturbed by Bromo-Complex-Forming Metal Ions, _J. Phys. Chem._, _89_, 1019.

LOSING AMPLITUDE AND SAVING PHASE

Bard Ermentrout
Department of Mathematics
University of Pittsburgh
Pittsburgh, PA 15260

1. Introduction

Many chemical and biological phenomena are modeled as systems of coupled limit cycle oscillators. These models are inherently complex in that they involve often large numbers of coupled ordinary and partial differential equations. To understand any of the behavior of these systems, simplifying assumptions are made. One such assumption is that the individual oscillators are nearly identical and weakly coupled. In this case only the phase of the individual oscillators matters and so the coupled system becomes a smaller system on a k-torus. This technique has been applied to discrete systems [1-3] as well as to reaction-diffusion equation [4,5]. Many interesting aspects of chemical and biological systems can be understood by studying the simple phase-models [6-8]. For example, see the paper by Kopell in this volume.

On the other hand, by restricting the oscillators to lie near the uncoupled limit cycle, a great deal of interesting phenomena may be missed such as "phase trapping", multiple stable periodic states, loss of periodicity. and chaos. These effects depend on "strong" coupling of the oscillators and force the system off of the torus of phases. That is, strong enough coupling can effect major changes on the amplitude and form of the oscillation. Bar-Eli [9] has recently used numerical techniques to show that for a large class of oscillators, coupling can lead to stable non-oscillating steady states. Phase models can never exhibit this behavior. Schreiber and Marek [10] have numerically found that two identical oscillators coupled with nonscalar diffusion can undergo a variety of complex dynamic phenomena including transitions to invariant tori, period doubling, and chaos. In continuous systems (i.e., reaction-diffusion) with oscillatory kinetics, many of these same phenomena occur. When diffusion is non-scalar, stationary patterns may emerge from an oscillatory medium [11]. It is also possible for inhomogeneous systems with strong diffusion to lose periodicity and come to a homogeneous resting level.

In this article, we will consider some of these phenomena for a model oscillator with very simple kinetics:

$$C_t = \begin{pmatrix} \lambda & -\omega \\ \omega & \lambda \end{pmatrix} C, \qquad (1.1)$$

where $\lambda = 1 - |C|^2$, $\omega = \omega_0 + q(1-|C|^2)$, $C \in \mathbb{R}^2$. Equation (1.1) has a unique asymptotically stable limit cycle, $(C_1,C_2) = (\cos \omega_0 t, \sin \omega_0 t)$. Equation (1.1) arises generally from a general Hopf bifurcation. We consider two types of coupling; discrete-diffusion:

$$\dot{C}_1 = \begin{pmatrix} \lambda_1 & -\tilde{\omega}_1 \\ \tilde{\omega}_1 & \lambda_1 \end{pmatrix} C_1 + \begin{pmatrix} \gamma & -\mu \\ \mu & \gamma \end{pmatrix} (C_2-C_1)$$

$$\dot{C}_2 = \begin{pmatrix} \lambda_2 & -\tilde{\omega}_2 \\ \tilde{\omega}_2 & \lambda_2 \end{pmatrix} C_2 + \begin{pmatrix} \gamma & -\mu \\ \mu & \gamma \end{pmatrix} (C_1-C_2), \tag{1.2}$$

and continuous diffusion:

$$\frac{\partial C}{\partial t} = \begin{pmatrix} \lambda & -\omega(x) \\ \omega(x) & \lambda \end{pmatrix} C + \begin{pmatrix} \gamma & 0 \\ 0 & \gamma \end{pmatrix} C_{xx}, \quad 0 < x < 1 \tag{1.3}$$

$$C_x = 0 \text{ at } x = 0 \text{ and } x = 1.$$

In Section 2 we determine the behavior of (1.2). We also discuss the restabilization of a homogeneous steady state for a pair of general coupled systems. Section 3 is an analysis of Eq. (1.3) when $\omega(x)$ is a linear function of x. A formal perturbation procedure is also developed which allows us to determine when phase-locking is lost.

2. Two Coupled Oscillators

Here we describe the behavior of Eq. (1.2) under two sets of different conditions. First we suppose diffusion is scalar ($\mu = 0$) but that the oscillators are different. This is essentially the situation that Bar-Eli has analyzed numerically for a pair of coupled Brusselators [9]. Secondly, we let the diffusion be nonscalar but assume identical oscillators. Schreiber and Marek [10] discuss this for coupled Brusselators.

Before analyzing (1.2), I will state a simple general result which concerns the restabilization of a rest state via coupling.

PROPOSITION 2.1: Consider the pair of coupled linear eigenvalue equations:

$$\nu x = Ax + \gamma(y-x)$$
$$\nu y = \alpha Ax + \gamma(x-y) \tag{2.1}$$

where $x,y \in \mathbb{R}^n$, A is an $n \times n$ matrix and $\alpha,\gamma \in \mathbb{R}$. Let λ be an eigenvalue of A. Then:

(a) $\{\nu - (\alpha\lambda-\gamma)\}\{\nu - (\lambda-\gamma)\} = \gamma^2$.

(b) If $\text{Re}\,\lambda < 0$ then both solutions to (a) have negative real parts.

(c) If $\lambda > 0$, then one solution to (a) is positive.

(d) If $\lambda = a + bi$, $a > 0$ and a/b is sufficiently small, then there are values of α and γ such that both solutions to (a) have negative real parts.

The proof of (a-c) is simple linear algebra. To prove part (d) let $a/b = \varepsilon$. Pick $\gamma \sim \frac{\alpha-1}{2}\sqrt{1-\varepsilon^2}\, b$. Then:

$$\text{Re}\,\nu^+ \sim b[\tfrac{\alpha-1}{2}(\sqrt{\varepsilon} - \sqrt{1-\varepsilon^2}) + \tfrac{\alpha+1}{2}\varepsilon],$$

where ν^+ is the solution to (a) with maximal real part. For ε small enough, and $\alpha > 1$, ν^+ has a negative real part.

This proposition has a useful implication. Suppose and oscillator has the equation:

$$\dot{x} = F(x) \tag{2.2}$$

and suppose $F(0) = 0$. Consider two such oscillators coupled with scalar diffusion:

$$\dot{x} = F(x) + \gamma(y-x)$$
$$\dot{y} = \alpha F(y) + \gamma(x-y).$$

The parameter α changes the frequency of the y-oscillator when $\gamma \equiv 0$. Proposition 2.1 implies that if all of the eigenvalues of the Jacobian of F have either negative real parts or large imaginary parts, then we can choose γ, α for which the homogeneous solution $x = y \equiv 0$ is linearly stable. This means that even if the origin $x = 0$ is an unstable equilibrium to (2.2), by coupling these systems we can stop the oscillation.

We turn now to Eq. (1.2):

$$\frac{d}{dt}\begin{pmatrix} u_1 \\ v_1 \end{pmatrix} = \begin{pmatrix} \lambda_1 & -\bar{\omega}_1 \\ \bar{\omega}_1 & \lambda_1 \end{pmatrix}\begin{pmatrix} u_1 \\ v_1 \end{pmatrix} + \begin{pmatrix} \gamma & -\mu \\ \mu & \gamma \end{pmatrix}\begin{pmatrix} u_2-u_1 \\ v_2-v_1 \end{pmatrix}$$

$$\frac{d}{dt}\begin{pmatrix} u_2 \\ v_2 \end{pmatrix} = \begin{pmatrix} \lambda_2 & -\bar{\omega}_2 \\ \bar{\omega}_2 & \lambda_2 \end{pmatrix}\begin{pmatrix} u_2 \\ v_2 \end{pmatrix} + \begin{pmatrix} \gamma & -\mu \\ \mu & \gamma \end{pmatrix}\begin{pmatrix} u_1-u_2 \\ v_1-v_2 \end{pmatrix} \tag{2.3}$$

with $\lambda_j = 1 - u_j^2 - v_j^2$ and $\bar{\omega}_j = \omega_j + q_j(1-u_j^2-v_j^2)$. Eq. (2.3) can be formally derived from a pair of coupled systems near a Hopf bifurcation [12]:

$$\frac{dx}{dt} = Ax + \lambda B_1 x + \lambda D(y-x) + O(|x|^2)$$

$$\frac{dy}{dt} = Ay + \lambda B_2 y + \lambda D(x-y) + O(|y|^2). \tag{2.4}$$

Here λ is the bifurcation parameter and A, B_1, B_2, and D are $n \times n$ matrices. A has a unique pair of conjugate imaginary eigenvalues, with left and right eigenvectors Φ_*, Φ, satisfying $\bar{\Phi}_* \cdot \Phi = 1$. Hagan [12] and others [13] have shown that:

$$\gamma + i\mu = \bar{\Phi}_* D \,\Phi \;.$$

In particular if D is scalar then $\mu \equiv 0$. In absence of coupling $(\gamma + i\mu \equiv 0)$, the solutions to (2.3) are:

$$u_j(t) = \cos(\omega_j t)$$

$$v_j(t) = \sin(\omega_j t), \quad j = 1,2. \tag{2.5}$$

Hence (2.3) represents a pair of nonidentical asymptotically stable limit cycles coupled together. We will consider two separate cases in order to relate our results with those of Bar-Eli and Schreiber et. al. We first suppose that the diffusion is scalar but that the oscillators are not identical. This resembles the case considered by Bar-Eli.

We introduce new polar coordinates:

$$u_1 + iv_1 = r\exp(i\theta_1)$$

$$u_2 + iv_2 = s\exp(i\theta_2) \qquad (2.6)$$

$$\phi = \theta_2 - \theta_1.$$

Then we can write (2.3) in terms of r, s, and ϕ:

$$\dot{r} = r(1-\gamma-r^2) + \gamma s \cos\phi$$

$$\dot{s} = s(1-\gamma-s^2) + \gamma r \cos\phi \qquad (2.7)$$

$$\dot{\phi} = \omega_2-\omega_1 + q_1 r^2 - q_2 s^2 - \gamma\left(\frac{r}{s}+\frac{s}{r}\right)\sin\phi.$$

Nonzero equilibria of (2.7) correspond to phaselocked limit cycles. We let $\Delta = \omega_2 - \omega_1$ denote the frequency difference and for simplicity set $q_1 = q_2 = 0$. We remark that if $q_j \neq 0$, the behavior of Eq. (2.7) does not differ substantially from $q_j \equiv 0$. The main effect is to alter the symmetry. For more details, a discussion of the "equi-amplitude" solutions ($r = s$) when $q_j \neq 0$ is given in [14]. No novel behavior is missed by making this assumption.

We first state a lemma concerning the stability of the trivial solution to (2.3), $u_j = v_j \equiv 0$, when $\mu \equiv 0$.

<u>LEMMA 3.1</u>: The homogeneous zero rest state of (2.3) is stable if and only if

$$1 < \gamma < \frac{1}{2}\left(1 + \frac{\Delta^2}{4}\right). \qquad (2.8)$$

The proof of this is contained in [14] and simply involves finding the roots of the eigenvalue equation for the linearization. Eq. (2.8) implies that the coupling must be sufficiently strong and furthermore, that there be a sufficiently large difference in frequencies between the two oscillators. We note the qualitative similarity between (2.8) and Proposition 2.1.

An analysis of the phaselocked behavior of (2.7) is easily made. We set the right hand sides equal to zero and this results in a set of algebraic equations for r, s, and a transcendental equation for ϕ. There are two types of equilibria:

$$\text{(a)}\quad r^2 = s^2 \equiv \rho^2$$

$$\text{(b)}\quad r^2 = 1 - \gamma - s^2 \equiv \rho^2 \qquad (2.9)$$

symmetric and asymmetric solutions. Substitution of (2.9) into (2.7) and use of some elementary trigonometry leads to

$$(1-\gamma-\rho^2) + \Delta^2 = \gamma^2. \qquad (2.10)$$

An analysis of (2.10) along with the concomitant linear stability analysis yields the bifurcation picture shown in Figures 2.1 a,b for $|\Delta| > 2$ and $|\Delta| < 2$ respectively.

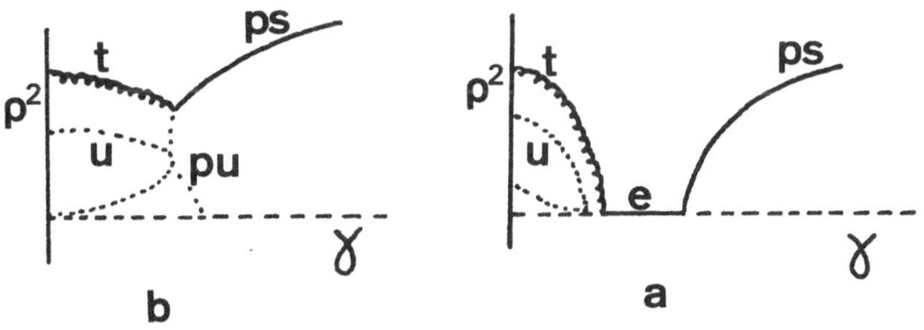

Figure 2.1

Amplitude versus coupling strength for
(a) $|\Delta| > 2$,
(b) $|\Delta| < 2$.
PS = stable phaselocked, PU = unstable phaselocked, U = unstable asymmetric phaselocked, E = stable equilibrium, T = torus.

We have also depicted the unstable asymmetric solutions defined by (2.9b). If we fix $|\Delta| > 2$ and vary the strength of coupling, we find the following. For strong coupling there is a stable phaselocked solution. As coupling decreases, the amplitude of this decreases and eventually the oscillation vanishes to a single point. Until the coupling reaches $\gamma = 1$, the zero solution is stable. At $\gamma = 1$ there is a complex bifurcation to a symmetric torus of the form:

$$r(t) = s(t) = \rho(t) = \rho(t+T)$$

$$\phi(t+T) = \phi(t) + 2\pi,$$

We contrast this behavior with the situation when $|\Delta| < 2$. Instead of a bifurcation to a "phase-drift" torus from a stable zero state, there is a direct transition from a phase-locked solution to a torus. As shown in Fig. 2.1 b. This bifurcation occurs through infinite period at $\gamma = |\Delta/2|$. This type of transition from phaselocking to tori is the type associated with phase models (see e.g. [3]) and in fact since (Δ) and (γ) can both be small for this to occur, we expect that the phase models will be valid approximations of the full system, (2.3).

We have shown that two sufficiently different oscillators can, when coupled, "anihilate" each other and leave only a stable rest state in their wake. It is crucial that the oscillators be different for this phenomena to occur; for if two identical oscillators are coupled with scalar diffusion, it is easy to show the homogeneous (bulk) oscillation is always stable no matter what the coupling strength [15].

Bar-Eli [9] describes a pair of identical, diffusively coupled oscillators with scalar diffusion and numerically discovers a new inhomogeneous stable non-oscillatory state. This bifurcation depends on the fact that the uniform rest state has a positive (not complex!) eigenvalue. Our simple model can never have positive eigenvalues around the unstable rest state; by construction they are always complex. Nevertheless, we are able to find regions where a nonoscillatory (time-independent) state can become stabilized. We remark that in this simple paradigm, phase-trapping is never observed; neither are stable anti-phase oscillations.

We turn our attention to some recent results on (2.3) when diffusion is not scalar but the oscillators are otherwise identical. This is an attempt to understand some of the complex dynamical behavior uncovered by Schreiber and Marek [10] through extensive numerical calculations for the Brusselator. We once again use the coordinates (2.6) and obtain the equations for r, s, and ϕ:

$$\begin{aligned} \dot{r} &= r(1-\gamma-r^2) + \gamma s \cos\phi - \mu s \sin\phi \\ \dot{s} &= s(1-\gamma-s^2) = \gamma r \cos\phi + \mu s \sin\phi \qquad (2.11) \\ \dot{\phi} &= q(s^2-r^2) - \gamma\left(\frac{r}{s} + \frac{s}{r}\right)\sin\phi + \mu\left(\frac{r}{s} - \frac{s}{r}\right)\cos\phi. \end{aligned}$$

We first examine the stability of the homogeneous inphase solution, $r = s = 1$, $\gamma = 0$. Linearization of (2.11) about this solution leads to the following stability requirement:

$$\gamma(\gamma+1) + \mu(\mu-q) > 0. \qquad (2.12)$$

We see that the inphase solution is always stable if μ or q is zero. When (2.12) is violated, then there is a transition from 3 negative eigenvalues to a positive eigenvalue and 2 negative eigenvalues. This is a symmetry breaking bifurcation and we expect the phase-locked solution to break up into phase-locked solutions which are not symmetric. In fact, we find numerically (the analysis of this bifurcation can be done exactly and we hope to do this in the future) that two new stable solutions exist with ϕ close to 0 and r,s close to 1. The symmetry of (2.11) under switching between r and s and changing $\phi \to -\phi$ requires us to calculate only half of the bifurcation picture. The remainder follows from symmetry. In Fig. 2.2 we show this transition to new asymmetric "in-phase" solutions.

Next we examine the out-of-phase solutions, $r^2 = s^2 = 1 - 2\gamma$, $\phi = \pi$. These solutions only exist for $\gamma < \frac{1}{2}$. A stability analysis yields two conditions:

$$\begin{aligned} &(a) \quad \gamma < \frac{1}{4} \\ &(b) \quad \mu[q(1-2\gamma) + \mu] > \gamma(1-3\gamma). \end{aligned} \tag{2.13}$$

If (2.13 a) is violated, there is a Hopf bifurcation; when (2.13 b) is violated stability is lost through a zero eigenvalue. The former case is interesting since it implies that "phase-trapped" solutions are possible for this problem. Numerical simulations of (2.11) indicate that there are in fact oscillatory solutions bifurcating from the anti-phase solution which correspond to invariant tori for Eq. (2.3). In Fig. 2.2, we sketch a rought bifurcation diagram for fixed values of (μ,q) letting γ, the coupling strength vary.

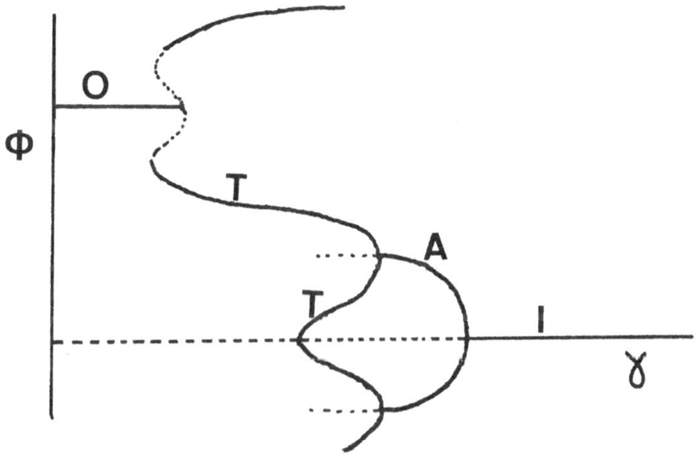

Figure 2.2

Phase versus coupling strength γ for $\mu = 1$, $q = 2$.
I = inphase solution, O = anti-phase solution, A = asymmetric solution, T = phase trapped solution. Solid lines are stable, dotted are unstable.

This is not a complete picture and will change if (μ,q) are in some other domain. It is illustrated to show the complex dynamical behavior possible with even a simple model. For large values of γ, the inphase symmetric solution is stable. As γ decreases, a pair of asymmetric phase-locked solutions bifurcates. There are two stable limit cycles for Eq. (2.3) at this point. Next, the asymmetric phase-locked solutions lose stability at an imaginary eigenvalue. A pair of "asymmetric-tori" bifurcate. These correspond to phase-trapped solutions; the phase is "trapped" around two phases, $\pm\phi_0$, $|\phi_0| < \frac{\pi}{2}$. As γ is decreased these two oscillations of (2.7) or tori of (2.3) merge into one large phase trapped oscillation. This merges with an unstable phase-trapped solution about $\phi = \pi$. Finally, the anti-phase solution is stabilized and remains so for all smaller values of γ.

The two stability conditions for the inphase and out-of-phase solutions, Eq. (2.12) and Eq. (2.13 a,b) show that it is possible to find parameter values for which both the in-phase and the anti-phase solutions are stable. In particular if we choose $q \equiv 0$, $\gamma < \frac{1}{4}$ and $\mu^2 > 1/16$, both of these solutions will be stable. Even in this simple system, there are multiple stable oscillating solutions. When the symmetry of this system is broken, we can expect to see other types of complicated behavior. It is clear that this type of behavior as well as complex phenomena such as phase-trapping cannot occur in simple phase models. The "forgotten" coordinate, amplitude, enters into these systems in a nontrivial manner. We are still far away from understanding the complicated numerical results of Schreiber and Marek, but, these simple oscillating systems can shed some light on the complex behavior of coupled limit cycle oscillators.

3. Continuum Models

Many authors have considered the behavior of forced and coupled oscillators, numerically and analytically. Generally, only one or two oscillators are considered in their analysis. Recently some advances have been made in understanding large numbers of coupled oscillators (see for example Kopell's paper in these proceedings).

In spite of these advances, little is known of the behavior of a continuum of oscillations with inhomogeneities. Kopell and Howard [16] and others have studied wave phenomena in oscillatory reaction-diffusion systems. More recently Hagan [5] using formal methods and Kopell [17] have studied the behavior of oscillatory systems with weak diffusion and weak inhomogeneities:

$$u_t = F(u) + \varepsilon[D\nabla^2 u + G(u,x)]. \tag{3.1}$$

Since ε is small, $u(x,t)$ remains close to the homogeneous limit cycle and as is the case with discrete systems, a phase model can be derived:

$$\phi_\tau = g(x) + d(\nabla^2\phi + a|\nabla\phi|^2) \tag{3.2}$$

where $\phi(x,\tau)$ is a slowly varying scalar phase. The terms d, a, and $g(x)$ depend on the matrix D, the kinetics, and the inhomogeneity $G(a,x)$. The one-dimensional version of (3.2) can be easily solved by converting to a second order linear equation through the Cole-Hopf transformation (see [5] for details). A phase-locked solution to (3.2) is one for which $\phi(x,\tau) \equiv \omega\tau + \int_0^x \psi(s)ds$. The function ψ satisfies:

$$\omega = g(x) + d(\psi_x + a\psi^2) \tag{3.3}$$

with appropriate boundary conditions. A solution can be found for most functions g(x), thus, the phase models predict phaselocking in all cases. This contrasts with discrete phase models which lose phaselocking through the appearance of phase-drift.

Numerical observations, on reaction-diffusion equations with a continuous frequency gradient [18] and an experiment performed by Winfree [19] indicate that phase models cannot tell the whole story. Winfree has designed an ingenious experiment which he calls a "continuum intestine". A strip of paper coated with the components of the oscillatory Belousov-Zhabotinski reagent is placed over a copper bar along which there is a continuous temperature gradient. Because of differences in the temperature of the bar, different regions of the medium can oscillate at different frequencies. The diffusion tends to synchronize these oscillations. The end result is "frequency plateaus"; regions of the medium oscillating at the same frequency connected to other regions at different frequencies. This same phenomena was observed numerically by Meinhardt in a model for sea shell pattern formation; a linear gradient of parameters was embedded in an oscillating reaction-diffusion system.

In this section, we describe some of the complex behavior possible in a reaction-diffusion system with a linear gradient of frequencies. We consider the λ-ω system:

$$\begin{pmatrix} u_t \\ v_t \end{pmatrix} = \begin{pmatrix} \lambda & -\omega(x) \\ \omega(x) & \lambda \end{pmatrix} \begin{pmatrix} u \\ v \end{pmatrix} + \gamma \begin{pmatrix} u_{xx} \\ v_{xx} \end{pmatrix} \tag{3.4}$$

$$u_x(0,t) = u_x(1,t) = 0$$

$$v_x(0,t) = v_x(1,t) = 0$$

where $\lambda = 1 - u^2 - v^2$, $\omega = \Omega_0 - \Delta x + q(u^2+v^2)$. This study concerns the case of $q \equiv 0$. For $q \neq 0$, the analysis is technically much more difficult; our numerics indicate that the major effect is to destroy symmetry of the solutions. Here, Δ is the depth of the frequency gradient and γ is the strength of the diffusion. The frequencies of this chain of oscillators are $\Omega_0 - \Delta x$ in absence of diffusion. Thus the net frequency difference from one end to the other is Δ. We will show that if $\gamma \geq 1$, then it is possible to phaselock at arbitrarily small amplitudes $r(x,t)$. We look for phaselocked solutions of (3.4) of the form:

$$\begin{aligned} u(x,t) &= r(x)\cos(\bar{\omega}t + \textstyle\int_0^x \phi(s)ds) \\ v(x,t) &= r(x)\sin(\bar{\omega}t + \textstyle\int_0^x \phi(s)ds) \end{aligned} \tag{3.5}$$

which when substituted into (3.4) yield:

$$\begin{aligned} 0 &= r(1-r^2) + \gamma(r''-r\phi^2) \\ 0 &= -\bar{\omega} + \Omega_0 - \Delta x + \gamma\left(\frac{2r'}{r}\phi+\phi'\right) \\ &\phi(0) = 0, \quad \phi(1) = 0 \\ &r'(0) = 0, \quad r'(1) = 0, \quad ' = \frac{d}{dx} \end{aligned} \tag{3.6}$$

We can utilize the symmetry of (3.6) to show that $\bar{\omega} = \Omega_0 - \Delta/2$, the mean frequency. Eq. (3.6) is then a third-order nonlinear two-point boundary value problem. Further symmetry arguments and a simple transformation leads to the following problem:

$$\begin{aligned} &(a)\quad r' = rs \\ &(b)\quad s' = \alpha(r^2-1) + \phi^2 - s^2 \\ &(c)\quad \phi' = \alpha\Delta(x-\tfrac{1}{2}) - 2s\phi \\ &(d)\quad \phi(0) = 0, \quad s(0) = 0, \quad s(\tfrac{1}{2}) = 0 \end{aligned} \tag{3.7}$$

where $\alpha = 1/\gamma$. The boundary conditions follow from symmetry and the definition of $s = r'/r$. We have recently used a shooting argument to prove the following theorem.

THEOREM 3.1: [20] For each $\alpha \leq 1$, if $0 \leq r(0) \leq 1$, there exists a $\Delta > 0$ such that (3.7) has a solution.

This theorem is not as strong as one would like, there is no statement about uniqueness, nor is the relationship between amplitude, $r(0)$ and Δ described. For this, we must turn to numerical calculations. First, we describe the general form of the solutions. In Fig. 3.1, we have drawn $r(x)$ and $\phi(x)$ on the interval $[0,1]$.

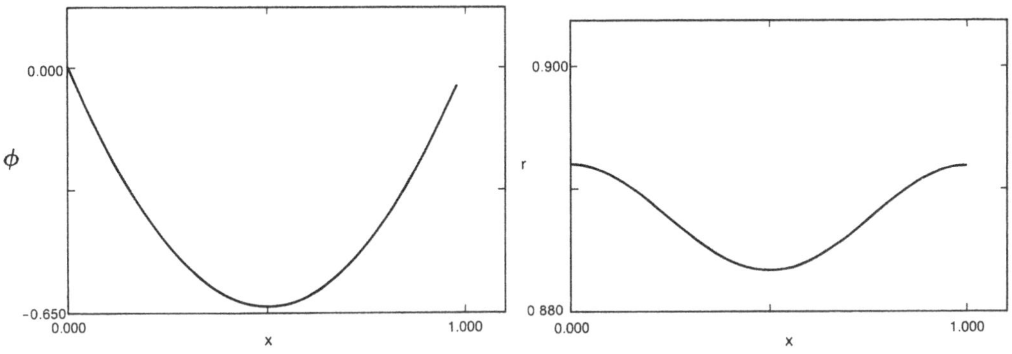

Figure 3.1

r(x) and $\phi(x)$ for $\Delta = 5$.

Note the generally parabolic shape of $\phi(x)$. r(x) is almost constant, taking its maximum values on the boundaries. Thus r(0) is the maximum amplitude of the coupled system. Because r(x) is almost constant, $s = r'/r$ is close to zero. Thus, a "phase-model" may seem warranted. Using the approximation $s = 0$, we see that (3.7 c) becomes a simple phase model which is readily solved:

$$\phi(x) = \frac{\alpha\Delta x}{2}(x-1). \tag{3.8}$$

This is a parabola. In Fig. 3.2, we plot $\phi(x)$ from (3.8) versus $\phi(x)$ as calculated numerically for $\alpha = 1$ $(\gamma = 1)$ and $\Delta = 10$. They are remarkably close. One might be led to believe that the whole story can be told with the phase model. Figure 3.3 shows why this is not the case. The maximal amplitude r(0) is plotted against the frequency difference Δ, for $\gamma = 1$. It is clear that r(0) is decreasing monotonically to zero as Δ increases. Furthermore, there is a critical value Δ^*, here $\Delta^* \approx 10.47$, above which oscillations disappear. Numerical simulations of the full PDE, Eq. (3.4), show that the uniform zero rest state is stable for all $\Delta > \Delta^*$. Thus, while the phase remains "unbothered" by the increasing frequency gradient, there is a drastic change in the amplitude; the rest state is actually restabilized.

The proof of Theorem 3.1 depends on the assumption that $\gamma \geq 1$ (i.e., $\alpha \leq 1$). A natural question is what happens for $\gamma < 1$. We have been unable to answer this question rigorously but numerical simulations imply a certain similarity to the discrete coupled pair of oscillators. In Fig. 3.4, we depict the behavior of Eq. (3.4) for $\gamma = 0.05$ and for two different values of Δ. For $\Delta < \Delta_p \approx 1.125$,

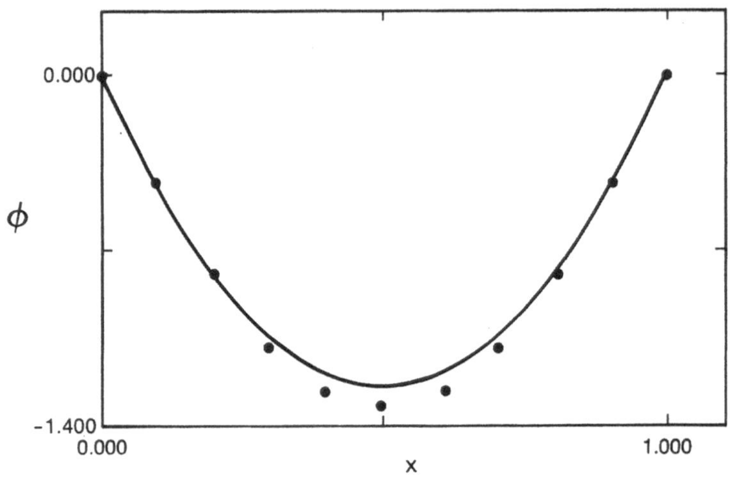

Figure 3.2

Comparison with true solution, $\phi(x)$ and approximation (3.8) for $\Delta = 10$. Dashed line is (3.8), solid line is numerical solution of (3.6).

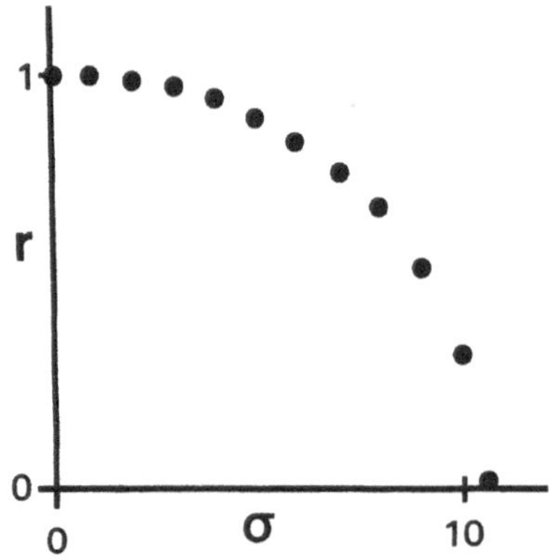

Figure 3.3

$r(0)$ versus Δ for $\gamma = 1$.

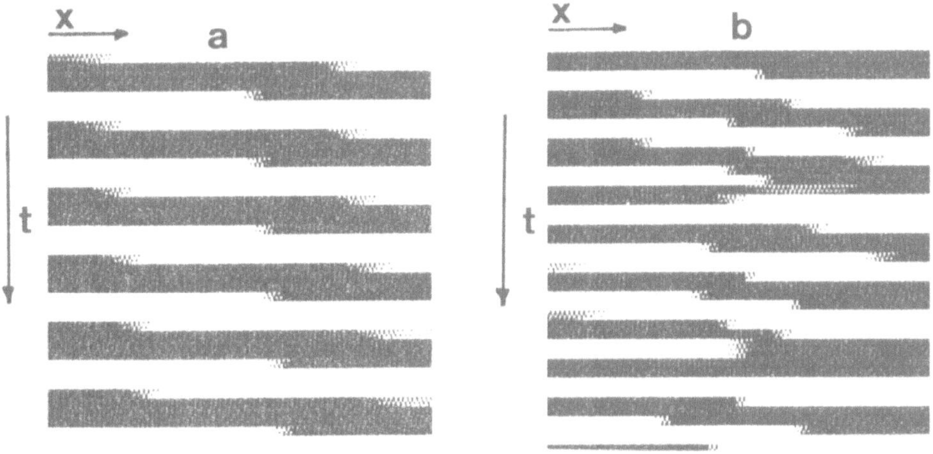

Figure 3.4

Positions of waves as a function of space and time, $\gamma = 0.05$
(a) $\Delta = 1$, phaselocked; (b) $\Delta = 2$, "plateaus".

there is a phase-locked solution; all points in the medium oscillate at the same frequency. For $\Delta > \Delta_p$, phaselocking is lost; "frequency plateaus" develop. The right hand side has synchronized at approximately 0.25 cycles per time unit while the left hand side is synchronized at 0.33 cycles per time unit. It is tempting to conjecture that this general behavior occurs for all $\gamma < 1$.

There is a strong similarity between the results in this section and those in the first part of Section 2. First there is a dependence of behavior on the absolute coupling strength, γ, and not the ratio of γ and D. This differs from the behavior of pure phase models where only Δ/γ matters (see [3]). The reason for this comes from the appearance of the amplitude term. If we divide by γ and rescale time, a term of the form:

$$r(1-r^2)/\gamma \tag{3.10}$$

appears in the equation for r in both Eq. (3.4) and Eq. (2.3). In a phase model where r is assumed to be constant, the term (3.10) is irrelevant. But when amplitude is important, it is clear that the coupling strength interacts with the rate of attraction to the limit cycle. In particular, for γ small, (3.10) implies r must be forced towards 1.

We conclude with a formal calculation which correctly predicts the important scalings for weak diffusion. That is, we describe the conditions under which phase-locking can be lost. We recall that in most analyses of locking in reaction-diffusion equations, both the diffusion and the inhomogeneity are of the same order of magnitude [4 , 5] (also, see (3.1)-(3.3)). Here, we consider a model problem in which the inhomogeneity is much larger than the diffusion. Our numerics indicate that phaselocking persists for quite large inhomogeneities; for example when $\gamma = 0.05$, it is necessary for the frequency gradient to exceed 1.125 in order to lose locking. We turn to our model system (3.4) which in polar coordinates becomes:

$$\begin{aligned}
&\text{(a)}\quad r_t = r(1-r^2) + \varepsilon^2(r_{xx} - r\theta_x^2)\\
&\text{(b)}\quad \theta_t = 1 + \varepsilon[\sigma x + q(r^2-1)] + \varepsilon^2\left(\frac{2r_x\theta_x}{r} + \theta_{xx}\right)\\
&\text{(c)}\quad \theta_x(0,t) = \theta_x(1,t) = 0\\
&\text{(d)}\quad r_x(0,t) = r_x(1,t) = 0.
\end{aligned} \tag{3.11}$$

the diffusion strength, ε^2 is small compared to the inhomogeneity, $\varepsilon\sigma x$, for $\varepsilon << 1$. We allow for the symmetry breaking "twist" term, q. The ratio of the gradient to the diffusion strength is σ/ε which is large for small ε. We let $\theta(x,t) = t + \phi(x,\tau)$, where $\tau = \varepsilon^2 t$. Then (3.11) becomes:

$$\begin{aligned}
&\text{(a)}\quad r_\tau = \frac{r(1-r^2)}{\varepsilon^2} + r_{xx} - r\phi_x^2\\
&\text{(b)}\quad \phi_\tau = \frac{\sigma x + q(r^2-1)}{\varepsilon} + \frac{2r_x}{r}\phi_x + \phi_{xx},
\end{aligned} \tag{3.12}$$

plus the associated boundary conditions. One might guess that (3.12 a) could be dispensed with by setting $r^2 = 1$. This would result in $\phi(x,\tau)$ satisfying $\phi_x = \frac{\sigma}{2\varepsilon}x(1-x)$. Thus ϕ_x^2 is also $O(1/\varepsilon^2)$ and so we find that $r^2 \sim 1$ is incorrect. Clearly the amplitude plays a role. Let $\varepsilon\phi = \psi$ be the scaled phase. Eq. (3.12) can be rewritten as

$$\begin{aligned}
&\text{(a)}\quad r_\tau = \frac{r(1-r^2-\psi_x^2)}{\varepsilon^2} + r_{xx}\\
&\text{(b)}\quad \psi_\tau = \sigma x + q(r^2-1) + \frac{2r_x\psi_x}{r} + \psi_{xx}.
\end{aligned} \tag{3.13}$$

Now we can let $\varepsilon \to 0$ and set $r^2 = 1 - \psi_x^2$ so (3.13) is reduced to a phase model:

$$\psi_\tau = \sigma x - q\psi_x^2 + \left(\frac{1-3\psi_x^2}{1-\psi_x^2}\right)\psi_{xx}, \tag{3.14}$$

with boundary conditions, $\psi_x = 0$ at $x = 0,1$. We seek phase-locked solutions to (3.14) of the form:

$$\psi(x,\tau) = \bar{\omega}\tau + \int_0^x v(s)ds,$$

where $v(x)$ satisfies

$$\bar{\omega} = \sigma x - qv^2 + \left(\frac{1-3v^2}{1-v^2}\right) v_x \tag{3.15}$$

$$v(0) = v(1) = 0.$$

This equation can be solved by numerical shooting. An exact solution can be found for $q \equiv 0$, which illustrates conditions for loss of phaselocking. $v(x)$ satisfies

$$\frac{\sigma}{2} x(1-x) = 3v - \log\left(\frac{1+v}{1-v}\right) \equiv f(v). \tag{3.16}$$

In order for phaselocking to occur the left hand side cannot exceed the local maxima of the right hand side. The local maxima and minima of f occur at $r = \pm 1/\sqrt{3}$ and the maximum value of the left hand side is $|\sigma/8|$. Hence, phaselocking occurs if and only if:

$$|\sigma| < 8\ f(1/\sqrt{3}) \approx 3.32074 \equiv \sigma_*. \tag{3.17}$$

We predict phaselocking will be lost if the frequency gradient is $O(\sqrt{\gamma})$. We have checked this procedure against numerical calculations and it appears to be correct.

References

[1] N. Levinson, "Small periodic perturbations of an autonomous system with a stable orbit", Ann. Math. 52 (1950), 727-738.

[2] G. B. Ermentrout and N. Kopell, "Frequency plateaus in a chain of weakly coupled oscillators, I.", SIAM J. Math. Anal. 15 (1984), 215-237.

[3] J. Neu, "Coupled chemical oscillators", SIAM J. Appl. Math. 37 (1979), 307-315.

[4] J. Neu, "Nonlinear oscillators in discrete and continuous systems", Ph.D. Thesis, Cal. Tech (1978), Chapter 5.

[5] P. Hagan, "Target patterns in reaction-diffusion systems", Adv. Appl. Math. 2 (1981), 400-416.

[6] S. Daan and C. Berde, "Two coupled oscillators: Simulations of the circadian pacemaker in mammalian activity rhythms", J. Theor. Biol. 70 (1978), 297-314.

[7] A. H. Cohen, P. J. Holmes and R. J. Rand, "The nature of coupling between segmental oscillators of the lamprey spinal generator", J. Math. Biol. 13 (1982), 345-369.

[8] N. Kopell and G. B. Ermentrout, "Symmetry and phaselocking in a chain of weakly coupled oscillators, preprint.

[9] K. Bar-Eli, "On the stability of coupled chemical oscillators", Physica 14D (1985), 242-252.

[10] I. Schreiber and M. Marek, "Strange attractors in coupled reaction-diffusion cells", Physica 15D (1982), 258-272.

[11] G. B. Ermentrout, S. P. Hastings and W. C. Troy, "Large amplitude stationary waves in an excitable lateral-inhibitory medium", SIAM J. Appl. Math. 44 (1984), 1133-1149.

[12] P. S. Hagan, "Spiral waves in reaction-diffusion equations", SIAM J. Appl. Math. 42 (1983), 762-786.

[13] G. B. Ermentrout, "Stable small-amplitude solutions in reaction-diffusion systems", Quart. Appl. Math., April (1981), 61-86.

[14] G. B. Ermentrout, D. Aronson and N. Kopell, (in preparation).

[15] V. Torre, "A theory of synchronization of two heart pace-maker cells", J. Theor. Biol. 61 (1976), 55-71.

[16] N. Kopell and L. N. Howard, "Plane wave solutions to reaction-diffusion equations", Stud. Appl. Math. 52 (1973), 291-328.

[17] N. Kopell, "Target pattern solutions to reaction-diffusion equations in the presence of imparities", Adv. Appl. Math. 2 (1981), 389-399.

[18] H. Meinhart, personal communication.

[19] A. J. Winfree, The Geometry of Biological Time, Springer-Verlag, New York (1980), 328-329.

[20] G. B. Ermentrout and W. C. Troy, "Phaselocking in a reaction-diffusion system with a linear frequency gradient", preprint (1985).

Spiral Waves in Excitable Media

James P. Keener
Department of Mathematics
University of Utah
Salt Lake City, Utah 84112

§1. Introduction

The beautiful spiral waves of oxidation in the Belousov-Zhabotinskii reaction [20] are the source of many interesting questions about periodic structures in excitable media. Because they are easy to produce and photograph, pictures of these spirals have appeared in a number of popular science oriented magazines. The study of spirals takes on a more personal interest when one realizes that fibrillation and sudden cardiac arrest from heart attacks may also be due to the appearance of rotating spiral waves of electrical activity on the ventricular myocardium [1], [14]. Immediately questions like "How do spirals form?" and "Can Spirals be prevented? " or "Can one predict if a heart attack will be fatal?" spring to mind. Going beyond questions of self preservation, we may also ask about the properties of spirals, such as their wavelength and frequency, or the conditions necessary to sustain spiral activity in a given medium.

We are not yet able to answer all of these questions although progress is being made in that direction. In this paper we will describe the basic ingredients of a theory which we believe will provide answers to many of these questions. We will show how the theory can be used to answer basic questions about the structure of rotating spirals in a homogeneous excitable medium and defer other problems to forthcoming works. Here we will use the theory to calculate the wavelength and frequency for FitzHugh-Nagumo dynamics (for which numerical verification is possible) and for the Oregonator model of the Belousov-Zhabotinskii reagent.

The main ingredient of our approach is a geometrical theory for wave propagation in excitable media [8]. This theory is an asymptotic theory which applies in the limit that the dynamics of the excitable medium evidence multiple time scales and that wave fronts occur as sharp transitions (boundary layers) between slowly varying regions. The advantage of this theory is that we can get useful approximate information by solving equations that are substantially easier than the full system of governing equations.

In fact, as the time scales of the dynamics become more disparate, the method becomes better. In contrast, exact solutions for these problems are unknown and numerical solutions are both difficult and costly to obtain when the problem is very stiff.

In the next section we describe the derivation of the theory from the governing equations. In section 3 we show how to apply the theory to the case of steady rotating waves in a two dimensional media, and in the final section we give the results of this analysis for the FitzHugh-Nagumo equations and the Belousov-Zhabotinskii reagent and for the latter show that the calculated values of wavelength and frequency agree reasonably well with observed data.

§2. Wave Propagation, Dispersion and Geometrical Wavefronts.

The model of excitable media that we use is given by the equations

$$\begin{aligned} \varepsilon \frac{\partial u}{\partial t} &= \varepsilon^2 \nabla^2 u + f(u,v) \\ \frac{\partial v}{\partial t} &= \varepsilon D\, \nabla^2 v + g(u,v) \end{aligned} \qquad (2.1)$$

where $0 < \varepsilon \ll 1$, $D > 0$. The number ε represents the ratio of temporal scales of the rates of reaction for the two species u and v. Diffusion in u is scaled to ε by a simple change of spatial scale.

The functions f and g have dynamics typical of excitable media. In particular, the null cline $f(u,v) = 0$ is "N-shaped" with $f(u,v) < 0$ for all v sufficiently large. The null cline $g(u,v) = 0$ is assumed to have one intersection with the null cline of $f(u,v)$ and we take $g(u,v) > 0$ for u sufficiently large. A sketch of the null clines $f(u,v) = 0$, $g(u,v) = 0$ is shown in figure 1.

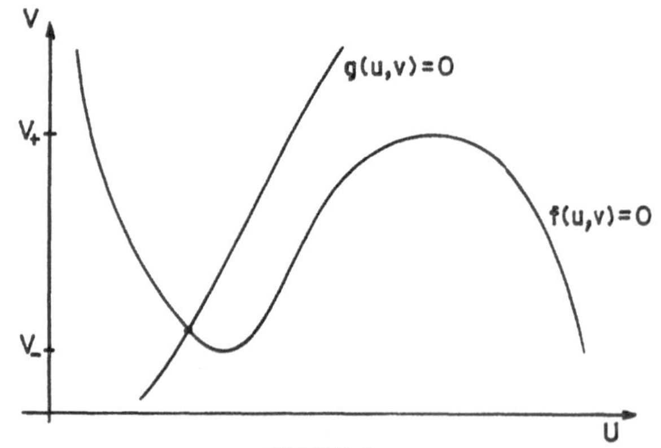

FIGURE 1

Three specific examples which we mention are the well-known Fitzhugh-Nagumo dynamics [6], [7]

$$f(u,v) = u(u - \alpha)(1 - u) - v \qquad 0 < \alpha < \frac{1}{2}$$
$$g(u,v) = u - \gamma v, \tag{2.2}$$

the piecewise linear caricature of Fitzhugh-Nagumo dynamics [13], [15]

$$f(u,v) = -u - v + H(u)$$
$$g(u,v) = u + \alpha \qquad 0 < \alpha < \frac{1}{2} \tag{2.3}$$

where H(u) is the Heaviside function, and a simplification of the Oregonator equations describing the Belousov-Zhabotinskii reaction [17]

$$f(u,v) = u - u^2 - \hat{f}\, v \left(\frac{u - q}{u + q}\right)$$
$$g(u,v) = u - v. \tag{2.4}$$

In Section 4, we calculate the wavelength and frequency for spirals in each of these media.

An important consequence of our assumptions about f(u,v) is that the null cline f(u,v) = 0 has three branches denoted $u = U_{\pm}(v)$ or $u = U_0(v)$ where $U_-(v)$ exists only for $v > v_-$ and $U_+(v)$ exists only for $v < v_+$, and $U_-(v) \leq U_0(v) \leq U_+(v)$ on the interval $v_- \leq v \leq v_+$.

The system (2.1) is oscillatory or excitable (non oscillatory) depending on the location of the unique critical point. In the limit $\varepsilon \to 0$, the system is self-oscillatory if the critical point lies on the branch $U_0(v)$ whereas it is excitable and non oscillatory if the critical point lies on one of the branches $U_{\pm}(v)$. For this discussion, it does not matter which case occurs.

To understand the behavior of system (2.1) in one spatial dimension we employ singular perturbation arguments [2], [4], [9]. These arguments reveal that waves consist of two typical regions. There are slowly varying regions in which diffusion has little effect and to lowest order in ε the variables u,v satisfy

$$\frac{dv}{dt} = g(u,v), \qquad f(u,v) = 0.$$

Since the middle branch $U_0(v)$ is unstable, u and v must satisfy $u = U_{\pm}(v)$ and on these branches the slow dynamics are

$$\frac{dv}{dt} = g(U_{\pm}(v),v) \equiv G_{\pm}(v). \tag{2.5}$$

Since u may lie on either the upper or lower solution branch of $f(u,v) = 0$, spatial discontinuities in u are possible. At such discontinuities diffusion becomes important and regions corresponding to different solution branches are patched together by a moving boundary layer. The boundary layer is found by making the exact change of variables $x = \varepsilon\zeta + y(\tau)$, $t = \tau$ and seeking translation invariant solutions $\left(\frac{\partial}{\partial\tau} \equiv 0\right)$ of the resulting equations. To lowest order in ε we obtain

$$u'' + cu' + f(u,v_0) = 0, \quad c = y'(\tau), \quad v_0 = \text{constant} \tag{2.6}$$

where to match the boundary layer to the jump discontinuity we require

$$\lim_{\zeta\to\pm\infty} f(u,v_0) = 0.$$

The boundary layer must have the correct orientation so that, for example, $\lim_{\zeta\to\infty} u(\zeta) = U_+(v_0)$ whenever the region immediately to the right of the boundary layer has $u = U_+(v)$.

The equation (2.6) is an eigenvalue problem to determine the speed c of the moving boundary layer as a function of the value v_0 in the boundary layer. The role of the boundary layer is to provide a transition between the two slow branches $U_{\pm}(v)$ keeping v fixed and continuous in the boundary layer.

Using these two ingredients we can describe wave propagation in one spatial dimension. We suppose that in some region the local dynamics are given by $\dot{v} = G_-(v)$, $u = U_-(v)$. At $x = x_1(t)$ there is a boundary layer traveling with the speed $\frac{dx_1}{dt} = c(v_1)$ where v_1 is the value of v specified from the dynamics at the point x_1. Behind the boundary layer the dynamics are switched to the upper branch $\dot{v} = G_+(v)$ with v continuous through the transition. At some distance behind the first transition there is a second transition at $x = x_2(t)$ traveling with velocity $\frac{dx_2}{dt} = -c(v_2)$ where v_2 is specified by the local dynamics and behind the layer the slow dynamics are once again switched back to the branch $\dot{v} = G_-(v)$. In both boundary layers the speed $c = c(v)$ is determined by the eigenvalue problem

$$u'' + cu' + f(u,v_0) = 0$$

$$\lim_{\zeta\to\infty} u(\zeta) = U_-(v_0) \qquad (2.7)$$

$$\lim_{\zeta\to-\infty} u(\zeta) = U_+(v_0)$$

This description of wave propagation is simplified in two special cases. If we seek periodic wave trains or solitary pulse propagation into a medium at rest, the speed of the leading wave front must be matched by the speed of the trailing wave back

$$c(v_1) = -c(v_2).$$

At any point in space, the time spent on the upper dynamics $u = U_+(v)$ is ,therefore,

$$T_1 = \int_{v_1}^{v_2} \frac{dv}{G_+(v)}$$

and if the waves are periodic, the time spent on the lower branch awaiting the next stimulus is

$$T_2 = \int_{v_2}^{v_1} \frac{dv}{G_-(v)} .$$

Thus, there is a curve parameterized by v_1

$$c = c(v_1), \qquad T = T_1(v_1) + T_2(v_1)$$

called the dispersion curve, which relates the period T of temporal oscillation to the speed of propagation. The dispersion curve as derived here is invalid for periods of order $O(\varepsilon)$ if $D = 0$ or of order $O(\varepsilon^{2/3})$ if $D > 0$.

To extend this one dimensional theory into two dimensional media we need to understand how boundary layers propagate. Between boundary layers the slowly varying region has the same dynamics, specifically

$$\frac{dv}{dt} = G_\pm(v) \qquad u = U_\pm(v).$$

To find the location of the boundary layer in $\mathbb{R}^2$ we introduce a coordinate system which is stretched in the direction normal to the boundary layer and which moves with the boundary layer. We make the change of variables

$$x = X(\varepsilon\zeta, \eta, \tau), \; y = Y(\varepsilon\zeta, \eta, \tau), \; t = \tau$$

where ζ and η are locally orthogonal coordinates and seek translation invariant solutions whose coordinate curves ζ = constant are level curves of u. In this moving coordinate system we find that

$$\frac{\partial^2 u}{d\zeta^2} + (N + \varepsilon K)\frac{\partial u}{\partial \zeta} + f(u,v) = 0, \; v = v_0 \tag{2.8}$$

to leading order, where

$$N = \frac{X_\tau Y_\eta - Y_\tau X_\eta}{(X_\eta^2 + Y_\eta^2)^{1/2}}$$

is the normal velocity and

$$K = \frac{X_\eta Y_{\eta\eta} - Y_\eta X_{\eta\eta}}{(X_\eta^2 + Y_\eta^2)^{3/2}}$$

is the curvature of ζ = constant coordinate curves. The sign convention chosen here is that the positive ζ axis is oriented at 90° clockwise from the positive η axis. One can show that $N + \varepsilon K = N + \varepsilon K|_{\zeta=0} + O(\varepsilon^2)$ where $\zeta = 0$ represents the center-line of the boundary layer. Since $N + \varepsilon K|_{\zeta=0}$ is independent of ζ we identify the boundary layer equation (2.8) with the one dimensional boundary layer equation (2.7) and require

$$N = c(v_0) - \varepsilon K \tag{2.9}$$

where N and K are the normal velocity and curvature of the center line of the wave front. A more detailed derivation of this equation is given in [8].

The equation (2.9) is analagous to the eikonal equation used in geometrical optics [11], [12]. Its interpretation is that waves travel with the plane wave velocity prescribed by v_0, incremented or decremented by a term proportional to the local curvature. The curvature correction makes physical sense when we realize that curvature can have the effect of focusing or de-focusing the flow of stimulating currents across the front.

A geometrical theory for wave front propagation in two dimensional excitable media can now be described. In slowly varying regions the dynamics are

$$\dot{v} = G_{\pm}(v), \quad u = U_{\pm}(v) \tag{2.10}$$

and boundaries between different regions move according to the eikonal equation

$$N = c(v) - \varepsilon K \tag{2.11}$$

where c(v) is the speed of plane waves determined by the value v at the boundary (continuous across the boundary).

In general, these equations are quite difficult to solve, even though they represent a major simplification of the original system. If, however, we seek periodic waves, then the two equations (2.10), (2.11) simplify further, namely we must solve

$$N = c(v) - \varepsilon K \tag{2.12}$$

where

$$T = T(v)$$

is the period between subsequent wave fronts determined by the dispersion relation.

§3. Spiral Waves.

To understand the implications of this geometrical theory to spiral waves we try to find k-armed spiral wave fronts of the form

$$\begin{aligned} X_j &= r\cos\left(\theta(r) - \omega t + \tfrac{2\pi j}{k}\right) \\ Y_j &= r\sin\left(\theta(r) - \omega t + \tfrac{2\pi j}{k}\right) \end{aligned} \qquad j = 1, 2, ..k \tag{3.1}$$

These arcs represent the center line $\zeta = 0$ of each localized coordinate system, and the radial variable r is the coordinate of the wave front.

We seek solutions of (2.12) in the form (3.1) on an annulus $r_0 < r < r_1$ to no flux boundary conditions $\theta'(r_0) = \theta'(r_1) = 0$. If $r_1 = \infty$, we want $\theta(r)$ to be asymptotically linear, and ideally, we want to solve this problem for all r_0, r_1, $0 < r_0 < r_1 < \infty$. For temporal consistency, we must require $\omega = \frac{2\pi}{kT}$ where T is the temporal periodicity of the waves.

When we calculate the normal velocity and curvature of the wave fronts (3.1) we find that

$$N = \frac{\omega r}{(1+\Psi^2)^{1/2}}, \quad K = \frac{\Psi'}{(1+\Psi^2)^{3/2}} + \frac{\Psi}{r(1+\Psi^2)^{1/2}} \qquad (3.2)$$

where $\Psi = r\theta'(r)$.

If we solve the zeroth order eikonal equation $N = c(v)$, we find by simple quadrature that

$$\theta(r) = \left(\frac{r^2}{r_0^2} - 1\right)^{1/2} - \tan^{-1}\left(\frac{r^2}{r_0^2} - 1\right)^{1/2}, \quad r_0 = \frac{c}{\omega}. \qquad (3.3)$$

The change of variables $s = \left(\frac{r^2}{r_0^2} - 1\right)^{1/2}$ yields wave fronts of the form

$$X(s) = r_0 \cos s + r_0 s \sin s$$
$$Y(s) = r_0 \sin s - r_0 s \cos s \qquad (3.4)$$

which is the parametric representation of an involute of a circle of radius r_0. This solution, which has been known for some time [19], has serious difficulties. Even with the consistency conditions $r_0 = \frac{c}{\omega}$, $\omega = \frac{2\pi}{kT}$, the involutes provide a family of spiral solutions, with no way to identify a unique member of this family as the physically realized solution. If we try to let r_0, the inner core size, approach zero, the spirals become degenerate. When r_0 is finite the involute has arbitrarily large curvature near the boundary, so that the implicit assumption that curvature is small fails to hold. Thus, (as is well known) the involute is not an adequate solution of the spiral problem.

When we include the curvature correction in (2.12) we find that $\theta(r)$ must satisfy

$$r\frac{d\Psi}{dr} = (1+\Psi^2)\left(\frac{rc}{\varepsilon}(1+\Psi^2)^{1/2} - \frac{\omega r^2}{\varepsilon} - \Psi\right), \quad \Psi = r\theta'(r) \qquad (3.5)$$

subject to the boundary conditions $\Psi(r_0) = \Psi(r_1) = 0$. Notice that the trajectories of (3.5) are monotone increasing in the parameter $c/_\varepsilon$ and monotone decreasing in the parameter $^\omega/_\varepsilon$. It follows that for every r_0, r_1 and $^c/_\varepsilon$ there is a unique value of $^\omega/_\varepsilon$,

$$^\omega/_\varepsilon = \Omega\left(\frac{c}{\varepsilon}, r_0, r_1\right)$$

for which a trajectory of (3.5) has $\Psi(r_0) = \Psi(r_1) = 0$. The function Ω is invariant under a change of scale, that is, for any scalar α

$$\omega = \varepsilon\alpha^2 \, \Omega\left(\frac{c}{\varepsilon\alpha}, \frac{r_0}{\alpha}, \frac{r_1}{\alpha}\right)$$

or, choosing $\alpha = \frac{1}{r_1}$

$$\omega = \frac{\varepsilon}{r_1^2} \, \Omega\left(\frac{r_1 c}{\varepsilon}, \frac{r_0}{r_1}\right) \qquad (3.6)$$

where the third argument of Ω is taken to be 1 and no longer indicated. Using numerical calculations, we have determined that for $x > 10$

$$\Omega(x,y) = m^* x^2 - \beta x^4 y^2 + O(y^4)$$

where $m^* = 0.330958$, $\beta = .097$.

The solution of our problem is finally completed by locating the intersection of the dispersion curve

$$c = c(v), \; \omega = \frac{2\pi}{kT(v)}$$

and the critical eigenvalue curve

$$\omega = \frac{\varepsilon}{r_1^2} \, \Omega\left(\frac{r_1 c}{\varepsilon}, \frac{r_0}{r_1}\right).$$

Although this can be solved numerically, we can estimate the solution of these equations for small ε and $r_0 = 0$ analytically. With $r_0 = 0$ the critical eigenvalue curve is

$$\omega = \frac{m^* c^2}{\varepsilon}$$

independent of outer radius r_1 provided $\frac{r_1 c}{\varepsilon} > 10$. The dispersion curve is approximated for small T by

$$c = a_0 T$$

where

$$a_0 = \frac{1}{2} \left| \frac{c'(v^*) G_-(v^*) G_+(v^*)}{G_-(v^*) - G_+(v^*)} \right| ,$$

$c(v)$ is the eigenvalue of equation (2.7), and v^* is that value of v for which $c(v^*) = 0$. With this estimate we find that the frequency, period, speed and wavelength of k-armed spirals are

$$\omega = \left(\frac{2\pi a_0}{k}\right)^{2/3} \left(\frac{m^*}{\varepsilon}\right)^{1/3}$$

$$T = \left(\frac{2\pi\varepsilon}{km^*}\right)^{1/3} a_0^{-2/3}$$

$$c = \left(\frac{2\pi a_0 \varepsilon}{km^*}\right)^{1/3} \qquad (3.7)$$

$$\Lambda = \left(\frac{2\pi\varepsilon}{km^*}\right)^{2/3} a_0^{-1/3}$$

respectively. Notice that these formulae are influenced by the actual details of the excitable dynamics only through the number a_0.

§4. Results and Conclusions

Equations (3.7) represent theoretical predictions of the properties of k-armed spirals in excitable media. We can make these predictions more explicit in the case of the FitzHugh-Nagumo equations and the Oregonator model, and these can then be checked against numerical and/or experimental data.

In the case of the simple piecewise linear dynamics (2.3) one can calculate

$$c\,(v) = \frac{1-2v}{2\left(v(1-v)\right)^{1/2}}$$

$$G_+(v) = 1 - v + \alpha, \qquad G_-(v) = \alpha - v$$

and $v^* = \frac{1}{2}$. It follows that $a_0 = \frac{1}{4} - \alpha^2$.

The case of cubic Fitzhugh-Nagumo dynamics involves more tedious calculations. However, one can show that

$$v^* = \frac{1}{27}(2 - \alpha)(1 - 2\alpha)(1 + \alpha), \quad |c'(v^*)| = \frac{9}{2} \frac{1}{\alpha^2 - \alpha + 1}$$

$$U_\pm(v^*) = \pm \left(\frac{\alpha^2 - \alpha + 1}{3}\right)^{1/2} + \frac{1}{3}(\alpha + 1)$$

from which the (complicated) expression for a_0 can be written down.

It should be possible to obtain numerical data to verify these predictions although to date we have not done so. Another check of this theory can be made by comparing with experimental data for Belousov-Zhabotinskii reagent. For this comparison we use the Tyson-Fife dynamics (2.4) [17] with parameter values chosen as

$$\hat{f} = 3, \; q = 10^{-3}, \; \varepsilon = 10^{-3}.$$

Using these values we calculate that $a_0 = .76$. To obtain this number we used numerical calculations of the eigenvalue c(v) and its derivative c'(v*) [18].

To compare our results with experimental data notice that the critical curve $\omega = \frac{m^* c^2}{\varepsilon}$ is independent of the chemistry of reaction. This expression can be written in terms of dimensional variables as

$$\frac{(\text{wavelength})^2}{\text{period}} = \frac{2\pi Du}{m^* k}$$

where D is the diffusion coefficient of the species u, in this case $u = [HBrO^{4+}]$. We take $D_u = 5 \times 10^{-5}$ cm^2/s from which we estimate

$$\frac{(\text{wavelength})^2}{\text{period}} = 10^{-3} \ cm^2/s$$

independent of the chemistry of the reaction. In fact, a collection of 5 different experiments reported in [20], [21] and [22] have $(\text{wavelength})^2/\text{period}$ ranging from 5×10^{-4} to 30×10^{-4} cm^2/s which shows relatively good agreement with our prediction.

Comparison with the dispersion curve requires knowledge of the details of the chemistry. In fact, most descriptions of spirals do not state the details of the chemistry with sufficient accuracy to allow a careful comparison with our theory. Nonetheless, using "reasonable" estimates of initial concentrations and rate constants we predict that the number $\text{wavelength}/(\text{period})^2$ should be about 2×10^{-4} cm/s^2. The collected data show $\text{wavelength}/(\text{period})^2$ ranging from 5×10^{-5} to 40×10^{-5} cm/s^2. Although these numbers show that our predictions have the correct order of magnitude, there remain discrepancies that need to be addressed. A more detailed description of this comparison with experimental data is given in [10].

The solution we suggest for spiral waves is, strictly speaking, valid only for spirals on an annulus with a central core $r_0 > 0$, even though we have used the approximation to obtain a solution for $r_0 = 0$. For small or nonexistent cores the approximation is invalid because diffusion of species v which we have ignored becomes important [5]. At this time, however, the numerical effects of these errors are not known nor is it known if correcting these errors will adequately explain the current discrepancies between theory and experiment.

In summary, with this asymptotic theory we are able to predict frequency, wavelength and period of rotating spiral patterns in excitable media. The geometric theory we use is derived in the limit that the governing dynamics act on different time scales, the ratio of which, ε , is small.

Many other questions about wave phenomena are answerable using this approximate geometric theory. These topics will be the subject of forthcoming papers.

Acknowledgement: Portions of this work were performed at the Centre for Mathematical Biology, Oxford University, supported by the Sciences and Engineering Research Council of Great Britain grant GR/C/63595 and also at the Mathematical Research Branch of the National Institute of Arthritis, Diabetes and Digestive and Kidney Diseases, National Institutes of Health, Bethesda, Md. This work was supported in part by NSF grant MCS 83-01881 and benefited greatly from the collaboration of J. Tyson as well as invaluable discussion with J. Rinzel and C. Peskin.

References

1. M. A. Allessie, F. I. M. Bonke, and F. J. G. Schopman, Circus movement in rabbit atrial muscle as a mechanism of tachycardia, Circ. Res. 41 9-18(1977).

2. R. G. Casten, H. Cohen, P. A. Lagerstrom, Perturbation analysis of an approximation to the Hodgkin-Huxley theory, Quart. Appl. Math., 32, 365-402(1975).

3. N. El-Sherif, R. Mehra, W. B. Gough, and R. H. Zeiler, Ventricular Activation Patterns of Spontaneous and Induced Ventricular Rhythms in Canine One-Day-Old Myocardial Infraction, Circ. Res. 51, pp. 152-166(1982).

4. P. C. Fife, Singular perturbation and wave front techniques in reaction-diffusion problems, Proc. AMS-SIAM Symp. on Asymptotic Methods and Singular Perturbation, New York, 1976.

5. P. C. Fife, Propagator - Controller Systems and Chemical Patterns, in Nonequilibrium Dynamics in Chemical Systems, C. Vidal and A. Pacault, eds., Springer-Verlag, 1984.

6. R. FitzHugh, Impulse and physiological states in models of nerve membrane, Biophysics J., 1, pp. 445-466(1961).

7. R. FitzHugh, Mathematical Models of Excitation and Propagation in Nerve, in Biological Engineering (ed. H. P. Schwan) pp. 1-85, McGraw-Hill, 1969.

8. J. P. Keener, Geometrical theory for spirals in excitable media, to appear.

9. J. P. Keener, Waves in Excitable Media, SIAM J. Appl. Math. 39, pp. 528-548(1980).

10. J. P. Keener and J. J. Tyson, Spiral Waves in the Belousov-Zhabotinskii Reaction, to appear.

11. J. B. Keller, Geometrical Acoustics I, The Theory of Weak Shocks, J. Appl. Phys. 25, 1938-947(1954).

12. J. B. Keller, A Geometrical Theory of Diffraction, Proc. Symp. Appl. Math., 8 27-52(1958).

13. H. P. McKean, Nagumo's equation, Adv. in Math. 4, 209-223(1970).

14. G. R. Mines, On Circulating Excitation on Heart Muscles and their possible relation to tachycardia and fibrillation, Trans. Roy. Soc. Can. 4, 43-53(1914).

15. J. Rinzel and J. B. Keller, Traveling wave solutions of a nerve conduction equation, Biophysics J. 13, 1313-1337(1973).

16. J. J. Tyson, On scaling the Oregonator equations, In: Nonlinear Phenomena in Chemical Dynamics, eds. C. Vidal and A. Pacault, Springer-Verlag, Berlin, 1981.

17. J. J. Tyson and P. C. Fife, Target Patterns in a realistic model of the Belousov-Zhabotinskii reaction, J. Chem. Phys. 73, (5) 2224-2237(1980).

18. J. J. Tyson and V. Manoranjan, The speed of propagation of oxidizing and reducing wave fronts in the Belousov-Zhabotinskii reaction, In: Nonequilibrium Dynamics in Chemical Systems, eds. C. Vidal and A. Pacault, Springer-Verlag, Berlin, 1984.

19. A. T. Winfree, The Geometry of Biological Time, Springer-Verlag, Berlin, 1980, p. 245.

20. A. T. Winfree, Rotating chemical reactions, Scientific American, June 1974.

21. A. T. Winfree, Spiral waves of chemical activity, Science 175 (1972) 634-636.

22. A. T. Winfree, Two kinds of waves in an oscillating chemical solution, Faraday Symp. Chem. Soc. 9 (1974) 38-46.

NEUROPHYSIOLOGY

BIOMECHANICAL AND NEUROMOTOR FACTORS IN MAMMALIAN LOCOMOTOR-RESPIRATORY COUPLING

D. M. Bramble
Department of Biology
University of Utah
Salt Lake City, Utah 84112/USA

Introduction

As a non-mathematical biologist I imagine that one of the great joys of those persons engaged in applied mathematics might be that of describing and predicting the behavior of complex living systems through abstractions that incorporate a relatively few key parameters. When this works, we can be a bit more confident about the rules underlying that behavior. Even when the biology does not conform, there may still be much of value in interesting, clever or otherwise novel mathematical models. But I presume (perhaps naively) that a principal objective in such ventures is that of uncovering the rules, interactions and constraints that govern the behavior of real biological systems. To the extent that this is so, the actual behavior of such systems will serve to both test and to set limits on such models.

In this essay I wish to discuss some issues and ideas concerning the phenomenon of locomotor-respiratory coupling (LRC) in mammals. This is partly because the topic is of current interest in our lab [7,8]. More important, however, is the strong suggestion that mammalian LRC is one of those phenomena that is especially amenable to analysis through a spectrum of disciplines, including mathematical modeling. Among the virtues of the topic are the strength and diversity of coupling patterns, the relative ease with which such patterns may be recorded in the intact animal [8] as well as the fact that a significant body of related neurophysiological information is already at hand. For these reasons and others, mammalian locomotor-respiratory coupling could well become one of the better understood examples of physiological and neural coupling in vertebrate animals. Indeed, preliminary attempts at mathematical modeling have already begun (Hoppensteadt, unpublished) and some excellent studies of related interest (i.e., phase locking of breathing to respirator pump cycle) have recently appeared [27,28].

What I shall attempt, then, is to provide some background on the topic and then consider especially the roles of morphology and mechanics in mammalian LRC as well as some possibilities concerning associated neuromuscular mechanisms. In so doing I will focus on quadrupedal species rather than humans. This is because in the former the role of mechanical constraint is more easily addressed. In addition, most of the relevant neurophysiological data have been obtained from quadrupedal mammals. My ultimate hope, however, is that this discussion may increase awareness of mammalian locomotor-respiratory coupling, and that it will make clearer the sorts of parameters involved in this phenomenon and why these might need to be incorporated into future realistic models of this mechanism, mathematical or otherwise.

Morphology, mechanics and coupling

By way of introduction it will be well to quickly review some basics concerning the locomotor behavior of mammals. Much of what is important and relevant is captured in Figure 1. The illustration shows how three key parameters change with increasing locomotor speed as a quadrupedal mammal progresses from its slowest gait, the walk, through its intermediate speed gait, the trot, to its fastest gait, the gallop. Oxygen consumption, which is a measure of the rate of energy expenditure, increases linearly with speed up to the point of maximum aerobic capacity. This point, which marks the transition from aerobic to substantially anaerobic metabolism, usually lies well above the speed at which the animal switches from trotting to galloping. The speed at which a given mammal breaks from a fast trot to its slowest galloping speed can be closely predicted on the basis of its body size alone [19] (see below).

In a running mammal speed is merely the product of stride frequency and stride length. The stride interval or period is defined as the time required for a given limb to execute one complete cycle of movement. The distance traversed during one full locomotor cycle is the stride length. Both stride frequency and stride length increase in a regular manner in the two slower mammalian gaits, but frequency then becomes nearly fixed at the trot-gallop transition. Thus, even though a mammal may greatly increase its speed between its slowest and fastest gallop, the rate at which it cycles its limbs changes relatively little (< 10%). It follows that nearly all increases in running speed within the gallop result from increases in stride length.

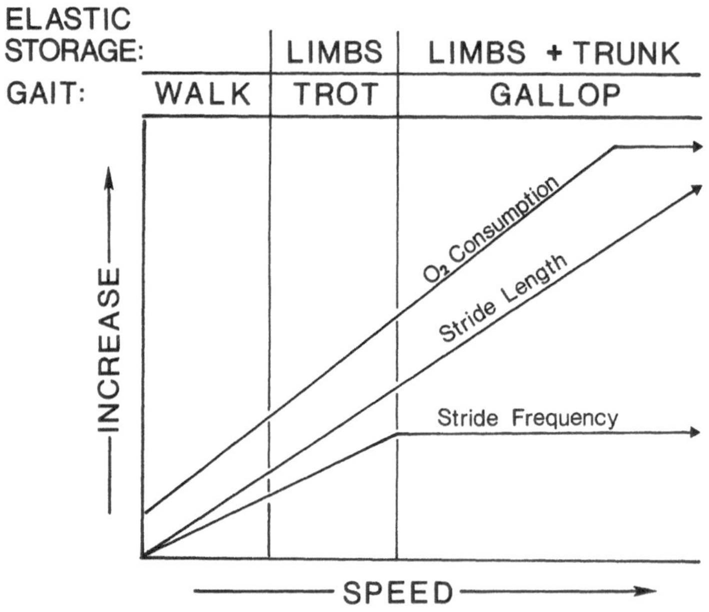

Figure 1. Relationships among physiological and gait parameters in quadrupedal mammals.

Figure 1 further indicates that the body of a running mammal possesses the ability to store mechanical energy in an elastic fashion. How this is done is gait dependent. In the trot, for example, nearly all of the storage of elastic strain energy appears to take place in the muscles of the limbs and especially in their associated tendons and ligaments [2,4]. Still more capacity for elastic storage seems to exist in the gallop and here it is thought that the oscillating trunk itself becomes a kind of large, complex spring (or series of springs) [37]. The ability to alternately store and retrieve such mechanical energy is of great physiologic importance to running mammals since it may appreciably reduce the overall metabolic cost of body transport [6,9,11]. The elastic properties of running mammals have been investigated experimentally and mathematically by several authors [1,3,24,26]. It is suggested below that the elastic properties of a running mammal may also be germane to the issue of locomotor-respiratory coupling.

Just why a mammal should attain its maximum stride frequency at the trot-gallop transition and well below its top running speed is not certainly known. Heglund et al. [19] have proposed perhaps the most attractive hypothesis to date. They suggest that the stride frequency attained at the initiation of the gallop may correspond to the natural or resonant frequency of vibration of the animal's body. Viewed as an interconnected system of masses and springs, the theory of elastic similarity [23,24,26], predicts that the natural frequency (f_n) of the body will bear the following relationship to body mass (M):

$$f_n \propto M^{.125}$$

This and all other scaling relationships discussed in the text are derived from the power function or allometric equation:

$$y = ax^b$$

wherein the parameter y is equal to the constant, a, times the parameter x raised to the power b. In all cases x will be taken as full body mass (M) in kilograms.

Experiments with mammalian species ranging in size from mice to horses have shown that there is indeed a regular scaling relationship between stride frequency at the trot-gallop transition and body size [19]. The relationship is

$$\text{Stride frequency}_{T-G} = 4.48\, M^{-0.14}$$

where frequency is in Hz. This empirical finding is in close agreement with scaling predictions based on elastic similarity criteria. The same experiments have revealed the following empirical relationships between body size and maximum running speed and stride length at the trot-gallop transition:

$$\text{Speed}_{T-G} = 1.53\, M^{0.24}$$

$$\text{Stride length}_{T-G} = 0.34\, M^{0.38}$$

where stride length is in meters and speed in m s^{-1}. (It should be mentioned here that such scaling comparisons are normally drawn at the trot-gallop boundary because this is considered to be an easily identifiable and physiologically equivalent running speed in mammals of very different body size [36].) Thus, the locomotor behavior of mammals at an equivalent running speed are now well established. What remains to be determined is what implications these scaling relationships might have for the coupling of the locomotor and respiratory cycles.

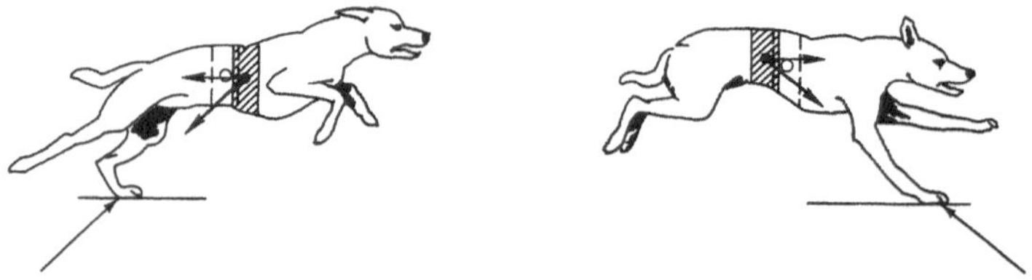

Figure 2. Mechanics of the stance (= limb supported) phases in the gallop of a medium sized mammal. Animal pushes off from hindlimbs (left) and lands on forelimbs (right). Accelerations and decelerations result in inertial displacements (from hatched position to unshaded position) of the "visceral piston." See text.

Previous studies have documented that the locomotor and respiratory cycles are tightly synchronized or phase locked in running quadrupedal mammals [5,8,21,22] as well as in bipedal humans [7,8]. The need for such synchronization in quadrupedal species is readily apparent when one considers the kinematics of the gallop (Fig. 2). The normal pattern in this gait requires that the animal break its fall and support its body on the forelimbs once each locomotor cycle (= stride). The peak ground reaction forces acting on the forelimbs are roughly equivalent to a 2 G acceleration at the speeds encountered at the trot-gallop transition [25]. Because the forelimbs are joined to the trunk by muscles, much of this load is passed directly to the ribcage [5,8]. The alignment of the load is such that it tends to force the ribcage medially and also to drive it rearwards. Both motions will reduce thoracic volume. In the alternate phase of the locomotor cycle, the animal pushes off with the rear limbs. At this time the backbone is extended and the volumes of both the thoracic and abdominal chambers increase.

Figure 2 suggests yet another important reason why mechanical constraints favor a close integration of locomotion and respiration. The abdominal contents represent a substantial inertial mass that is very loosely joined to the surrounding musculo-skeletal framework.

Because of the weak mechanical coupling, this mass must experience cyclic displacements within the body cavity. Specifically, the visceral mass should shift forward into the ribcage during the forelimb braking and stance phase of the gallop and should slide rearward during the propulsive thrust from the hindlimbs. The abdominal contents thus represent a potential "visceral piston" whose inertial displacements could either facilitate or impede the filling and emptying of the lungs depending upon their phase relationships relative to that of the ventilatory cycle. This suggestion is reinforced by the fact that the liver, which is a major component of the visceral piston, is directly attached to the respiratory diaphragm.

Morphological organization, locomotor kinematics and related mechanical considerations all imply that we should expect 1:1 coupling of the locomotor and respiratory cycles in a galloping mammal. In fact, all quadrupedal species so far examined exhibit this pattern [8]. As would be expected, the onset of exhalation is coincident with or shortly follows the impact of the forelimbs. At this time the ribcage begins to experience compressive loading and the visceral piston surges forward, thereby simultaneously compressing the thoracic chamber from behind. The inhalation half-cycle corresponds to that portion of the stride cycle when the forelimbs are free of the substrate (or nearly so) and the ribcage is therefore not under compressive load. Much of inhalation occurs while the animal is extending the trunk and driving it forward with the hindlimbs.

Although 1:1 coupling is the rule for galloping mammals, a more complex situation accompanies the trot. Both horses and dogs demonstrate the 1:1 coupling ratio while trotting but in horses, at least, several other coupling ratios are known to exist [22]. The kinematic and mechanical profile of the trot is different from that of the gallop and may permit considerably greater flexibility in the entrainment of breathing. It is worth noting that although bipedal, the normal gait of humans is a trot. Humans exhibit up to seven distinct coupling ratios while running, but the 1:1 pattern so common among quadrupeds is utilized only under exceptional circumstances [7]. In order to simplify matters, we will restrict attention to the gallop in which 1:1 coupling is the rule.

If the visceral mass of a galloping mammal were actually to operate as an oscillating piston whose displacements were an integral part of the ventilation cycle, then it is reasonable to ask whether such motions might not influence the maximum viable stride frequency. The elastic components of the musculo-skeletal framework of the body might conceivably be made more or less stiff by varying the level of contractile activity in the working musculature (although there is recent evidence that reflex mechanisms may act to maintain constant stiffness in limb muscles over a wide range of imposed loading [17,20]). The ability to alter stiffness also implies the ability to adjust the resonant frequency. It is difficult to see, however, by what means a mammal could effect comparable adjustments in the frequency response of its visceral mass. Because 1:1 coupling in the gallop requires that the visceral piston vibrate at the same frequency as the body frame, it is at least conceivable that the upper limit on sustained stride frequency in a galloping mammal is set, not exclusively by the elastic properties of the musculo-skeletal system, but also by the frequency limits of the visceral piston. If so, galloping mammals must choose locomotor frequencies close to the resonant frequency of their guts.

As a starting point we might think of the visceral piston as a simple mass and spring system that approximates harmonic motion as the animal breaks into a gallop. The advantage of such a model is that empirically derived scaling data then permit specific predictions about the dynamic behavior of the piston. This model requires several basic assumptions. The first is that the liver is our visceral piston. This is a reasonable assumption since the liver is the largest and heaviest of the abdominal organs and it is always attached directly to the muscular diaphragm. Liver mass has been found to scale as $M^{.89}$ [29]. Likewise, it is assumed (for reasons given above) that the scaling of the natural frequency of the visceral piston is the same as that for stride frequency at the trot-gallop transition, or $M^{-.14}$. Finally, it is also assumed that there is a periodic displacing force (at the same frequency). This derives from the ground reaction force and passes from the limbs to the trunk as the animal alternately lands on the forelimbs and pushes off with the hindlimbs. Force platform studies indicate that the peak ground reaction force on the forelimbs of galloping mammals is equivalent to a nearly constant 2.1 G load [25]. The load appears to be the same for both large and small species, thus indicating that peak ground reaction forces scale as $M^{1.00}$.

Using the basic equation relating natural frequency (f_n) to mass (m) and spring modulus (k), it is possible to obtain the expected scaling value for the latter:

$$f_n = \frac{1}{2\pi}\sqrt{\frac{k}{m}} \quad \text{or} \quad M^{-.14} = \frac{1}{2\pi}\sqrt{\frac{k}{M^{.89}}} \quad \text{or} \quad k \propto M^{.61}$$

Given a spring whose stiffness scales as $M^{.61}$ and a displacing force that goes as $M^{1.00}$, it is then possible to calculate the scaling relationships of several dynamic properties of the visceral piston. These are listed in Table 1, together with several potentially correlated gait parameters.

TABLE 1: Scaling predictions for dynamics of mammalian visceral piston and possible locomotor correlates at trot-gallop transition

Visceral Piston[a]		Locomotor Parameter	
Natural frequency (f_n)	$\propto M^{-.14}$	Stride frequency	$\propto M^{-.14}$
Maximum velocity (v_o)	$\propto M^{.24}$	Running speed	$\propto M^{.24}$
Stroke length (s)	$\propto M^{.39}$	Stride length	$\propto M^{.38}$
Acceleration (a)	$\propto M^{.10}$	Limb excursion angle	$\propto M^{.10}$
Spring modulus (k)	$\propto M^{.61}$		

[a]The piston is assumed to experience simple harmonic motion, but with a periodic displacing force ($\propto M^{1.00}$) acting at a frequency ratio (r) close to 1.00. Piston mass scales as $M^{.89}$. See text.

As the Table indicates, maximum amplitude of the visceral piston displacement is predicted to scale as $M^{.39}$. This value is not meaningfully different from that relating stride length to body mass ($M^{.38}$) and it is not difficult to imagine that piston displacement might well be functionally related to the displacement of the surrounding body frame (i.e., to stride length). Similarly, maximum velocity (v_o) of the piston at the midpoint of its cycle is expected to scale as $M^{.24}$ which is exactly that of maximum trunk velocity or running speed ($M^{.24}$) of mammals at the trot-gallop transition. Finally, the acceleration of the visceral piston increases as the .10 power of body mass. This is also the observed scaling factor for limb excursion angle as mammals achieve their lowest galloping speed [25]. The limb excursion angle determines the fraction of the total stride cycle over which the running mammal must alternately brake and accelerate its center of mass.

It is necessary to note that the close match between the predicted behavior of the visceral piston and empirically determined gait parameters is critically dependent upon the scaling of liver mass to overall body mass in mammals (i.e., $M^{.89}$). This parameter in the model is nominally independent of such specifically gait dependent parameters as stride and resonant frequency. Were the liver to scale with the value generally expected for the mass of internal organs ($\sim M^{1.00}$), there would be no special correspondence between piston and locomotor dynamics.

Admittedly, this is a very rough and overly simplistic model. It is vulnerable on several counts. It includes, for example, no specific provision for damping of any sort. A true forced vibration model with viscous damping would, in many ways, be more realistic. The practical difficulty with such realistic models, however, is that they incorporate additional parameters for which no empirical values exist and for which there are no ready estimates. The simple vibrational model employed here is one for which all the necessary real values are known. That it yields predictions which make some intuitive sense may be purely fortuitous. On the other hand, the explicit predictions of this or possible alternative models are subject to direct testing with current technologies (e.g., high speed cineradiography).

Neuromotor mechanisms and coupling

There is considerable evidence that rhythmogenesis in both locomotion [18,40] and respiration [10,16,42] involves central pattern generators (CPGs). Accordingly, locomotor-respiratory integration must require a coupling of these pattern generators or neural oscillators. This implies some means of communication between the two types of pattern generators so that, at minimum, information on frequency, phase and possibly intensity can be exchanged. Since virtually all available information indicates that locomotion entrains breathing and not the reverse [8], the locomotor pattern generator (LPG) must be regarded as the driving or dominant oscillator which then entrains the respiratory pattern generator (RPG) through phasic input signals of some sort. There would appear to be two basic means of achieving this (Fig. 3).

The first possibility is a direct linkage in which the LPG communicates with the RPG via central afferent pathways. The second option is that of a more indirect and distal linkage through peripheral mechanoreceptors stationed in skeletal muscles that are stimulated phasically during locomotion. Here two possible pathways exist for linking LPG and RPG. The simpler would send phasic afferent signals from stretch receptors (SRs) in the working locomotor muscles (in limbs, back) back to the respiratory pattern generator in the brainstem. The other mechanism would utilize signals from SRs situated within the respiratory musculature itself. However, these receptors would report locomotor pattern to the RPG by virtue of mechanical linkage between locomotor and respiratory structures. The basis of peripheral rather than central communication between the two pattern generators will be addressed first, since several lines of evidence suggest that such a system could and very possibly does operate in the intact animal.

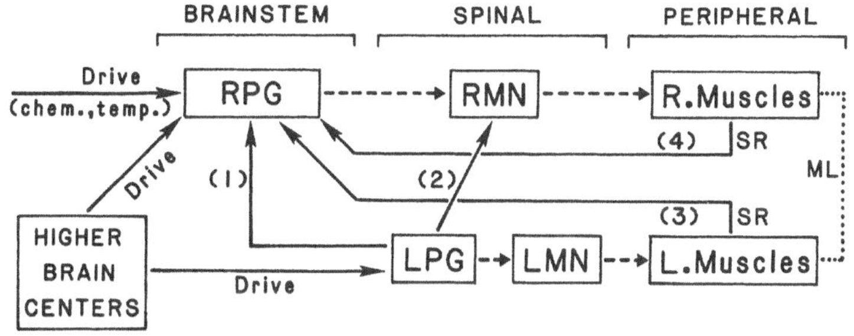

Figure 3. Flow diagram showing possible mechanisms for functionally linking locomotor and respiratory pattern generators (LPG, RPG) through afferent pathways (1-4). A direct central linkage (path 1) is possible as are two peripheral pathways (3,4) involving stretch receptors (SR) within the locomotor or respiratory muscles (L., R. Muscles). Mechanical linkages (ML) between locomotor and respiratory muscles permit the latter to report phasic locomotor events. Pattern generators drive muscles through intermediate locomotor and respiratory motoneuron pools (LMN, RMN). Pathway 2 depicts a potential entrainment linkage between LPG and respiratory motoneurons at the spinal level [see ref. 38]. Various "drive" influences also affect the pattern generators.

There is little solid evidence that phasic input from peripheral mechanoreceptors of working muscle can alone effect the entrainment of breathing to limb motion. Several studies involving bicycle exercise in humans have purported to show evidence of respiratory phase locking to leg motion, but it now appears that such evidence must be viewed with suspicion [41]. While not denying a possible role for locomotor-based afferents in mammalian LRC, I wish here to concentrate on another possibility which has so far gone unnoticed, i.e., that chest wall receptors play a direct role in the synchronization of gait and lung ventilation.

Experiments on several species of mammals have demonstrated that afferent signals from intercostal muscle spindles can affect the timing of the ventilation cycle [12,31,33,34]. The receptors are located in both external and intercostal muscles but appear to be more numerous in the former. Sufficient mechanical or electrical stimulation of such receptors within those intercostal muscles residing in the anterior or mid-region of the ribcage produces reflexive inhibition of the inspiratory phase of the respiratory cycle [31,33]. The threshold for this reflex diminishes with the elapsed time since the onset of the inspiratory activity and is therefore inversely related to lung volume [35]. Hence, a much stronger stimulus is required to effect a phase switch (i.e., inspiration to expiration) early in the process of lung filling than later when inspiratory duration is approaching the normal limit of its period. The inhibitory reflex from the anterior and mid-intercostal muscles is suprapinal and involves ascending afferent pathways which impinge ultimately on the medullary RPG [30,31]. It should also be mentioned that both RPG and LPG are subject to modulation through various "drive" inputs. The respiratory generator is particularly sensitive to chemical stimulation and both pattern generators can be strongly affected (or overridden) by input from higher brain centers (e.g., motivational influences).

Some detail on the structure of the RPG and its connections to peripheral mechanoreceptors is given in Figure 4. The structure of the RPG complex is based largely on the model developed by von Euler [16]. Three essential components are depicted. There is a central inspiratory activity generator (CIA). This motoneuron pool provides the excitatory drive to lower level motoneuron pools (i.e., phrenic, PMN; intercostal, ICMN) serving the two principal sets of respiratory muscles (diaphragmatic, DM; intercostal, ICM). The two additional components of the pattern generator are the inspiratory off-switch (I O-S) and an integrating and summating center (*S*). The latter receives an excitatory corollary discharge from the CIA as well as excitatory input from various peripheral mechanoreceptors, including the spindles of the anterior intercostal muscles. The integration center itself sends excitatory stimuli to the inspiratory off-switch. As the name implies, the I O-S will, at threshold, inhibit the CIA, thereby terminating contractile activity in the inspiratory muscles.

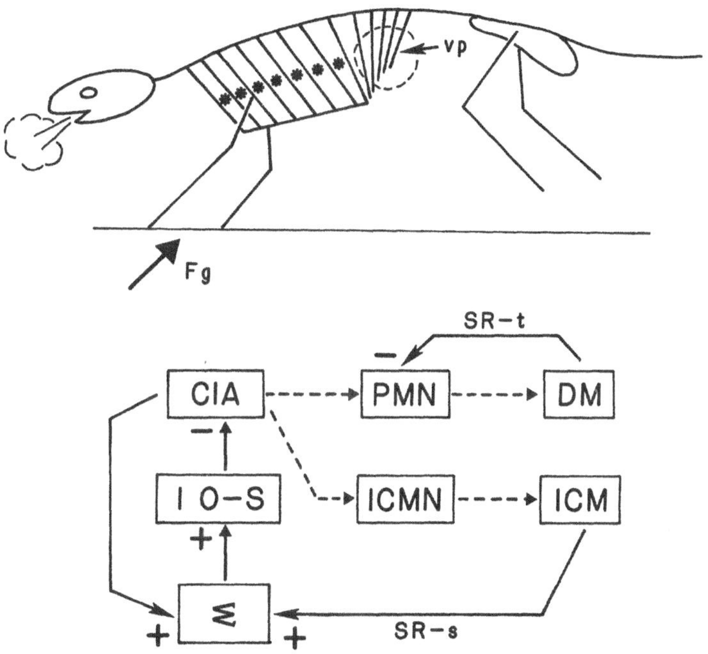

Figure 4. Model for synchronization of exhalation with forelimb stance phase of gallop. Associated mechanical (upper) and neuromotor (lower) events are depicted. Ground reaction forces (Fg) compress ribcage, thereby stimulating spindle stretch receptors (SR-s) in the anterior and middle intercostal muscles (ICM; asterisks). Afferent signals from spindles have an excitatory (+) input to integration center (Σ) of respiratory pattern generator. This helps to excite the inspiratory off-switch (I O-S) which then has an inhibitory effect (-) on the central inspiratory activity center (CIA). Inhibition of CIA terminates motor activity in the diaphragmatic and intercostal muscles (DM, ICM) and the associated phrenic and intercostal motoneuron pools (PMN, ICMN). Phrenic motoneuron activity is further inhibited by stimulation of Golgi tendon organ receptors (SR-t) in the diaphragm caused by forward surge of visceral piston (vp).

The corollary discharge of the CIA onto the integration center assures that inspiratory activity is self-limiting even in the absence of afferent feedback from peripheral stretch receptors. Still, it is clear that additional excitatory input to the integration center will trigger an earlier activation of the inspiratory off-switch. In this way peripheral mechanoreceptors can, to a considerable extent, regulate the timing of the respiratory cycle.

The biomechanical and neurophysiological factors that could produce a synchronization of respiratory and locomotor events in a galloping mammal are depicted in Figures 4 and 5. As previously noted, there are sound mechanical reasons why exhalation should be coincident with impulsive loading of the forelimbs and thorax (Fig. 4). Such loading will tend to collapse the ribcage and force it posteriorly about its articulations with the vertebral column. This will subject the anterior intercostal muscles (especially external intercostals) to a sudden and large stretching force, thus stimulating the associated spindle organs. These mechanoreceptors will, in turn, send excitatory signals to the integration center of the respiratory pattern generator. Inspiratory off-switch will then be pushed to threshold, thereby inhibiting the CIA and hence the principal inspiratory muscles.

As Figure 4 also suggests, one additional mechanical factor and related neuromotor reflex may also have a role in synchronizing the onset of exhalation with forelimb impact. The forward surge of the visceral piston will cause it to load the central, tendinous region of the diaphragmatic muscle to which the liver is joined. Although the diaphragm has few muscle spindles, it is well endowed with stretch receptors in the form of Golgi tendon organs. These receptors are known to have an inhibitory affect on phrenic motor activity when appropriately stimulated [14]. Because the diaphragm will be taut (i.e., near the end of the inspiratory phase) when the forelimbs strike the ground, anterior displacement of the liver (and associated viscera) will be very effective in stretching its tendon organs. The inhibitory stimuli from the tendon organs are shown as converging on the phrenic motoneuron pool, but it is also possible that they also act at the level of the RPG.

The mechanics and circuitry for coupling inhalation with the appropriate phase of the locomotor cycle are less complex (Fig. 5). As the animal thrusts from the hindlegs, the ribcage and abdominal cavities are expanded. There is no external loading on the anterior intercostal muscles and hence no inhibitory signal is sent to the RPG. Likewise, the visceral mass will shift posteriorly and will not subject the tendon organs of the diaphragm to tensil stress. On the other hand, the posterior (floating) ribs of the thorax will be spread as the back is extended. This will stretch the associated intercostal muscles and stimulate their spindle organs. These spindles, in distinct contrast to those of the anterior ribcage, facilitate inspiratory motor activity in the diaphragm [13,14,31]. This reflex pathway is apparently spinal rather than supraspinal.

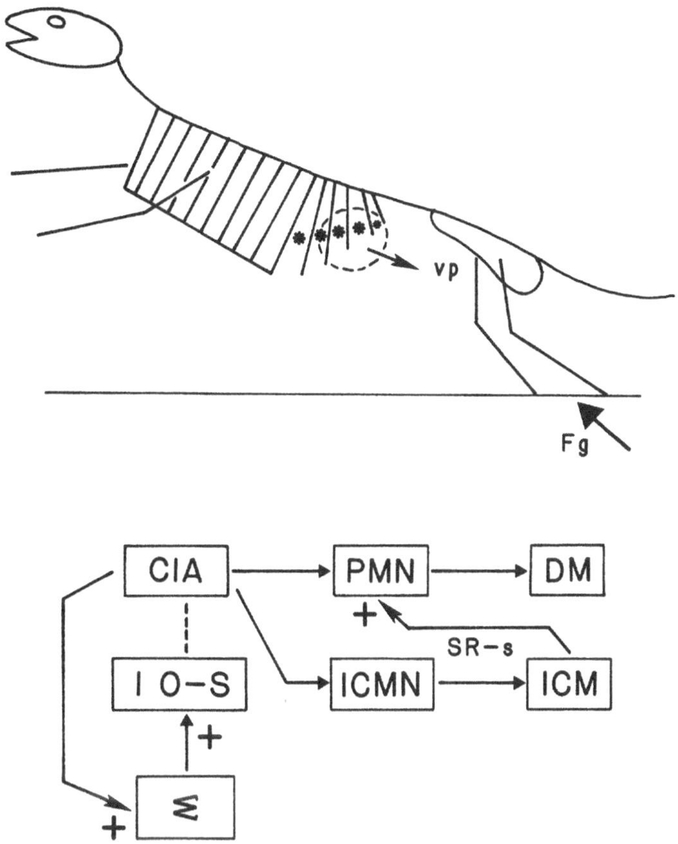

Figure 5. Model for synchronization of inspiration with hindlimb stance phase of gallop cycle (cf. Fig. 4). I O-S has not reached threshold, thereby allowing CIA to provide motor drive to inspiratory musculature. Inspiratory activity in the diaphragm is enhanced by excitatory input from posterior intercostal muscle spindles to phrenic motoneurons. Visceral piston shifts rearward. Abbreviations as in Fig. 4.

Though experimentally well documented, the actual functional significance of the intercostal reflexes to ventilatory control in the intact animal has remained decidedly enigmatic [32]. Their real function, as suggested here, may be related to the close coupling of locomotor and breathing cycles in naturally exercising mammals. Such a function, vital as it may be to the exercising individual, is not likely to reveal itself in traditional physiologic animals (i.e., anesthesized, spinalized or decerebrate). Nonetheless, exactly such preparations have now provided the first solid suggestion that there may be locomotor-respiratory integration through direct interaction of central pattern generators.

Recent studies by French workers [38,39] have shown that pharmacologically induced fictive locomotion in anesthesized rabbits not only produces rhythmic locomotor-like bursting in the motor nerves to hindlimb and forelimb muscles, but also elicits synchronized bursting in the phrenic nerve to the diaphragm. This pattern, generally at a 1:1 coupling ratio, persists even after high spinal transection in which all connections with the traditional brainstem RPG are presumably severed. Still lower transection of the spinal cord (thoracolumbar junction) terminates the coupling of hindlimb and respiratory motor activity. It therefore seems that lumbar LPGs entrain cervical respiratory motoneuron pools serving the diaphragm. Since slower but nonetheless rhythmic bursting is seen in the phrenic nerve of rabbits with both high and low spinal transections, there is also reason to believe that CPGs for respiratory activity may exist at the spinal level [38]. At the moment too little is known about these "spinal respiratory oscillators" to assess their possible role in locomotor-respiratory integration in the intact animal. Still, as Figure 3 indicates, we must now entertain the prospect of direct linkages among locomotor and respiratory oscillators at a level not previously anticipated.

One additional piece of evidence points to the possibility of direct central integration and coupling of the locomotor and ventilation cycles. Electrical stimulation of subthalamic locomotor centers has recently been shown to elicit parallel and correlated motor output to both limb and respiratory muscles (= diaphragm) during fictive locomotion in cats [15]. Since afferent traffic to the brain was blocked in these animals, peripheral mechanoreceptors could presumably play no coordinating role in linking the breathing and locomotor cycles. The thrust of this study was to demonstrate that "feedforward" mechanisms might account for much of the observed close matching of ventilatory response to exercise level and corresponding metabolic demand in mammals. The same data hint, but do not prove, that the periodic locomotor and respiratory bursts in such animals may be phase locked.

Certainly the anatomic distribution, biomechanical milieu, as well as the specific physiological effects of the intercostal stretch receptors make them an exceptionally attractive mechanism for assuring the entrainment of breathing and locomotor cycles in running mammals. Yet clearly there are also reasons for suspecting that such coupling could involve direct linkages between appropriate CPGs. If both peripheral (stretch receptor) and central mechanisms are responsible, then the control system is to some extent redundant.

Assuming that both central and peripheral components are involved, it may be that the central components, acting alone, provide for only an approximate and relatively soft coupling, and that this is then strengthened and modulated by peripheral sensory input of the types advocated here. Thus, the very rigorous phase locking seen in naturally exercising mammals may require this modulatory component. Such peripheral input, conveying as it must, detailed information on the associated mechanics of the chestwall, trunk, and limbs, may be essential in establishing specific phase relations and coupling ratios.

Two lines of evidence provide some support for this hypothesis. First, peripheral input from mechanoreceptors located in the lungs and major airways appear to have a dominant role in the entrainment of the breathing cycle of anesthesized cats to a respirator pump cycle, and also in determining viable coupling ratios [27]. Secondly, bicycle exercise experiments designed to look for the possible entrainment of human breathing pattern to leg motion also suggest that peripheral mechanoreceptive input may be essential to strict phase locking. Close approximations to coupling (at several different coupling ratios) are observed in many subjects during such experiments, but the most carefully controlled studies indicate that precise, sustained phase locking of leg and breathing cycles is not common and generally occurs for only brief intervals at relatively high work loads [41]. This result contrasts sharply with the persistent and strong coupling observed between the limb and respiratory cycles in humans running at all speeds [7].

A simple but essential difference, however, exists between bicycle exercise and natural locomotion. In the former, the upper body is relatively stationary and supported by the seat. It does not therefore experience the periodic impulsive loadings which are inevitably associated with running. Such loadings, with their attendant deformations of the ribcage and displacements of the viscera, are a likely source of stimulation for chestwall (and diaphragmatic) stretch receptors. However, this mechanical climate may be approximated in bicycle exercise when riders lift off the seat and shift their weight to the legs as may happen, for example, during high levels of exertion.

Acknowledgments

I thank two former students, David R. Carrier and Joel E. Russel, for many stimulating discussions (and arguments) about thoracic mechanics and LRC in mammals. Russel also produced the illustration in Figure 2. This work has been supported in part by NIH Biomedical Research Support Grant RR07092 administered by the University of Utah.

Literature Cited

[1] Alexander, R. McN.; Mechanics and scaling of terrestrial locomotion, In: *Scale Effects in Animal Locomotion* (Pedley, T. J., ed.), Academic Press, London (1977), pp. 93-110.

[2] Alexander, R. McN. and Bennet-Clark, H. C.; Storage of elastic strain energy in muscle and other tissues, Nature 265 (1977), 114-117.

[3] Alexander, R. McN. and Vernon, A.; The mechanics of hopping by kangaroos (Macropodidae), J. Zool., Lond. 194 (1975), 265-303.

[4] Alexander, R. McN., Maloiy, G. M. O., Ker, R. F., Jayes, A. S. and Warni, C. N.; The role of tendon elasticity in the locomotion of the camel (*Camelus dromedarius*), J. Zool., Lond. 198 (1982), 293-313.

[5] Attenburrow, D. P.; Time relationships between the respiratory cycle and limb cycle in the horse, Equine vet. J. 14 (1982), 69-72.

[6] Biewener, A., Alexander R. McN., Heglund, N. C.; Elastic storage in the hopping of kangaroo rats (*Dipodomys spectabilis*), J. Zool, Lond. 195 (1981), 369-383.

[7] Bramble, D. M.; Respiratory patterns and control during unrestrained human running, In: *Modelling and Control of Breathing* (Whipp, B. J. and Wiberg, D. M., eds.), Elsevier, New York (1983), pp. 213-220.

[8] Bramble, D. M. and Carrier, D. R.; Running and breathing in mammals, Science 219 (1983), 251-256.

[9] Cavagna, G. A., Heglund, N. C. and Taylor, C. R.; Walking, running and galloping: mechanical similarities between different animals, In: *Scale Effects in Animal Locomotion* (Pedley, T. J., ed.), Academic Press, London (1977), pp. 111-125.

[10] Cohen, M. I.; Neurogenesis of respiratory rhythm in the animal, Physiol. Rev. 59 (1979), 1105-1173.

[11] Dawson, T. J. and Taylor, C. R.; Energetic cost of locomotion in kangaroos, Nature 246 (1973), 313-314.

[12] Decima, E. E. and Euler, C. von; Intercostal and cerebellar influences on afferent phrenic activity in the decerebrate cat, Acta. Physiol. Scand. 76 (1969), 148-158.

[13] Decima, E. E., Euler, C. von and Thoden, U.; Intercostal to phrenic reflexes in the spinal cat. Acta. Physiol. Scand. 75 (1969), 568-579.

[14] Duron, B.; Intercostal and diaphragmatic muscle endings and afferents, In: *Regulation of Breathing* (Pt. 1), (Hornbein, T. F., ed.), Marcel Dekker, New York (1981), pp. 473-540.

[15] Eldridge, F. L., Milhorn, D. E. and Waldrop, T. G.; Exercise hyperpnea and locomotion: parallel activation from the hypothalamus, Science 211 (1981), 844-846.

[16] Euler, C. von.; Central pattern generation during breathing, Trends in Neurosci., Nov. (1980), 275-277.

[17] Greene, P. R. and McMahon, T. A.; Reflex stiffness of man's antigravity muscles during knee bends while carrying extra weights, J. Biomech. 12 (1979), 881-891.

[18] Grillner, S.; Control of locomotion in bipeds, tetrapods, and fish, In: *Handbook of Physiology* (Brooks, V. B., ed.), Sec. I, *The nervous system* (Brookhart, J. M. and Mountcastle, V. B., eds.), Vol. 2, *Motor control*, American Physiological Society, Bethesda (1981).

[19] Heglund, N. C., Taylor, C. R. and McMahon, T. A.; Scaling stride frequency and gait to animal size: mice to horses, Science 186 (1974), 1112-1113.

[20] Hoffer, J. A. and Andreassen, S.; Regulation of soleus muscle stiffness in premammillary cats: intrinsic and reflex components, J. Neurophysiol. 45 (1981), 267-285.

[21] Hörnicke, H. and Meixner, R.; Depth and frequency of breathing in exercising horses, Proc. Int. Union Physiol. Sci. 13 (1977), 332.

[22] Hörnicke, H. and Meixner, Pollman, U.; Respiration in exercising horses, In: *Equine Exercise Physiology* (Snow, D. H., Persson, S. G. B., Rose, R. J., eds.), Burlington Press, Cambridge (1983), pp. 7-16.

[23] McMahon, T. A.; Size and shape in biology, Science 179 (1973), 1201-1204.

[24] McMahon, T. A.; Using body size to understand the structural design of animals: quadrupedal locomotion, J. Appl. Physiol. 39 (1975), 619-627.

[25] McMahon, T. A.; Scaling quadrupedal galloping: frequencies, stresses, and joint angles, In: *Scale Effects in Animal Locomotion* (Pedley, T. J., ed.), Academic Press, London (1977), pp. 143-151.

[26] McMahon, T. A.; *Muscles, Reflexes, and Locomotion*, Princeton University Press, Princeton (1984).

[27] Petrillo, G. A. and Glass, L.; A theory for phase locking of respiration in cats to a mechanical ventilator, Amer. J. Physiol. 246 (Regulatory Integrative Comp. Physiol. 15) (1984), R311-R320.

[28] Petrillo, G. A., Glass, L. and Trippenbach, T.; Phase locking of respiratory rhythm in cats to a mechanical ventilator, Can. J. Physiol. Pharmacol. 61 (1983), 599-607.

[29] Prothero, J. W.; Organ scaling in mammals: the liver, Comp. Biochem. Physiol. 71A (1982), 567-577.

[30] Remmers, J. E.; Inhibition of inspiratory activity by intercostal muscle afferents, Resp. Physiol. 10 (1970), 358-383.

[31] Remmers, J. E.; Extra-segmental reflexes derived from intercostal afferents: phrenic and laryngeal responses, J. Physiol. (London) 233 (1973), 45-62.

[32] Remmers, J. E.; Functional role of the inspiratory terminating reflex from the intercostal muscle spindles, In: *Respiratory Centres and Afferent Systems* (Duron, B., ed.). Editions de l'Institut National de la Santé et de la Recherche Medicale, Paris (1976), pp. 183-191.

[33] Remmers, J. E. and Martila, I.; Action of intercostal muscle afferents on the respiratory rhythm of anesthetized cats, Resp. Physiol. 24 (1975), 31-41.

[34] Shannon, R. and Freeman, D.; Nucleus retroambigualis respiratory neurons: responses to intercostal and abdominal muscle afferents, Resp. Physiol. 45 (1981), 357-375.

[35] Speck, D. F. and Webber, C. L., Jr.; Time course of intercostal afferent termination of the inspiratory process, Resp. Physiol. 43 (1981), 133-145.

[36] Taylor, C. R.; The energetics of terrestrial locomotion and body size in vertebrates, In: *Scale Effects in Animal Locomotion* (Pedley, T. J., ed.), Academic Press, London (1977), pp. 127-141.

[37] Taylor, C. R.; Why change gaits? Recruitment of muscles and muscle fibers as a function of speed and gait, Amer. Zool. 18 (1978), 153-161.

[38] Viala, D. and Frenton, E.; Evidence for respiratory and locomotor pattern generators in the rabbit cervico-thoracic cord and their interactions, Exp. Brain Res. 49 (1983), 247-256.

[39] Viala, D., Viala, C. and Frenton, E.; Coordinated rhythmic bursting in respiratory and locomotor muscle nerves in the spinal rabbit, Neurosci. Letters 11 (1979), 155-159.

[40] Wetzel, M. and Stuart, D. G.; Ensemble characteristics of cat locomotion and its neural control, Progr. Neurobiol. 7 (1976), 1-98.

[41] Yonge, R. P. and Petersen, E. S.; Entrainment of breathing in rhythmic exercise, In: *Modelling and Control of Breathing* (Whipp, B. J. and Wiberg, D. M., eds.), Elsevier Biomedical, New York (1983), pp. 197-204.

[42] Younges, M. K. and Remmers, J. E.; Control of tidal volume and respiratory frequency, In: *Regulation of Breathing* (Hornbein, T. F., ed.), Pt. 1, Marcel Dekker, New York (1981), pp. 621-671.

Analysis of a VCON Neuromime

by

F. C. Hoppensteadt
Department of Mathematics
University of Utah
Salt Lake City, UT 84112

The voltage-controlled oscillator neuromime (VCON) described below is motivated by three things: First, van der Pol's equation [1] and similar "flush-and-fill" models have been used since the 1920's to study neural activity. Subsequent work on van der Pol's equation resulted in a map of parameter space that describes phase-locking of the oscillator to external forcing [2,3]. Second, R. Guttman, et. al., [See 4] used experimental procedures developed by Hodgkin and Huxley for studying squid axon membranes, and they obtained a similar phase-locking portrait for these membranes. This showed that neuron membranes have rich phase-locking, or synchronization properties. Third, I developed and studied a model of a neuron that emphasizes frequency encoded information in [5]. This model is based on a voltage controlled oscillator circuit, and it is called **VCON**. A VCON provides a straightforward method for building circuit analogs for neural networks, and its associated mathematical model is in phase-amplitude coordinates, so it avoids a major difficulty in dealing with nonlinear oscillators.

The work in [5] describes first-order VCONs and networks of them. The work here investigates second order VCONs. Second order VCONs are quite similar to models of pendula and cryogenic electronic devices, they can generate relaxation oscillations, and they exhibit interesting hysteresis. Both the free behavior of these circuits and their response to oscillatory forcing are described here.

1. Neurons and VCONs.

A ***neuron***, or nerve cell, includes a network of ***dendrites*** that collect incoming signals; a ***cell body*** that processes inputs and is capable of generating output signals in the form of voltage pulses called ***action potentials***, an ***axon*** that carries these action potentials away from the cell body; and a ***synapse*** that converts action potentials into chemical signals that are released from the cell to drift away and interact with other dendritic structures. The principal parts of a neuron are shown in Figure 1.

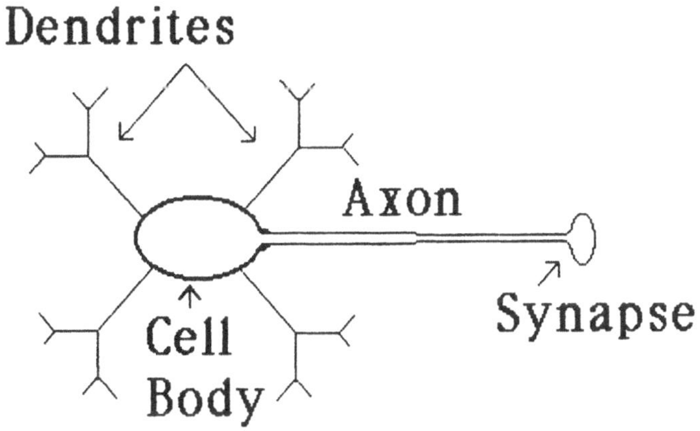

Figure 1. Principal parts of a neuron.

The VCON model simulates each of these processes using integrated circuits. The ***cell body*** is modelled by a voltage controlled oscillator, denoted by VCO here. VCOs are devices whose output is a constant or oscillatory voltage with its frequency being controlled by an input voltage. A neuron cell body behaves like a VCO in several ways: First, experimental evidence [4] suggests that neuron membranes can generate action potentials at a rate that depends on a stimulating voltage. In addition, the standard models in mathematical neurobiology are the Hodgkin-Huxley, FitzHugh-Nagumo and van der Pol models, and each acts like a voltage controlled oscillator since each oscillates with a frequency that is related to the strength of stimulation. For example, an increased applied voltage in the FitzHugh-Nagumo model causes the firing frequency to increase from zero to a maximum.

Although an ***axon*** can be modelled using a sequence of VCONs [See 5], we will ignore axon propagation times here. In addition, we restrict attention here to synapses impinging directly on a cell body, and so we will ignore the cell's *dendritic* network. The *synapse* is modelled by a rectifying diode followed by a low pass filter. The diode passes only the positive part of the cell body voltage to reflect a threshold that must be exceeded by the membrane potential to cause release of neurotransmitters. The low-pass filter simulates first-order chemical kinetics of neurotransmitter binding with the post-synaptic membrane and diffusion out of the synaptic gap. The output, U, is the signal received by the post-synaptic cell. The circuit is described in Figure 2.

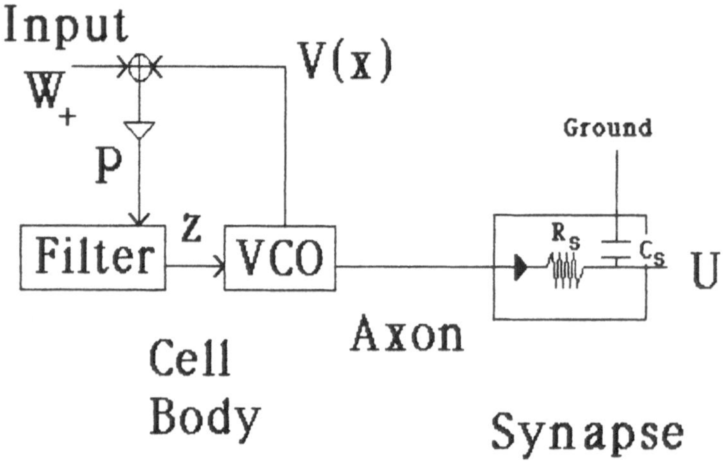

Figure 2: VCON.

In Figure 2, a voltage W coming from an impinging synapse is combined with the post-synaptic potential V in a voltage adder. The sum is then passed through a linear amplifier to clip its amplitude to physiological limits. The result is $P(V + W_+)$, where P is an increasing sigmoidal function. We suppose that $P(0) = 0$, $P(-v) = -P(v)$, $dP/dv > 0$ and $P^* = \max_V P(v) < \infty$. P is then filtered to clip its frequency to physiological limits, and the filter output, z, is used as the controlling voltage in the VCO. This feedback of the cell's membrane potential is similar to the Hodgkin-Huxley model where feedback simulates the opening and closing of membrane channels, but we make here no direct connection between the VCON circuit and the ionic channels in a neuron's membrane. The circuit described here is a form of phase-locked loop, and it is similar to ones commonly used in communications devices.

The cell body's membrane potential is modelled by the VCO output. It passes through a rectifying diode, which simulates a threshold for release of neurotransmitters, and then through a low pass filter to simulate diffusion of neurotransmitter across a synaptic gap and chemical binding with the post-synaptic membrane. The voltage that would be seen by a receiving membrane is U, and it plays the role of W in this circuit.

The mathematical formulation is as follows: The unknowns in this model are the filter output $z(t)$ and the VCO phase $x(t)$. The wave form V is fixed, and in commercially available VCOs it can be a square wave, a triangular wave or a sinusoidal wave. We take $V(x)$ to be a 2π periodic function of the phase x. Also, the linear amplifier characteristic P is fixed, and it is an increasing sigmoidal function. The model for this circuit is

$$dx/dt = \omega + z$$

$$R_f C_f \, dz/dt = -z + P(V(x) + W_+)$$

$$R_s C_s \, dU/dt + U = V(x)_+ .$$

The first equation describes the VCO: ω is the VCO center frequency and z is its controlling voltage. The second equation describes the output z of a low-pass filter into which P is put: $R_f C_f$ gives the time constant of this filter. Finally, U is the output of the synapse filter, and its time constant is $R_s C_s$. With these choices, the model is quite similar to a pendulum with an oscillating support point and an applied torque [See 6], and to a model for a point Josephson junction [See 7].

2. **Analysis of a free VCON.**

We can ignore the third equation, since it determines U once x is known. Setting $y = z + \omega$ and $r = R_f C_f$ in the first two equations gives

$$dx/dt = y$$

$$r \, dy/dt = -y + \omega + P(V(x) + W_+)$$

In this section, we will study this system when $W = 0$ (the free case). To fix ideas, we take $V(x) = \cos(x)$ and $P(v) = P^* \tanh(v)$, where P^* is a constant.

The free problem results when no external forcing is present (i.e., $W = 0$). In this case, the problem can be completely analyzed. First, the strip

$$S = \{ (x, y) : \omega - P^* \leq y \leq \omega + P^* \}$$

is globally asymptotically stable. In fact, we see directly from the second equation that solutions approach this strip with rate $1/r$. It is appropriate to view this system as being one on the cylinder $|y| < \infty$, $0 \leq x < 2\pi$, since the equations are invariant under the translation $x \to x + 2\pi$.

$\omega > P^*$. When $\omega > P^*$, there are no static states since the isoclines $dx/dt = 0$ and $dy/dt = 0$ never cross. The fact that the strip S is invariant shows that the mapping defined from the cross section of this strip at $x = 0$ to the one at $x = 2\pi$ has a fixed point in S, and so there is a periodic solution, say $y = Y(x)$. This periodic solution can be viewed as being a limit cycle on the cylinder. In fact, this limit cycle is unique, and it is globally asymptotically stable. [See 7]. The relation between x and t can be determined on this limit cycle by solving the equation

$$dx/Y(x) = dt.$$

For example, the period of the free oscillation can be approximated when $r \ll 1$. In this case, there is an outer solution of the problem [See 8],

$$y \sim \omega + P(\cos(x)).$$

Thus, $dx/dt \sim \omega + P(\cos(x))$, and so $dt \sim dx/[\omega + P(\cos(x))]$. Integrating this quantity for $x = 0$ to $x = \pi$ gives half a period of the limit cycle:

$$T(\omega) \sim 2\pi / \sqrt{(\omega^2 - P^{*2})}.$$

This rough calculation is useful in computer simulations.

$\omega < P^*$. In this case, there are two equilibria on the cylinder; namely, when $P(\cos(x)) = -\omega$. One is left of $x = \pi$, and it is denoted by x_L. The other one is denoted by x_R. The linearization of the system about these states has the form

$$r\, d^2x/dt^2 + dx/dt - A(x^*)\, x = 0$$

where $A(x^*) = -P'(\cos(x^*)) \sin(x^*)$. The characteristic polynomial is

$$r\lambda^2 + \lambda - A(x^*) = 0,$$

and its roots are $\lambda = [-1 + \sqrt{(1 + 4\, r\, A(x^*))}]/2r$. Thus, we see that for $x^* = x_L$, $A < 0$, and so this is a stable sink. It is a stable spiral point if ω is small, and it is a stable node if ω is near P^*. Similarly, we see that x_R is a saddle point.

For each value of r, there is a number $\omega^*(r)$ such that

i) If $\omega < \omega^*(r)$, then the sink is globally stable, with the exception of the saddle point and its stable manifolds.

ii) If $\omega^*(r) < \omega < P^*$, a stable sink and a stable limit cycle both exist.

The value $\omega = \omega^*(r)$ corresponds to the value at which there is a saddle-saddle connection on the cylinder. These results are proved for the case $P(v) = v$ in [7], and the proofs for general P are similar, but they are not presented here.

The following figures show solutions of this system for $r = 2.0$, $P(v) = \tanh(v)$, $W = 0$ and $V(x) = \cos(x)$.

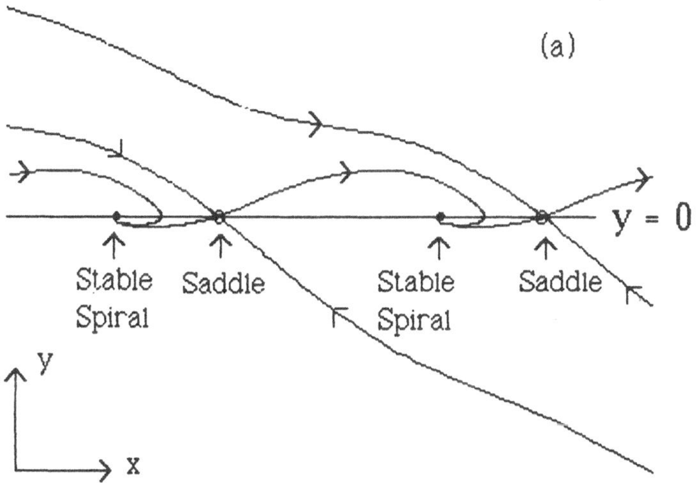

$x = -2\pi$ $x = 0$ $x = 2\pi$

Figure 3a: In this case, $\omega = 0.5$ and we see that all solutions approach stable nodes except for the stable manifolds of the saddle points. Thus, the VCON equilibrates to a stable voltage, $V(x_L)$.

Figure 3b: In this case, $\omega = 0.75$. We see here that there are stable nodes and there is a stable periodic solution, $y = Y(x)$, coexisting for this choice of parameters. Y is referred to as being a *running periodic solution* or *a limit cycle* on the cylinder x modulo 2π.

The running periodic solution corresponds to repetitive firing by the cell. Thus, the VCON has two stable modes of behavior in this case.

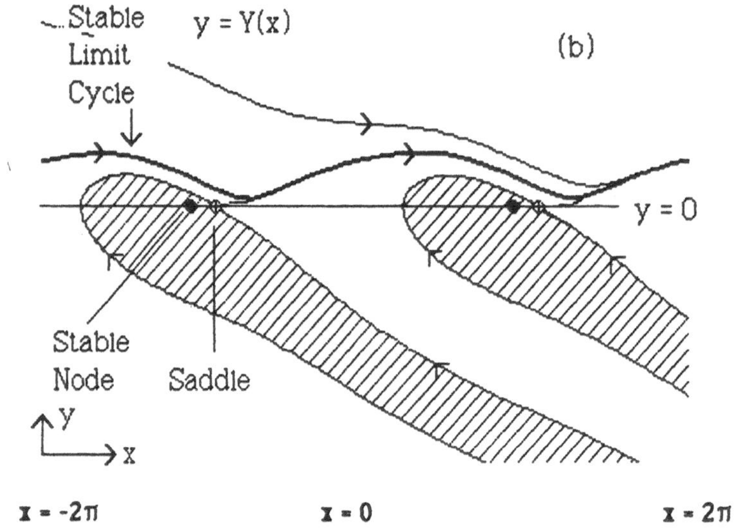

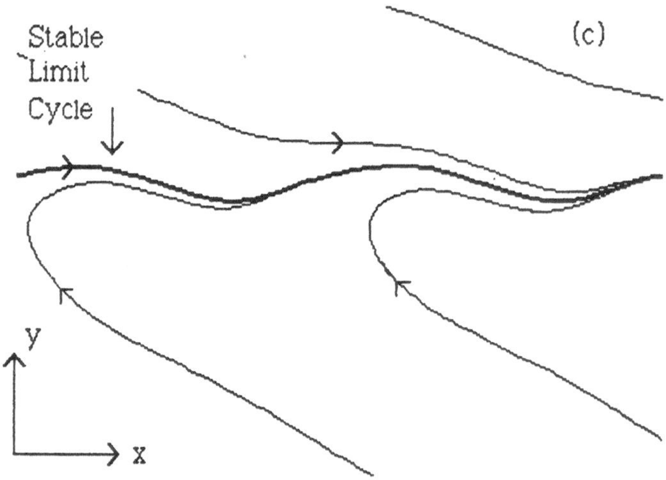

Figure 3c: In this case, $\omega = 1.0$. There are no equilibria, but there is a unique limit cycle, and it is globally stable.

We note that in the preceding figures, as ω increases through the value $\omega^*(2.0)$ the upper stable manifolds of the saddle points pass through saddle-saddle connections and they approach $z = -\infty$ rather than $z = +\infty$ as $t \to -\infty$. This system exhibits hysteresis as ω increases from zero to 1.0 and back to zero: For small values of ω, the system equilibrates to a stable node, where it remains until the saddles and nodes coalecse and disappear (at $\omega = P^*$), after this the running periodic solution is reached. When ω is decreased, the system remains on the limit cycle until the saddle-saddle connection is reached. Below this, the solution equilibrates to the sink.

A similar result follows for general P and V when ω is fixed and W is a constant parameter. In fact, a comparable system results when the mean value of $P(V(x) + W_+)$ is subtracted from z, but with ω replaced by a function of W.

3. **Response of a VCON to external forcing.**

We consider the case where $r \ll 1$. It follows that y approaches an outer solution of the form $y = Y(x,t,r)$ [See 8]. Thus, the system reduces on the outer solution to the form

$$dx/dt = Y^*(x,t,r) = \omega + P(V(x) + B\cos(\nu t)_+) + O(r)$$

where the error estimate holds uniformly for $0 \le t < \infty$. This is a problem to which Denjoy's theory applies [See 9], and so we are motivated to consider the rotation number

$$\rho = \lim_{t\to\infty} x(t)/2\pi t \ .$$

This number can be used to discover phase-locking behavior using numerical simulations. Figure 4 describes dominant phase locking regions for this reduced model. In Figure 4, the intervals over which ρ is constant are ω-intervals over which the oscillator is phase-locked onto the forcing function.

A detailed study of the response of the VCON to oscillatory forcing is beyond the scope of this presentation, and it will be carried out elsewhere. However, we summarize certain results and problems about the general case:

$\omega < \omega^*(r)$. In this case, the free system has a stable sink, so the forced system has a stable oscillation of period 2π. [See 6].

$\omega > P^*$. In this case, the problem is one of a forced limit cycle. Similar models

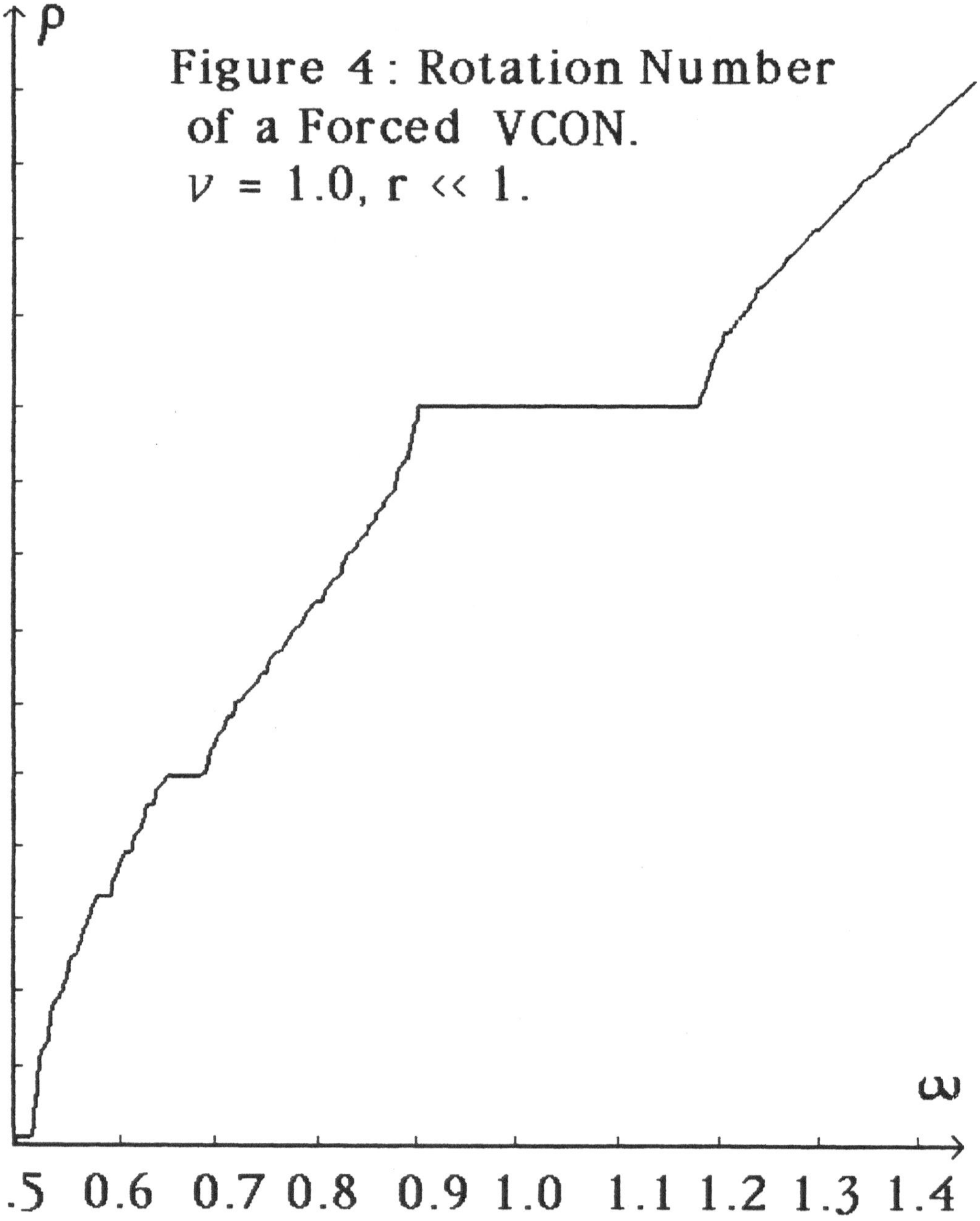

Figure 4: Rotation Number of a Forced VCON. $\nu = 1.0$, $r \ll 1$.

have been studied by methods of averaging and harmonic balance [10,11]. In particular, when the problem is non-resonant, a stable oscillation exists having a period that is a rational number multiple of the forcing period. In the case of resonance, the system behaves erratically.

$\omega^*(r) < \omega < P^*$. This is an overlap region in which two stable solutions of the free problem coexist. The forcing drives two stable oscillations having different periods in non-resonance cases, but the resonance case has not yet been investigated.

References:

1. B. Th. van der Pol, Phil. Mag. 2(1926)978; see also B. Th. van der Pol and J. van der Mark, ibid. 6(1928) 763, and Arch. Neerl. Physiol. 14(1929) 418. Also, K. F. Bonhoeffer, J. Gen. Physiol. 32(1948) 69, Naturwissensch. 40(1953)301. R. FitzHugh. In Biological Engineering, H. P. Schwann, ed., pp1-85, McGraw-Hill, 1962.

2. M. L. Cartwright, J. E. Littlewood, Ann. Math. 54(1951)1-37; Camb. Phil. Soc. Proc. 45(1949)495-501; Ann. Math. 48(1947)472-494; J. London. Math Soc. 20(1945)180-189.

3. J. E. Flaherty, F. C. Hoppensteadt, Frequency entrainment of a forced van der Pol oscillator, Studies in Appl. Math. 58(1978)5-15.

4. R. Guttman, L. Feldman, E. Jacobsson, Frequency entrainment of squid axon, J. Membr. Biol. 56(1980)9-18.

5. F. C. Hoppensteadt, An Introduction to the Mathematics of Neurons, Cambridge U. Press, 1985.

6. J. J. Stoker, Nonlinear Vibrations, Interscience, New York, 1950.

7. M. Levi, F. C. Hoppensteadt, W. Miranker, The dynamics of the Josephson junction, Quart. Appl. Math., July, (1978) 167-198.

8. F. C. Hoppensteadt, Properties of solutions of ordinary differential equations with small parameters, Comm. Pure Appl. Math. XXIV (1971) 807-840.

9. V. I. Arnol'd, Geometrical Methods in the Theory of Ordinary Differential Equations, Springer-Verlag, 1983.

10. W. Chester, The forced oscillations of a simple pendulum, J. Inst. Maths. Applics., 15(1975)289-306

11. C. Hayashi, Nonlinear Oscillations in Physical Systems, McGraw-Hill, New York, 1964.

Acknowledgement: This work was supported in part by NSF Grant #MCS82-08986.

COUPLED OSCILLATORS AND LOCOMOTION BY FISH

N. Kopell
Department of Mathematics
Northeastern University
Boston, MA 02115

1. Locomotor Central Pattern Generator

Fish of many species propel themselves through water by rhythmic undulations; fins are used for stabilization and change of direction, but not for stereotypic straight line movements [1]. The undulations are caused by contractions which pass down the muscles along the spinal cord (with muscles on the opposite sides of the fish 180° out of phase). These contractions are in turn directed by motoneurons which emerge from special positions in the spinal cord having a spatial periodicity that matches the segmentation of the backbone. Measurements from these positions ("ventral roots") show rhythmic voltage changes (bursts of activity) at each such point, with a uniform frequency and a phase lag between any two points that are proportional to the distance between the points; i.e., the neural activity is a constant speed travelling wave. For technical reasons, much of this data has been gathered for dogfish and lamprey [1,2]; some of the observations have been corroborated for other species.

The neural "program" which produces this behavior is called a central pattern generator (CPG) and is a special case of very general phenomena involving neural control of rhythmic processes such as walking, running, chewing and breathing. Such programs can be modified by feedback from muscles, joints, etc. [3,4,5] but are known to be able to operate in the absence of such feedback [6]. For vertebrates, many of the CPG's, including the fish locomotor CPG, are localized in the spinal cord; appropriate behavior can be elicited even in an animal in which descending control from supraspinal structures has been eliminated (as in "spinal" fish).

Except for very simple invertebrate systems [7,8], the circuitry of CPG's is mostly unknown. However, it is an accepted working hypothesis [9] that CPG's involve the interaction of coupled oscillators, each unit of which may be a single cell or a subnetwork. One major question is to understand how the oscillators interact in order to stably and spontaneously generate the appropriate spatial and temporal pattern of phases so that the muscles contract at the right moments. Since little is known, this may seem to be an impossible question. However, as we shall illustrate for fish locomotion, many observations are consequences of general properties of coupled oscillators, independent of the details of the physiology; the mathematics provides a tool for identifying which aspects of the physiology are significant. Indeed, the mathematics described below makes deep predictions about the physiology.

Before discussing the mathematics, we give some more observations about the neural activity corresponding to swimming in intact and spinal dogfish. The propagation speed of the neural wave is proportional to the swimming speed of the fish, which can vary over an order of magnitude [1]. Furthermore, it is also proportional to the frequency of oscillation at each position, so the wave-length of the wave is independent of the frequency. (This contrasts with the dispersion function for propagation down an axon in Hodgkin-Huxley models [10].) The ratio of the wave-length to the size of the fish is species dependent; for dogfish the wavelength is about twice the fish; for thinner fish, such as lamprey, the ratio is smaller.

Intact or spinal dogfish can swim backward in response to appropriate stimulation. When this happens, the wave of neural activity passes from the tail to the head (i.e., "caudal" to "rostral") instead of vica versa [2]. In forward swimming in spinal animals with the head fixed in place, the leading segment (the segment at which the phase is most advanced) need not be the head of the fish, but can be anywhere in the most rostral third; one then observes neural waves travelling outward in both directions from the leading segment [2]. The position of the leading segment is easily changed by experimental alterations [2]. Finally, there is nothing special about the ends of the spinal cord; small pieces of the cord, functionally isolated from the rest, have the ability to "fictively" swim forward or backward with appropriate phase relationships [2]. (The activity of the spinal cord which would give rise to normal swimming in an intact animal is called "fictive swimming", and is believed to correspond to neural activity in the intact swimming animal.)

The model to be presented is directed to the question: how are phase lags generated such that the above observations are easily understandable? The form of the answer, elaborated below, is that the waves depend on the oscillators being coupled more generally than diffusive coupling, in a way one might expect from cells interacting via synapses. In particular, the waves to be described are quite different from those studied in reaction-diffusion equations [11] or related parabolic P.D.E.'s [10].

2. The Model

Since little is known about the circuitry, it is desirable for the model to be robust, with as few assumptions as possible. It is assumed that the spinal cord can be treated as a chain of coupled oscillators. (Each oscillator may correspond to a vertebral segment; however, the mathematics is unchanged if each oscillator extends over many segments, or is much smaller than a segment, e.g. a single cell. See [12] for further comments and relevant data.) Since the two sides of the cord operate in antiphase, only one side is considered. Each oscillator is assumed to be described by an O.D.E. of arbitrary dimension

$$\underline{u}_k' = F_k(\underline{u}_k), \quad \underline{u}_k \in R^m, \quad k = 1,\ldots,N+1 \tag{1}$$

having a stable limit cycle solution. The major hypothesis is that the coupling is "weak". We also assume that oscillators interact only with their nearest neighbors along the chain; the latter hypothesis is not strictly necessary (see [13] for generalization). The full equations then have the form

$$\underline{u}_k' = F_k(\underline{u}_k) + \varepsilon[G^+(\underline{u}_{k+1},\underline{u}_k) + G^-(\underline{u}_{k-1},\underline{u}_k)], \tag{2}$$

where $\varepsilon \ll 1$ and G^+ need not equal G^- (i.e., the coupling may be isotropic or nonisotropic). The finiteness of the chain implies

$$G^+(\underline{u}_{N+2},\underline{u}_{N+1}) \equiv 0 \equiv G^-(\underline{u}_0,\underline{u}_1). \tag{3}$$

We assume that the uncoupled oscillators (1) are similar in dynamics, i.e., there is some F satisfying $F_k = F + O(\varepsilon)$ for each k. The most interesting case for dogfish locomotion has $F_k = F$. (This contrasts with a special case of equations (2), (3) analyzed in [14], motivated by "frequency plateaus" in intestinal peristalsis. There the coupling was simple - linear diffusion - but $F_k \not\equiv F$, and the interest was in the behavior for $F_{N+1} - F_1$ large enough to prevent phase-locking.) We shall consider the more general hypothesis partly to compare the results with those of Cohen et. al. [15] on a simple model for lamprey locomotion.

There is one more hypothesis on the coupling which is generic, but crucial, and which corresponds to assuming that the oscillators are not coupled in a form mathematically equivalent to diffusion (e.g. by electrotonic coupling). For diffusive coupling, $G^{\pm}(\underline{u},\underline{u}) \equiv 0$; the assumption ((7)) to be made below always fails for such coupling.

Before this last hypothesis is spelled out, we reduce (2), (3) to a set of equations of much lower dimension, one per oscillator. This can be done using invariant manifold theory: at $\varepsilon = 0$, the assumption that (1) has a limit cycle oscillation implies that (2), (3) has an asymptotically stable invariant torus of dimension (N+1). Such a torus persists under perturbations, as in (2), (3), $\varepsilon \neq 0$ [16]; the size of the perturbation ε is uniform in N [13]. The equations on this invariant torus turn out to have a special form: Coordinates θ_k, $\underline{Y}_k$ are chosen in a neighborhood of the limit cycle of (1), θ_k the phase around the limit cycle. Let $\phi_k \equiv \theta_{k+1} - \theta_k$, ω_k the frequency of the limit cycle of (1), and $\varepsilon\Delta_k \equiv \omega_{k+1} - \omega_k$. Then $\theta_1,\phi_1,\ldots,\phi_N$ satisfy

$$\theta_1' = O(1) \tag{4}$$

$$\phi_k' = \varepsilon[\Delta_k + H^+(\phi_{k+1}) - H^+(\phi_k) + H^-(-\phi_k) - H^-(-\phi_{k-1})] + O(\varepsilon^2) \tag{5}$$

and (3) becomes

$$H^+(\phi_{N+1}) = 0 = H^-(-\phi_o). \tag{6}$$

Here $H^\pm$ are 2π-periodic scalar functions which are explicitly computable from F and $G^\pm$ by, e.g. averaging techniques. (For simple "λ-ω" oscillators [11] with linear diffusive coupling, $H^\pm$ have the form $H^+ = H^- = A \sin \phi + B[\cos \phi - 1]$; if $G^\pm = a_\pm \underline{u}_{k\pm1}$, then $H^\pm = A^\pm \sin \phi + B^\pm \cos \phi$ [14].) The "synaptic coupling hypothesis" referred to above is

$$H^+(0) \neq 0 \neq H^-(0). \tag{7}$$

We note that for ordinary linear diffusion, $H^+ = H^- \equiv H$, and $H(0) = 0$.

3. Why "Synaptic Coupling" Makes a Difference

For fish locomotion, we are interested in solutions to (2), (3) that are "phase-locked", i.e., solutions in which all oscillators move at the same frequency, and with phase lags ϕ_k that are independent of time. The existence and stability of such solutions is not affected by the $O(\varepsilon^2)$ terms in (5) [17, p. 168], which are the only ones that couple the $\{\phi_k\}$ to θ_1; hence, the stable phase-locked solutions are precisely the stable critical points of (6) and

$$\dot{\phi}_k = \Delta_k + H^+(\phi_{k+1}) - H^+(\phi_k) + H^-(-\phi_k) - H^-(-\phi_{k-1}). \tag{8}$$

(Here $\dot{\phi}_k \equiv (d/d\tau)\phi_k$, where $\tau = \varepsilon t$ is slow time.) Suppose, now, that (7) is not satisfied, e.g. $H^+(0) = 0 = H^-(0)$. Then $\phi_k \equiv 0$ is a solution to (6), (8); this corresponds to oscillation in unison, with no phase lags. If, however, (7) is satisfied, then (6) implies that the first and last equations of (8) are not satisfied by $\phi_k \equiv 0$, so any phase-locked solution must involve phase lags and/or leads. The finiteness of the chain is crucial; it will be seen below that the edge effects force nontrivial ϕ_k for all k, uniformly in N.

4. A "Phony Continuum" Equation (P.C.E.)

If $H^+ = H^- \equiv H$, and H is an odd function of its argument, then it is easy to explicitly compute the critical points of (6), (8) [14,15]. In general, however, this is not possible. Instead, we shall exploit the structure of (8) to guess at the behavior of the solution for large N. (For dogfish, the number of vertebral segments is more than 65 [2].)

Let $H^{\pm} = H_e^{\pm} + H_o^{\pm}$ be the decomposition of $H^{\pm}$ into its even and odd components. The time independent version of (8) can be rewritten as

$$0 = \Delta_k + [f(\phi_{k+1}) - f(\phi_{k-1})] + [g(\phi_{k+1}) - 2g(\phi_k) + g(\phi_{k+1})] \tag{9}$$

where

$$f(\phi) = \frac{(H_e^+ + H_e^-)}{2} + \frac{(H_o^+ - H_o^-)}{2}$$

$$g(\phi) = \frac{(H_o^+ + H_o^-)}{2} + \frac{(H_e^+ - H_e^-)}{2} \quad .$$

Note that, in the isotropic case $(H^+ = H^-)$, $f(\phi) = H_e(\phi)$ and $g(\phi) = H_o(\phi)$. The boundary condition (6) can be written

$$f = g \ \text{ at } \ \phi_o, \quad f = -g \ \text{ at } \ \phi_{n+1}. \tag{10}$$

For N large, we may look for solutions to (9), (10) for which the graph of ϕ_k vs. k lies approximately on some smooth curve. This is facilitated by rewriting (9) as

$$0 = \frac{1}{N}\{\beta_k + \frac{2[f(\phi_{k+1}) - f(\phi_{k-1})]}{2/N} + \frac{1}{N}\frac{[g(\phi_{k+1}) - 2g(\phi_k) - g(\phi_{k-1})]}{1/N^2}\}. \tag{11}$$

(Here $\Delta_k \equiv \beta_k/N$; this scaling keeps the frequency difference from one end of the chain to the other independent of N.) Assume $\beta_k = \beta(k/(N+1))$ for some smooth $\beta(x)$; e.g. $\beta_k \equiv \beta$, which corresponds to a linear frequency gradient. Equation (11) then suggests the continuum equation

$$0 = \beta(x) + 2f(\phi)_x + \frac{1}{N}\, g(\phi)_{xx} \tag{12}$$

where $0 \leq x \leq 1$ and $\phi(x) = \phi_k$ for $x = k/(N+1)$. Boundary conditions (10) become

$$f = g \ \text{ at } \ x = 0; \quad f = -g \ \text{ at } \ x = 1. \tag{13}$$

For diffusive coupling, (13) implies that $\phi = 0$ at $x = 0$ and $x = 1$; for synaptic coupling, $\phi \neq 0$ at the end points.

Equation (12) is not a standard continuum limit associated with a physical continuum, as is, e.g. a reaction-diffusion equation. Discretization of the latter leads to coupling coefficients between adjacent elements whose size is unbounded as the mesh size decreases; by contrast, for (2), the coupling strength is $O(\varepsilon)$ for all N. However, as will be seen below, (12) provides the correct "diagnostic equation" for understanding the behavior of (2).

5. Formal Solutions to the P.C.E.

Equations (12), (13) form a singularly perturbed 2-point boundary value problem. It is singularly perturbed because the coefficient of the highest derivative ϕ_{xx} is multiplied by 1/N, which is small when N is large. Such equations can often be solved formally by asymptotic methods. Over much of the interval [0,1], the solution is close to some solution of the "outer equation"

$$0 = \beta(x) + 2f(\phi)_x. \tag{14}$$

Near $x = 0$ or $x = 1$, or sometimes in the interior, there can be a "shock layer" which satisfies the second order equation

$$0 = 2f(\phi)_X + g(\phi)_{XX} \tag{15}$$

where $X = Nx$ is a "stretched variable". Solutions to (12), (13) can be constructed by "patching together" solutions to (14), (15), each satisfying one of the boundary conditions of (13). For more information about the procedure, see [18]; more details for (12), (13) are found in [13].

We shall require some not terribly restrictive hypotheses on f and g in order that there be stable phase-locked solutions. These hypotheses are required to hold in an interval which includes the range of the appropriate solution to (14) and the boundary values $\phi = \phi_L$ at $x = 0$ and $\phi = \phi_R$ at $x = 1$. The first is that $g'(\phi) > 0$. This is necessary for temporal stability. (Note that, otherwise, in some part of its range, the time dependent version of (12) could behave like a backward heat equation!). For "λ-ω" oscillators, coupled by $G^{\pm} = D(\underline{u}_{k+1} - \underline{u}_k)$ or $D\underline{u}_{k\pm1}$, $g' > 0$ near $\phi = 0$ if D is close to being diagonal [14], but can otherwise be negative near $\phi = 0$; the hypothesis holds for simple models of "excitatory synaptic" coupling, but not "inhibitory synaptic coupling" (G. B. Ermentrout, pers. comm. See [12] for the equations.)

A more technical hypothesis is that $f'' \neq 0$. This is a "genuine nonlinearity" condition; it is possible that results can be proved in its absence. Finally, we assume a hypothesis that is always true if $|H^+ - H^-|$ is not too large: there is a value ϕ_T of ϕ for which $f'(\phi_T) = 0$, and ϕ_T is between ϕ_L and ϕ_R. (See Fig. 1. If $H^+ = H^-$ then $f = H_e$ and $\phi_T = 0$.)

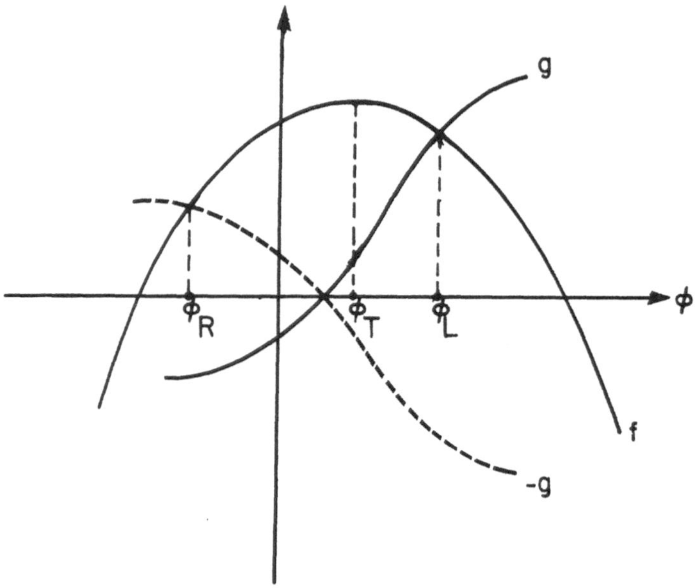

Figure 1

The functions f and g. The dashed curve is the graph of -g. ϕ_L satisfies $f(\phi) = g(\phi)$; ϕ_R satisfies $f(\phi) = -g(\phi)$. ϕ_T satisfies $f'(\phi) = 0$.

Under these hypotheses, (12), (13) can be uniquely solved, with a shock layer whose position depends on the sign of f'', and a balance between the size of the frequency gradient and the amount of anisotropy. The solution for $H^{\pm} = A \sin \phi + B \cos \phi$, $A,B > 0$, with $\beta(x) < 0$ (a decreasing frequency gradient) is given in Fig. 2. For frequency gradients that are sufficiently large, the hypothesis $g'(\phi) > 0$ cannot be satisfied since the range of the solution to (14), satisfying the appropriate one of the boundary conditions, is too large.

Of particular interest are the solutions to (12), (13) when $\beta(x) \equiv 0$. For these, the solutions to (14) are constants. There can be a boundary layer at either end depending on the sign of f'' and the direction of the stronger coupling; if $H^+ = H^-$, then the boundary layer is in the middle at $x_0 = 1/2$. The solution for $H = A \sin \phi + B \cos \phi$, $A,B > 0$ is shown in Fig. 3; the solution for $A > 0$, $B < 0$ is its mirror image. The position of the boundary layer is very sensitive. Any anisotropy or frequency gradient moves it (in the limit as $N \to \infty$) to one end or the other.

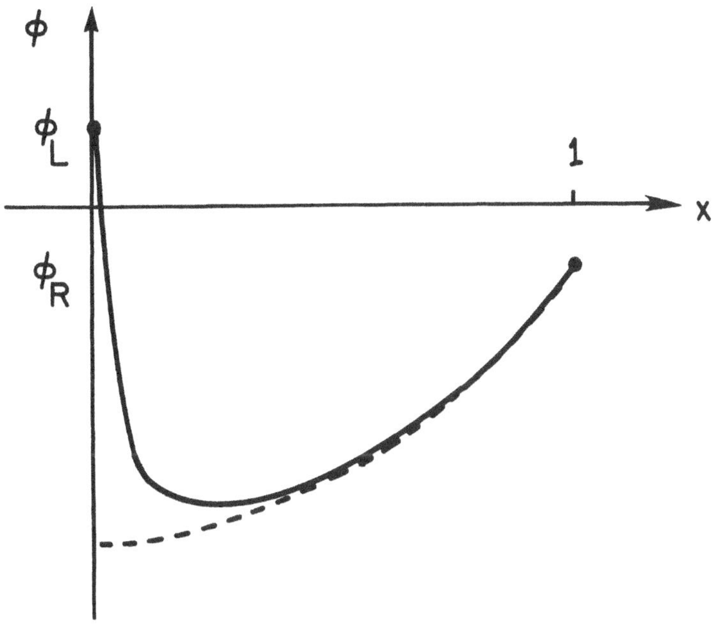

Figure 2

A schematic picture of the solution to (12), (13) for $H^{\pm} = A \sin \phi + B \cos \phi$, $A,B > 0$, $\beta(x) < 0$. The dashed curve is the solution to the outer equation (14) satisfying the R.H. boundary condition.

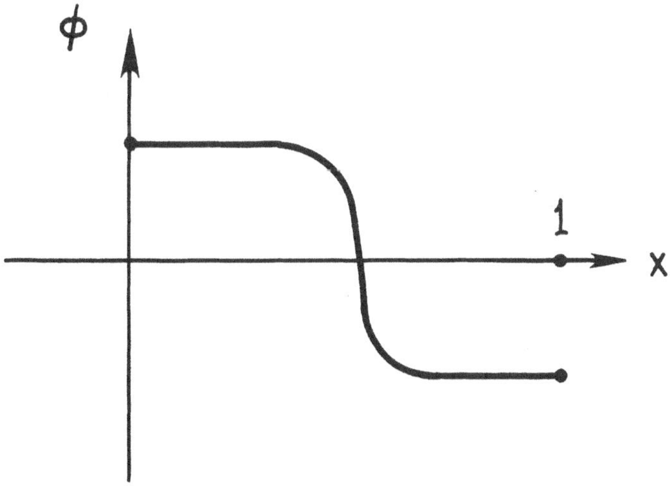

Figure 3

A schematic picture of the solution to (12), (13) for $H = A \sin \phi + B \cos \phi$, $A,B > 0$, $\beta(x) \equiv 0$.

6. P.C.E.'s and C.P.G.'s

We now interpret the meaning of these solutions for the fish C.P.G., and compare them with the solutions of a model by Cohen et. al. [15] for lamprey locomotion. Cohen et. al. considers an equation of the form (8), with $H = \sin\phi$ and $\Delta_k \neq 0$. Thus, they are analyzing a special case of what we are calling diffusive coupling, with a non-zero frequency gradient. (There is some evidence for a spread of natural frequencies in lamprey [15].) With $\Delta_k < 0$ (frequency higher at the rostral end), this equation has travelling wave solutions going rostral to caudal, i.e., forward swimming. To compare this with solutions to (12), (13), recall that $\phi_k \approx \phi(k/(N+1))$ and $\phi_k \equiv \theta_{k+1} - \theta_k$. Thus, $\phi_k > 0$ implies $\theta_{k+1} > \theta_k$, or the k-th oscillator lags its caudal (k+1) neighbor. Similarly, $\phi_k < 0$ implies that the k-th oscillator leads its caudal neighbor. Hence, a solution as in Fig. 2 corresponds to a wave that is rostral to caudal (except very close to the boundary $x = 0$). It is not, however, a wave of constant speed. The speed is inversely proportional to $|\phi_k|$, so the speed increases rostral to caudal (away from the boundary layer). For any $\beta(x) < 0$ and any f,g satisfying the above hypotheses, there is a wave going rostral to caudal, but the wave speed is never constant; neither is it for the model of Cohen et. al. [15].

There are solutions to (12), (13) corresponding to constant speed waves if $\beta(x) \equiv 0$ and the coupling is synaptic. The graph in Fig. 3 corresponds to a pair of waves moving outward from some point (as seen in dogfish); the mirror image (a solution when $f'' < 0$) corresponds to waves travelling inward toward some point. A small change in the forward or backward coupling strength moves the boundary layer to one end or the other (corresponding to forward or backward swimming), with ϕ constant away from the boundary layer. The physiological implications of these and other facts about the solutions will be discussed in Section 8, after a discussion of the rigorous relationship between (8) and (12).

7. Rigorous Results

The P.C.E. analogue is, indeed, a correct diagnostic equation for (6), (8), but only in the presence of an additional hypothesis, one related to a "numerical entropy" condition on numerical algorithms for weak solutions to quasilinear equations $u_t + \bar{f}(u)_x = 0$ [19]. The condition is

$$g' > |f'| \tag{16}$$

throughout the range of the relevant solution $\phi(x)$ to (14). Condition (16) can be interpreted as saying that the boundary layer of (12), (13), whose width is governed by the ratio of the coefficient of ϕ_{xx} to that of ϕ_x, is sufficiently large relative to the mesh size 1/N. We believe that this hypothesis is sharp; when it fails, there may be temporally stable solutions to (6), (8) (as found numerically), but these have spatial oscillations, and hence are not qualitatively like those of (12), (13).

Let $Q = f(\phi_L) - f(\phi_R) - (\Sigma\beta_j)/N$. Q measures the relevant balance between the amount of anisotropy and the frequency gradient; $Q \neq 0$ insures that the shock layer is near one boundary or the other, the choice determined by the sign of $f''Q$. (See [13] for more details.) The main result [13] is:

THEOREM: Suppose there is an interval J containing ϕ_L, ϕ_R and ϕ_T on which $QF'' < 0$ (resp. > 0) and (16) holds. Assume that J contains the range of the solution to (14) satisfying $\phi(1) = \phi_R$ (resp. $\phi(0) = \phi_L$). Then there is a time-independent solution to (6), (8) which converges, as $N \to \infty$, to the solution to (12), (13). (The convergence is nonuniform near the boundary layer.) The solution is locally unique and locally asymptotically stable.

The proof of the existence of a time-independent solution mimics the formal construction of the solution to the P.C.E.. The analogues of (14) and (15) are

$$0 = \frac{\beta_k}{2N} + f(\phi_{k+1}) - f(\phi_k) \tag{17}$$

$$0 = f(\phi_{k+1}) - f(\phi_{k-1}) + g(\phi_{k+1}) - 2g(\phi_k) + g(\phi_{k-1}). \tag{18}$$

The proof of the asymptotic stability of this solution uses ideas related to monotone methods for parabolic P.D.E.'s. Both parts of the proof use hypothesis (16).

8. Physiological Implications

According to Section 7, there are temporally stable solutions to (6), (8) which behave like the time-independent solutions to (12), (13). Hence, we can infer properties of solutions to the discrete equations from those of solutions to the P.C.E., and consider their physiological significance.

We have seen that constant speed travelling waves (corresponding to solutions to (12), (13) with constant "outer solutions") requires non-diffusive coupling between adjacent oscillators, and no significant frequency gradient. This mechanism does not require any inhomogeneities ("pacemakers") at the ends of the spinal cord, and, for sufficiently long chains, gives the same phase lag between adjacent oscillators, independent of the length of the chain. (It is determined by the boundary condition ϕ_L or ϕ_R satisfied by the outer solution.)

As observed in the dogfish C.P.G., the leading segment is not necessarily at one end or the other. However, the leading segment in the solution to (12), (13) or (6), (8) is at a considerable distance from the ends only if the coupling is close to isotropic; its position is then very sensitive to changes in parameters. Thus, the mathematics predicts that there is at most a small amount of anisotropy. This prediction is nontrivial since, in other systems which produce such travelling waves, the connections are thought to be almost entirely in one direction [20].

The mathematics suggests methods by which control systems can produce other behavior of swimming dogfish. For example, if the coupling is indeed close to isotropic, then small changes in the strength of forward or backward coupling can move the leading segment to one of the ends; movement to the caudal end produces backward swimming. For control systems in the lamprey see [21]. In order to have solutions to (12), (13) with waves going outward from (not inward to) some segment, the function f must satisfy $f'' < 0$. For this sign of f'', an increase in coupling from rostral to caudal oscillators (i.e., down the chain) moves the leading segment to the tail, and hence produces waves which go caudal to rostral.

A variation on this theme concerns electric fish Eigenmania and Gymnarchus which use their fins for locomotion. Although detailed EMG measurements have not been made of the muscular (and hence neural) activity, there are behavioral observations of travelling waves of muscular contractions in the fins [22]. For electric fish, the wavelength does not appear to be constant in time, and the leading segment moves rapidly back and forth along the fish. As explained above, this can be the consequence of close to isotropic coupling of the oscillator units governing the fin muscles, with uniform changes in forward or backward coupling strengths. For small changes, the wavelength is not perturbed much, but for larger such changes, the wave length (determined by ϕ_L or ϕ_R, which depend on H^- and H^+) can be altered.

The mathematics also suggests a simple physiological mechanism for producing a variety of swimming speeds, while keeping constant the phase lag per segment (as observed). If the oscillators undergo a uniform increase in frequency (e.g. by added uniform excitation), and the coupling is left alone, then the equations for the phase lags (expressed in angles, not time) are left untouched. However, the speed of the wave of neural activity (which is proportional to the swimming speed) increases with increased frequency. Thus, by independently manipulating coupling and frequencies, the fish control system can produce backward swimming or forward swimming at different speeds.

Finally, the mathematics has implications for the embryological development of motor behavior in fish. Early coordinated motion called "C-coils" in fish (and, in fact, in all vertebrates) involves simultaneous contractions of muscles on one side of the cord. Somewhat later in development, fish embryos display travelling waves (and S-waves, which involve antiphase coupling of units on opposite sides of the cord). According to the mathematics, C-coils are to be expected if the neural or muscular units are coordinated electrotonically. Evidence for such early electrotonic coupling has been found [23,24]. For neural units coupled "synaptically", the expected behavior is travelling waves. Thus the mathematics suggests that the change from C-coils to travelling waves corresponds to the development of appropriate functional synapses.

C-coil behavior persists in adults, e.g. in "escape swimming". This is usually mediated by a special cell known as the Mauthner cell, but also happens in the absence of a Mauthner cell [25]. The persistence of C-coils is compatible with synaptic connections; a relative increase in electrotonic coupling decreases the phase lag, and a large such increase could produce behavior with almost zero phase lag.

9. Related Mathematics

The mathematics necessary to understanding equations (6), (8) has further implications for the behavior of chains of coupled oscillators. It can be shown that if $H^+ = H^- \equiv H$ is an odd function (e.g. $H = \sin \phi$), then the size of the total frequency difference that can be sustained and still have phase-locking is much smaller $(O(1/N))$ than if H has an even component. Also, the frequency at which the oscillators lock is different for the two cases. For more details, see [13].

This paper deals with large chains of weakly coupled oscillators. Simpler but related methods can be used to investigate a pair of weakly coupled oscillators, or a weakly forced oscillator [12]. For a pair of identical such oscillators, synaptic coupling implies that the frequency of the coupled pair need not be the same as the uncoupled frequency, and the change in frequency is related to the phase lag or lead when one oscillator unilaterally forces the other.

Acknowledgments

The mathematics on which this work is based is joint with G. B. Ermentrout. A. Cohen got me interested in the mathematics of CPG's, and generously kept me afloat as I struggled to swim through the literature. C. Peskin asked the question that provoked this paper.

This work was supported in part by NSF Award No. MCS8301249, USAFSAM Research in Biotechnology, Contract F33615-82-D-0627 and the J. S. Guggenheim Memorial Foundation.

References

1. Grillner, S. and Kashin, S., "On the generation and performance of swimming in fish," in Neural Control of Locomotion, R. M. Herman, S. Grillner, P. S. G. Stein and D. G. Stuart, eds., Plenum, NY (1976).

2. Grillner, S., "On the generation of locomotion in the spinal dogfish," Exp. Brain Res. 20, 459-470 (1974).

3. Forssberg, H., Grillner, S., Rossignol, S. and Wallén, P., "Phasic control of reflexes during locomotion in vertebrates," in Neural Control of Locomotion, R. M. Herman, S. Grillner, P. S. G. Stein and D. G. Stuart, eds., Plenum, NY, 647-674 (1976).

4. Grillner, S. and Wallén, P., "On peripheral control mechanisms acting on the central pattern generators for swimming in the dogfish," J. Exp. Biol. 98, 1-22 (1982).

5. Grillner, S., "Mechanosensitive neurons in the spinal cord of the lamprey," Brain Res. 235, 169-173 (1982).

6. Grillner, S., "Locomotion in vertebrates: central mechanisms and reflex interactions," Physiol. Rev. 55, 247-304 (1975).

7. Eisen, J. S. and Marder, E., "A mechanism for production of phase shifts in a pattern generator," J. Neurophys. 51, 1375-1393 (1984).

8. Poon, M., Friesen, W. D. and Stent, G. S., "Neural control of swimming in the medicinal leech, V., connections between the oscillatory interneurons and the motor neurons," J. Exp. Biol. 75, 45-63 (1978).

9. Grillner, S., "Control of locomotion in bipeds, tetrapods and fish," in Handbook of Physiology, Section 1: The Nervous System, 2, V. B. Brooks, ed., Amer. Physiol. Soc., Bethesda, 1179-1236 (1981).

10. Rinzel, J., "Models in neurobiology," in Nonlinear Phenomena in Physics and Biology, R. H. Enns, B. L. Jones, R. M. Miura and S. S. Rangnekar, eds., Plenum, 345-367 (1981).

11. Kopell, N. and Howard, L. N., "Plane wave solutions to reaction-diffusion equations," Studies in Appl. Math. 52, 291-328 (1973).

12. Kopell, N., "Toward a theory of modelling central pattern generators," to appear in Neural Control of Rhythmic Movements, A. H. Cohen, S. Grillner and S. Rossignol, eds., J. Wiley, NY.

13. Kopell, N. and Ermentrout, G. B., "Symmetry and phase-locking in chains of weakly coupled oscillators," to appear.

14. Ermentrout, G. B. and Kopell, N., "Frequency plateaus in a chain of weakly coupled oscillators, I," SIAM J. Math. Anal. 15, 215-237 (1984).

15. Cohen, A. H., Holmes, P. J. and Rand, R. H., "The nature of the coupling between segmental oscillators of the lamprey spinal generator for locomotion: a mathematical model," J. Math. Biol. 13, 345-369 (1982).

16. Fenichel, N., "Persistence and smoothness of invariant manifolds for flows," Indiana Univ. Math. J. 21, 193-226 (1971).

17. Guckenheimer, J. and Holmes, P., Nonlinear Oscillations, Dynamical Systems and Bifurcations of Vector Fields, Springer, NY (1983).

18. Lin, C. C. and Segel, L. A., Mathematics Applied to Deterministic Problems in the Natural Sciences, MacMillen, NY (1974).

19. Le Roux, A. Y., "A numerical conception of entropy for quasi-linear equations," Math. of Comp. 31, 848-872 (1977).

20. Stein, P. S. G., "Mechanisms of interlimb phase control," in Neural Control of Locomotion, R. M. Herman, S. Grillner, P. S. G. Stein and D. G. Stuart, eds., Plenum, NY, 465-487 (1976).

21. Grillner, S. and Wallén, P., "How does the lamprey central nervous system make the lamprey swim," J. Exp. Biol. 112, 337-357 (1984).

22. Lissman, H. W., "Zoology, locomotory adaptations and the problem of electric fish," in The Cell and the Organism, J. A. Ramsey and V. B. Wigglesworth, eds., Cambridge Univ. Press, Cambridge, 301-317 (1961).

23. Furshpan, E. J. and Potter, D. D., "Low-resistance junctions between cells in embryos and tissue culture," in Current Topics in Developmental Biology 3, A. Moscona and A. Monroy, eds., Academic Press, NY, 95-127 (1968).

24. Blackshaw, S. E. and Warner, A. E., "Low resistance junctions between mesoderm cells during development of trunk muscles," J. Physiol. 255, 209-230 (1976).

25. Eaton, R. C., Lavender, W. A. and Wieland, C. M., "Alternative neural pathways initiate fast start responses following lesions of the Mauthner neuron in goldfish," J. Comp. Physiol. 145, 485-496 (1982).

Appendix: On the Assumption of Weak Coupling

At the meeting at which this paper was presented, two questions were raised about the plausibility of the assumption of weak coupling. The first was whether coupling between neural elements could indeed be weak. The second involved a mathematical consequence of weak coupling, that the equations are then reducible to ones involving only differences in phase, not the phases themselves. This was thought to be anti-intuitive, since a pair of neurons which are both depolarized do not in general interact with each other in the same way as when they are both hyperpolarized.

We start with the second objection. The theory of weakly coupled oscillators does not assert or require that coupling depends only on the phase differences. It says that, if the coupling is weak, coordinates can be chosen (using near-identity transformations) so that, in these coordinates, the coupling depends, to lowest order, on only the phase differences. The function H is determined by averaging the interactions over a θ-cycle [A1]. (This is in addition to the implicit averaging that is done, for bursting cells, over action potentials, to produce equations of the form (1) which involve smooth functions of the membrane potential.) According to the theory, for sufficiently small coupling it is necessary to know only the averaged part of the interactions in order to deduce the phase-locking properties.

The first question was asked by someone thinking in terms of systems for which phase response curve analysis is appropriate, i.e., systams in which a single pulse has a large effect on the forced oscillator. Whether such a description is a more accurate representation of the fish C.P.G. is a factual question for which the data does not (to my knowledge) exist. However, there are many observations on related systems which lend plausibility to the assumption of weak coupling. First of all, post-synaptic potentials can, in general, be very small. Secondly, some interneurons in C.P.G. systems are known to be non-spiking, and to operate in a continuous and graded way [A2]. Indeed, for some spiking cells as well, neurotransmitter release can be graded and sensitively dependent on membrane potential, even when the cells are hyperpolarized [A3]. Finally, weak coupling implies that it takes many cycles to reach locking; Grillner [pers. comm.] noticed such long transients in his work on the lamprey C.P.G.

References

[A1] Hale, J., Ordinary Differential Equations, John Wiley, NY (1969).

[A2] Pearson, K. G. and Fourtner, C. R., "Nonspiking interneurons in the walking system of the cockroach", J. Neurophysiol. 38, 33-51 (1975).

[A3] Graubard, K., Raper, J. A. and Hartline, D. K., "Graded synaptic transmission between identified spiking neurons", J. Neurophysiol. 50, 508-521 (1983).

EXPERIMENTAL STUDIES OF CHAOTIC NEURAL BEHAVIOR:

CELLULAR ACTIVITY AND ELECTROENCEPHALOGRAPHIC SIGNALS

P. E. Rapp

I. D. Zimmerman

Department of Physiology and Biochemistry
The Medical College of Pennsylvania
3300 Henry Avenue
Philadelphia, PA 19129

A. M. Albano

G. C. deGuzman

Department of Physics
Bryn Mawr College
Bryn Mawr, PA 19010

N. N. Greenbaun
Department of Mathematical Sciences
Trenton State College
Trenton, NJ 08625

and

T. R. Bashore
Department of Psychiatry
The Medical College of Pennsylvania
The Eastern Pennsylvania Psychiatric Institute
3200 Henry Avenue
Philadelphia, PA 19129

Abstract:

Deterministic systems can display a form of highly irregular, quasirandom behavior called chaos. Even though the observed behavior is very complex, the systems which generate it can be very simple. Thus, in at least some instances, irregular biological systems may obey a simple, potentially discoverable, deterministic dynamical law. These systems can undergo reversible transitions to and from chaotic dynamics in response to small changes in parameter values. As a long term goal, this form of analysis may suggest more effective responses to disordered behavior in physiological control systems.

This contribution is concerned with chaos in neural systems and its possible role in epileptogenesis. Calculation of information dimension provides a procedure for distinguishing between chaotic and random behavior. This technique is applied to experimental data from two preparations: spontaneous activity of cortical neurons in the pre- and post-central gyri of the squirrel monkey and human electroencephalographic signals. In each case the results suggest that these systems can display low dimensional chaotic behavior.

I. Chaotic dynamics and physiological systems

A. What is chaos?

It is often assumed that deterministic systems display well ordered behavior. It is now known that this is not always the case. The term deterministic chaos is used to describe the highly irregular, quasi-random behavior that has been observed in deterministic systems (65, 117, 60). No completely satisfactory, general definition of chaos exists. Though a proposal has been made for the specific case of discrete dynamical systems (77), no consensus for continuous systems has been found. Indeed, as knowledge of dynamical behavior of nonlinear systems increases, the term appears to be losing rather than gaining specificity.

A useful characterization of continuous systems that can lead to a qualitative understanding of chaotic dynamics can be obtained by examining their attractors. Consider the n-dimensional differential equation $dx/dt=f(x)$ where x is a member of R^n. The vector function f(x) is sufficiently well behaved to ensure existence and uniqueness of solutions. A set A in the n-dimensional state space is an attractor of the flow defined by that equation if there exists a set U, a neighborhood of A, which is attracted by A. Set A is said to attract set U if every point of U, when taken as an initial point of the differential equation, results in a solution that comes arbitrarily close to A as time increases. Two types of attractor dominated classical analysis of differential equations, the fixed point attractor (resulting in steady state behavior) and the periodic attractor (resulting in oscillations). These two structures remain central to most dynamical models of physical and biological behavior. Kronecker introduced a third structure, the irrationally wound torus. The simplest case is a two dimensional manifold that is periodic in both of its variables v and w with incommensurate frequencies. The simplest description of the flow on the torus itself is:

$$dv/dt = 1$$

$$dw/dt = k$$

where k is an irrational number. Because k is irrational, an integer number of rotations in v never equals an integer number of rotations in w, that is, the orbit never closes on itself to become periodic. The idea generalizes to higher dimensions; a q-torus is the q-dimensional analog and is constructed with q incommensurate frequencies. The behavior is termed quasiperiodic.

Recently it has been established that an important additional class of attractors exists. These attractors have such complex structures that they have been termed strange attractors. Remarkably, even very simple differential equations can provide examples of strange attractors. Ruelle (110, 111) has suggested that a successful definition of a strange attractor will incorporate the following elements:

(i.) A strange attractor is an attractor and is neither a fixed point nor a periodic solution.

(ii.) A strange attractor is indecomposable, that is, it cannot be split into two or more separate attractors.

(iii.) There is sensitive dependence on initial conditions for initial points in the set U.

It is the third element of the definition that introduces irregular dynamical behavior. Sensitive dependence on initial conditions is said to occur if any arbitrary infinitesimal change in initial values results in solutions that diverge exponentially for small values of time. Sensitive dependence as defined here does not require that two trajectories diverge, rather, with the possible exception of some unusual initial points (having zero relative measure on U), any pair of trajectories originating in U diverge. The divergence throughout a neighborhood of the attractor results in complex mixing of trajectories and suggests why the presence of strange attractors is linked to chaotic dynamics.

However, it should be noted that systems governed by strange attractors are not the only type of deterministic chaotic system. Mees and Sparrow (94) have argued that systems containing strange attractors constitute a small fraction of all chaotic systems, and that a second more common form of chaos is obtained as the consequence of the simultaneous existence of a weakly attracting periodic solution and a complex set (called a strange invariant set) which is not itself an attractor. The periodic solution may be very irregular and have a long period; the structure of the strange set is very like that of a strange attractor.

The potential importance of complex attractor topologies in studies of transitions to and from disordered behavior is established by recognizing that the attractors of a differential equation are not fixed by the form of the vector function f(x). Typically these functions contain parameters whose values are fixed for a given case but can vary from case to case. Dramatic changes in the number and in the qualitative form of the attractors occurs as the result of possibly very small changes in parameter values. For example, consider a set of differential equations modeling an electronic circuit. A small change in a parameter value specifying the resistance of a component can cause a stable oscillation in voltage to deteriorate to seemingly random chaos. Importantly, parameter-dependent transitions of attractor topology to chaotic behavior are not limited to mathematical systems. They can be observed experimentally in physical and chemical systems in response to changes in parameters that are subject to experimental control such as temperature, Reynolds number and flow rate. Examples include electronic circuits (41), Couette-Taylor flow (14, 15); Rayleigh-Bernard convection (81); spin wave instabilities in ferromagnetic resonance (36); laser instabilities (5) and chemical reactions (125).

B. Functional significance of chaotic physiological control systems

The growing successes of the parameter-dependent transition paradigm in identifying the underlying dynamical mechanism generating disordered behavior in many physical and chemical systems suggested that it might also prove responsible for failures of physiological control systems. Defects in physiological regulation resulting from these transitions, including transitions to chaos, have been termed dynamical diseases by Mackey and Glass (90). Several possible examples are presently being investigated. They include respiratory instabilities (38, 39), hematopoietic disorders (84, 85, 86, 87, 89), cardiac arrhythmias (3, 22, 24, 38, 51, 52, 67, 68, 108, 109, 121, 122) and seizure disorders (104, neurological disorders will be described in greater detail in a subsequent section).

To a first approximation, the functional significance of transitions between fixed point, periodic, toroidal and strange attractors can be reduced to two statements:

1. In at least some instances, irregular biological behavior is subject to deterministic analysis. A disordered system may be following a very simple, potentially discoverable, dynamical law.

2. This irregular behavior may be controllable. Chaotic activity is generated by deterministic systems, and deterministic systems respond to intervention in a deterministic fashion. Small changes in parameter values that could be effected pharmacologically could produce a reverse transition to ordered, physiologically acceptable behavior.

Thus, this form of analysis offers the prospect of rationally designed, minimal interventions of increased specificity and efficacy.

II. Chaos in neural systems

A. Positive instances

The preceding description of chaotic behavior and its possible role in dynamical diseases might suggest that chaotic neural activity is invariably deleterious. This need not be so. It is possible to speculate that in some instances chaotic neural activity could be benign or even beneficial.

The usefulness of irregular neural activity in searching memory and in the early stages of decision making processes seems plausible at an intuitive level. Recent results in optimization theory makes it possible to place these intuitive speculations on a more quantitative basis. It can be shown that some control systems can function successfully only by introducing noise into the system (74), and some numerical optimization processes can be greatly accelerated by annealing algorithms (17, 83) that use noise to prevent spurious convergence to local minima. Thus, neural computations that are dynamically equivalent to an optimization could be accelerated by appropriate irregular activity. Given the highly evolved state of the mammalian central nervous system, optimization might be a frequent operational mechanism. The deterministic nature of chaotic behavior again becomes potentially beneficial since annealing algorithms function by reducing the noise level as the optimization progresses. A model of a specific example of neural computation in which noise, rather than chaos, plays an important role is Sejnowski's (9, 118) model of image recognition in the visual cortex.

It should be recalled that the beneficial use of noise has a long history in control engineering. Dithering is the intentional injection of noise into a control network (35). It has the effect of smoothing nonlinearities and consequently suppressing periodic behavior. Dithering results in a system that displays steady state behavior subject to acceptable low amplitude random variation. Though systematic theoretical investigations have not been reported, it seems probable that chaotic input might have a similarly stabilizing effect that could prove useful in neural control networks.

B. Chaos in neural systems: Negative instances

Though it is possible to speculate on the possible benign or helpful effects of some instances of chaotic neural activity, most attention has centered on the possible relationship between neural chaos and failures in neural regulation. The effect of chaos on information processing in dynamical systems raises several fundamental questions (45, 119). However, these abstract issues have not yet influenced present thinking about neural control. Rather, attention has been directed to models of specific processes. For example, King, Barchas and Huberman (73) have constructed a model of dopamine metabolism in the central nervous system. This is of particular interest because abnormalities in brain dopamine transmission have been identified in one neurological disorder, Parkinson's disease, and have been implicated in a psychiatric disorder, schizophrenia. A number of dopamine systems have been isolated in brain, one of which, the nigro-neostriatal system, is known to have deficient levels in Parkinson's disease. Neurons in the substantia nigra pars compacta contain dopamine and project in large part to the neostriatum where the neurotransmitter acts primarily to inhibit postsynaptic activity. Parkinson's disease is related to the progressive loss of these dopamine containing neurons. The dopamine hypothesis of schizophrenia holds that either excessive receptor activation by supranormal levels of the transmitter or supersensitive dopamine receptors in the neostriatum cause the dramatic symptoms (e.g. hallucinations and labile affect) of certain types of schizophrenics. The King, Barchas and Huberman equations model the synthesis, release and degradation of dopamine. Varying the parameter quantifying the efficacy of dopamine at the postsynaptic receptor can result in transitions to chaos. These authors suggest that chaotic variation in dopamine levels could result in neural control failures of Parkinson's disease and in the unstable mental behaviors that are characteristic of schizophrenia.

Mackey and an der Heiden (88) have modelled CA3 neurons of the hippocampus. Mossy fibers of the granule cells in the dentate gyrus provide excitatory stimulation to CA3 pyramidal cells that in turn project to the fornix. The pyramidal cells also stimulate inhibitory interneurons, the basket cells. On stimulation, basket cells release GABA which inhibits pyramidal cell discharge. Mackey and an der Heiden modelled a specific process; the effect of penicillin on pyramidal cells. Penicillin binds competitively to the GABA receptors; an increase in penicillin results in a decrease in GABA receptor availability. In their model, increasing penicillin results in a transition from bursting behavior to irregular (possibly chaotic) firing to sustained high frequency firing. This theoretical result is suggestive because excessive concentrations of penicillin can induce seizures and direct topical application of penicillin crystals to the cortex is used to create epileptogenic foci in experimental animals.

The Mackey-an der Heiden result is one of a number that suggest a possible role for chaos in epileptogenesis. (See also: 70, 71, 114, 115, 116). It has been subsequently suggested that a seizure is not itself chaotic (104) indeed, a seizure is characterized by a decrease in the degree of disorder in brain electrical activity. However, a seizure might be the corrective resynchronizing response to the loss of coherence in brain activity that in turn, is the result of chaotic neural behavior. A cortical analogy to cardiac defibrillation has been made (104).

Though the possible clinical consequences of chaotic neural activity are presently a matter for speculation, these theoretical investigations do lead to the central motivating question of this study: is chaotic neural behavior possible?

III. Is chaotic neural behavior possible?

A. Theoretical evidence

The growing number of theoretical investigations of chaotic neural behavior includes: Aihara and Matsumoto (4), an der Heiden, Mackey and Walther (7), Carpenter (18, 19), Chay (20, 21), Colding-Jorgensen (25), Ermentrout (28), Guevara, Glass, Mackey and Shrier (55), Holden, Winlow and Haydon (62), Jensen, Christiansen, Scott and Skovgaard (69), King, Barchas and Huberman (73). Mackey and an der Heiden (88), Rapp (104, 105) and Sbitnev (114, 115, 116). The emerging conclusions from this mathematical activity are:

(i.) Because neurons are intrinsically complex nonlinear devices, they are capable of generating chaotic signals at the cellular level.

(ii.) Neurons are organized into networks that can display chaotic behavior as a collective network phenomenon.

It should be noted that the characterization of these models as chaotic is typically qualitative. Quantitative measures based on computation of attractor dimension, Lyapounov exponents and entropy, have not been made. Thus, while this very large body of results is suggestive; it is not definitive.

B. Experimental evidence

Depending on the amplitude and the frequency of stimulation, periodically forced nonlinear oscillators can display either periodic phase locking behavior, quasi-periodic behavior or chaotic activity. This is probably the most readily observed form of chaotic behavior. In experiments with invertebrates, periodically forced pacemaker neurons can fire chaotically (57, 58, 61, 93). Chaotic behavior in a neural system not subject to periodic stimulation has been reported by Holden, Winlow and Haydon (62) in a molluscan neuron in response to high concentrations of the convulsant 4-aminopyridine. However, the effect of the drug, irregular variations in amplitude, appeared only after prolonged exposure at high convulsant concentrations.

As with the theoretical models, the identification of chaos was qualitative. While this is probably a reliable procedure in the experiments with periodically forced neural oscillators, it may be less satisfactory in assessing spontaneous behavior. This suggests that the quantitative procedures for calculating the dimension of an attractor described in the next section will be important.

IV. Dimension: The quantitative characterization of dynamical behavior

A. The importance of attractor dimension

Informally stated, the dimension of a system (called the dimension of the state space or the behavior space) is its number of degrees of freedom. A planetary system consisting of a single planet confined to a plane orbiting about a fixed sun, has a two dimensional state space. A system consisting of point masses moving in a 3-space has a state space dimension equal to three times the number of particles (assuming that no conservation conditions are used to reduce the system). In chemical problems the dimension of the state space is equal to the number of distinct chemical species in the reaction system, and thus for biochemical cases the dimension is large. Similarly, the dimension of the state space of a neurophysiological system can be very large. However, the dimension of the attractors of systems with high dimensional state spaces can be much smaller than the dimension of the state space itself. The fixed point, corresponding to a chemical steady state, has dimension zero. This is true even if this point is an attractor within a very large dimensional state space. A periodic solution has dimension one, and a q-torus has dimension q. These finite values contrast with random signals which have infinite dimension. The examination of attractors is valuable in dynamical analysis because structurally complex systems, such as a neuron, may have very simple dynamical behavior that is characterized by its attractors. The degree to which the dynamics of a system can be adequately specified by the properties of its attractors depends upon the speed of convergence. If the attractors are only weakly attracting, behavior in their immediate neighborhood will not be representative of the behavior of the system. However, while it is possible to construct mathematical systems in which the attractors are only weakly attracting, chemical and physical systems observed in the laboratory typically converge rapidly to an attractor. For these systems behavior near attractors is of primary importance. Thus, characterizing attractors is an important step in dynamical analysis. The first step in the investigation of an attractor is determination of its dimension. This is particularly true because of the relationship between attractor dimension and chaotic behavior. This relationship will be described after a discussion of dimension and its measurement.

B. Different definitions of dimension

Historically, the intuitive concept of dimension was made more precise with the definition of topological dimension. It was subsequently generalized with the definition of the Hausdorff-Besicovitch and Renyi dimensions. These definitions are sketched in this section and followed in the next section by a description of how they can be measured.

The topological dimension of a set, D_T, is an integer that specifies the number of distinct directions in the set. The idea is classical and was elaborated by Poincare'. (An historical discussion appears in Mandelbrot (91). See also Hurewicz and Wallman (64). The failure of the intuitive definition was established at the end of the century by results of Cantor and Peano. This motivated a more precise definition of a topologically invariant dimension by Brouwer, Menger, and Urysohn (64). However, topological dimension has limited immediate applicability in the analysis of dynamical systems; its generalizations prove more useful.

The Kolmogorov capacity (29, 30, 75) is a measure of the structure of a set that reduces to the topological dimension for simple objects like fixed points and line segments. Let the set be covered (filled) by a collection of p-dimensional cubes of size ϵ^p and let $N(\epsilon)$ be the minimum number of cubes needed to fill it. The Kolmogorov capacity C of that set is

$$C = \lim_{\epsilon \to 0} [\ln N(\epsilon)/\ln(1/\epsilon)]$$

The capacity is closely related to the Hausdorff-Besicovitch dimension (56). However, they are not identical (32). The literature is complicated by a nonuniform nomenclature. Farmer (29) uses the term fractal dimension for the Kolmogorov capacity. Mandelbrot (91) uses the term similarity dimension for the capacity and refers to the Hausdorff dimension as the fractal dimension.) The relationship between capacity and dimension can be approached in a formal fashion by beginning with the expression for capacity and noting that for small ϵ, $N(\epsilon)\,\epsilon^C$, the volume of $N(\epsilon)$ elements of each volume ϵ^C, should be constant. Let $L_C(\epsilon) = N(\epsilon)\,\epsilon^C$. If there exists a number, D_H, such that

$$\lim_{\epsilon \to 0} L_C(\epsilon) \longrightarrow \infty \qquad \text{if } C < D_H$$

and

$$\lim_{\epsilon \to 0} L_C(\epsilon) \longrightarrow 0 \qquad \text{if } C > D_H$$

then that number D_H is the Hausdorff-Besicovitch dimension of the set.

The Szpilrajn inequality (64, page 107) states

$$D_T \leq D_H$$

For topologically simple objects these dimensions are the same. However, for complex sets such as a Cantor set or the Koch curve, the Hausdorff dimension is not an integer. When this is the case, a strict inequality is obtained. Mandelbrot (91, p. 15) defines a fractal as a set for which the Hausdorff-Besicovitch dimension strictly exceeds the topological dimension.

The Hausdorff dimension can be used to assess the geometric structure of an attractor, but it does so independently of the density of points on that set. The Renyi information dimension incorporates a density dependence. For attractors with uniform density, the Renyi dimension reduces to the Hausdorff dimension. The information dimension was originally defined by Renyi (8, 107) in an investigation of the probabilistic relationship between thermodynamics and information theory. Early applications of this concept in dynamics include those of Farmer (30, 31) and Alexander and Yorke (6). The description here follows Farmer (29). Let X have k different possible values with probabilities $P_1 \ldots\ldots P_k$. Then the information obtained by measurement of X is

$$I = \sum_{i=1}^{k} P_i \log(1/P_i)$$

As the number of possible values k increases, the discrimination of the measurement increases. Let the resolution, r, be defined as r=logk. For a given value of r, n(r) measurement values of the total of k different values will have nonzero probability. Let $P_i(r)$ be the probability of the i-th value. The set $P_i(r)$ is the probability distribution at resolution r. The information obtained by a measurement of resolution r is I(r).

$$I(r) = \sum_{i=1}^{n(r)} P_i(r) \log(1/P_i(r))$$

The rate at which information increases with increasing resolution in the limit of infinite resolution is the information dimension.

$$D_I = \lim_{r \to \infty} I(r)/r$$

A related hierarchy of dimensions, the order-q information dimensions, can be constructed (49) from I_q, the order-q information.

$$I_q(r) = [1/(1-q)] \log \sum_{i=1}^{n(r)} P_i(r)$$

$$D_q = \lim_{r \to \infty} I_q(r)/r$$

As with the Hausdorff dimension, the information dimension can assume noninteger values and obeys the inequality $D_T \leq D_I$ (29). It is also possible to construct an inequality relating information dimension to the Kolmogorov capacity

$$D_I \leq C$$

The proof follows from the concavity of xlog (1/x) for positive values of x. Equality is obtained only in the special case where all n(r) measurements of nonzero probability are equally probable, $P_i(r)=1/n(r)$. Dynamically this is equivalent to measuring the shape of the attractor while neglecting any differences in the distribution of points on the attractor.

C. Quantitative estimates of dimension

As observed by Farmer (24, p. 230) the definition of topological dimension is not readily reducible to a computational procedure. For this reason, calculations of dimension have of necessity centered on the Hausdorff and Renyi dimensions.

The Hausdorff dimension itself is also not readily computable but the related Kolmogorov capacity can be calculated. Box counting algorithms for estimating capacity were independently introduced by a number of researchers (34, 43, 100, 113, 123). Starting with the definition of capacity as a limit, it follows that for small ϵ,

$$\ln N(\epsilon) = C \ln(1/\epsilon) + \text{correction terms}$$

Thus a plot of $\ln N(\epsilon)$ against $\ln(1/\epsilon)$ for small ϵ should have a linear region with slope C. Consider a mathematical system, for example an ordinary differential equation. The dimension of the behavior space is the dimension of the equation and thus is known. Attractors are confined to a compact subset of the behavior space. The box counting algorithm proceeds by partitioning that subset into boxes of length ϵ on each side. At each box, the procedure determines if there is one or more points of the attractor in the box. If there are, $N(\epsilon)$ is increased by 1. Depending on the specific implementation of the algorithm, this procedure tends to overestimate $N(\epsilon)$. This can in part be compensated against by progressively decreasing ϵ. However, this results in a procedure that will stress the importance of calculations at small ϵ and thus be sensitive to noise.

This method would seem to work only for mathematical systems in which the dimension of the state space is determined. This is in contrast with experimental data where the total dimension of the behavior space is not known. Experimental time series data consists of measurements of one component of the behavior space vector. For example, in an experimental study of a neuron a single variable, such as membrane potential, is measured. All of the other equally important variables in the system, such as intracellular and extracellular ion concentrations, are not measured directly. An extension of the box counting algorithm for experimental data that resolves this difficulty was introduced by Shaw and his colleagues (34, 100) and Takens (123). The analysis begins with measurements of a single dynamical variable at a sequence of time points, x_1, x_2, ..., x_k, ..., x_N. This string of numbers is used to create a new set S. Elements of S will be points, X, in an n-dimensional space; n is referred to as the embedding dimension. There is more than one way to construct S. The simplest is to form n-vectors from the data serially.

$$X_1 = (x_1, x_2 \ldots x_n)$$

$$X_2 = (x_2, x_3, \ldots x_{n+1})$$

$$X_i = (x_i, x_{i+1}, \ldots x_{i+n-1})$$

The structure of the attractor is then inferred from the structure of set S. This inference is justified by a theorem of Takens' (123) which is derived from the Whitney embedding theorem (127). In the proof of the theorem, it is assumed that all data points are on the attractor, that is, it is assumed that transient initial behavior has died away. It is also assumed that an infinite number of measured values that are uncorrupted by noise are also available. Given these assumptions, the theorem shows that an embedding exists between the attractor and set S whenever the embedding dimension, n, is sufficiently large. (Specifically, n is to be twice the capacity of the attractor.). An embedding is a 1 to 1 differentiable map, whose inverse is also differentiable. Whenever an embedding exists between two sets, they are structurally very similar. Specifically, the capacity of the attractor is equal to the capacity of S. There are some technical qualifications to this statement in Takens, (123).

In practical applications, the conditions of the theorem are never satisfied because data are corrupted by noise and an infinite number of values are not measured. However, the structure of S provides the best estimate of the structure of the attractor that can be extracted from that data set. It is possible to relax the requirement on embedding dimension n. It is sufficient for n to be greater than the capacity of the attractor. The capacity of the attractor is not known _a priori_, but the theorem demonstrates that if n is large enough, the capacity of the set S will not change on increasing n. This establishes a procedure for estimating the dimension of the original system's attractor from experimental data. Pick a value of n and construct S. The capacity of set S is then calculated using the previously described box counting algorithm. Then, increase n, construct a new S and calculate the capacity of this set. If the value of the capacity of S is strictly less than n and does not change through several increments of n, then n is large enough.

While this box counting procedure is readily reduced to computable form, it is not practical. The data requirements, both in terms of the number of measurements and the accuracy, are very severe (53, 113). Explicit calculations of data requirements were performed by Greenside, _et al_ (50). They conclude that the procedure is not feasible for systems with noisy data and attractor capacities greater than two.

Guckenheimer (54) constructed a different procedure for estimating capacity from experimental data. Its mathematical foundation rests on an embedding argument. It differs from the previous method in using an alternative to box counting. An n-dimensional set S is constructed from the data in the same way as before. Let z be an element of S. Let $P_z(\epsilon)$ be the proportion of the elements of S lying within an ϵ of point z. The function $\ln P_z(\epsilon)$ is plotted against $\ln \epsilon$. Guckenheimer has shown that, in the linear region, the slope gives an estimate of capacity. Results available thus far suggest that this procedure may succeed in cases where attractor structure cannot be resolved by box counting.

A method for estimating the Renyi dimension based on a scaling property of the correlation integral was constructed by Grassberger, Procaccia and their colleagues (23, 44, 46, 47, 48, 49). As in box counting, when the procedure is to be used with experimental time series data, measured from a single dynamical variable, the data is embedded in a space of dimension n

$$X_i = (x_i, \ldots\ldots, x_{i+n-1}).$$

Suppose there are N measurements. K vectors, K = N-n+1, are formed serially from the data. The discrete form of the correlation integral is:

$$C_n(\epsilon) = (1/N_p) \sum_{i=1}^{K} \sum_{j=i+1}^{K} \Theta(\epsilon - |X_i - X_j|)$$

where N_p is the number of distinct pairs of vectors X_i and X_j and Θ is the Heaviside function.

For sufficiently large data sets, sufficiently large embedding dimension n, and ϵ small enough it can be shown (47) that

$$C_n(\epsilon) \propto \epsilon^{\nu}$$

where ν is the correlation exponent (it is also called the correlation dimension) and is an estimate of the order-2 Renyi information dimension, which in turn estimates the dimension according to the relation (47).

$$\nu \leq D_I \leq \text{Kolmogorov capacity}$$

The Kolmogorov capacity is equal to the information dimension in special cases where there is a uniform distribution of points on the attractor.

$C_n(\epsilon)$ is a step function increasing monotonically with positive ϵ from zero at $\epsilon=0$ to 1. If $C_n(\epsilon)$ is plotted, the stepwise nature of the function is usually not obvious. The number of steps between zero and one is N_p, which is order N^2, N = the number of data points. It follows that $\ln C_n(\epsilon)$ is monotone increasing in $\ln\epsilon$ and from the theorem we know that for an appropriate range of ϵ, $\ln C_n(\epsilon)$ versus $\ln\epsilon$ should have a linear region with slope ν. Figure 1 shows $\ln C_n(\epsilon)$ versus $\ln\epsilon$ for a neural data set described in the next section. The function is drawn for ten successive values of embedding dimension. For small values of the correlation distance ϵ, the distance is on the order of experimental noise, and $C_n(\epsilon)$ is disordered. For large values of ϵ, the function saturates at $C_n(\epsilon) = 1$.

As with box counting algorithms, the procedure can succeed only if the embedding dimension n is greater than the dimension of the attractor. Therefore the correlation integral is calculated for several values of n. If n is sufficiently large, the slope of the linear region should be equal to ν, and the value will not change with successively greater values of embedding dimension. If n is less than the dimension of the attractor, the slope should be equal to the embedding dimension. (A technical qualification to this point will be made presently.) Random signals have infinite dimension. Thus, if the method is applied to a random time series, the embedding dimension is always less than the attractor's dimension, and the value of the slope will increase with each increase in embedding dimension. Strictly, the method never distinguishes between a random signal and a chaotic signal with a dimension larger than the largest embedding dimension tested. The method only determines that, if there is a finite dimensional attractor, its dimension is greater than the maximum n tested.

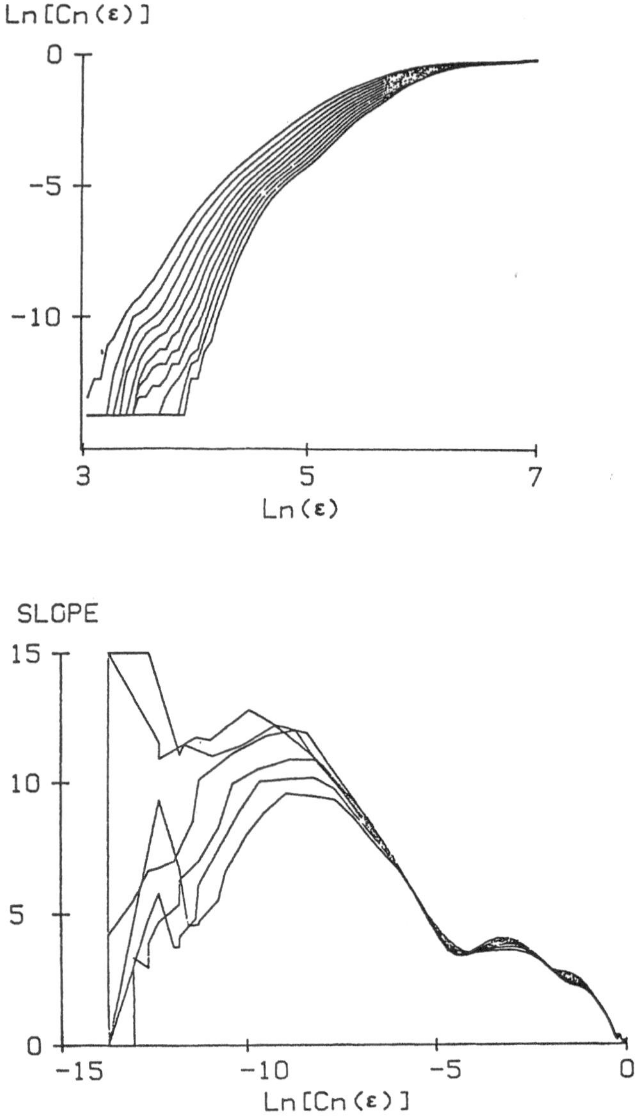

Figure 1:

Plot of $\ln C_n(\epsilon)$ versus $\ln\epsilon$ for embedding dimensions n=10 to n=20. The corresponding plot of $\mathbb{D}$ as a function of $\ln C_n(\epsilon)$ is shown for n=15 to n=20. The horizontal plateau exists between -4.5 and -2.5. The correlation exponent is 3.5.

In principle it should be possible to estimate ν directly from the slopes of linear region of the $\ln C_n(\epsilon)$ versus $\ln\epsilon$ curves. This is acceptable when the procedure is used to estimate the dimension of numerical examples where the data is noise-free to double precision accuracy. However, it is not always a satisfactory practice when using experimental data. A more reliable procedure results when the derivative $\mathbb{D} = d\ln C_n(\epsilon)/d\ln(\epsilon)$ is explicitly calculated and plotted against $\ln C_n(\epsilon)$. This is also shown in Figure 1. The linear region of the previous family of curves results in a horizontal region of $\mathbb{D}$ which corresponds to a constant value of the derivative. The horizontal region is called the scaling region. The motivation for plotting $\mathbb{D}$ against $\ln C_n(\epsilon)$ rather than the more natural, though mathematically equivalent, choice of $\ln\epsilon$ may require explanation. When $\mathbb{D}$ is plotted against $\ln C_n(\epsilon)$, the horizontal regions superimpose. When plotted against $\ln\epsilon$, they are displaced horizontally relative to each other. Requiring superposition for several values of embedding dimension is an additional self-consistency test that is particularly valuable in analyzing small sets of noisy data (1, 5).

The plot of $\mathbb{D}$ versus $\ln C_n(\epsilon)$ is irregular for small values of $\ln C_n(\epsilon)$. This is expected. The function $\ln C_n(\epsilon)$ is monotone increasing in ϵ. Therefore small values of $\ln C_n(\epsilon)$ correspond to small values of ϵ, and as previously noted when the noise level is order ϵ the correlation integral is irregular. $\mathbb{D}$ approaches zero as $\ln C_n(\epsilon)$ increases. This is also anticipated. As ϵ increases, the correlation integral approaches the saturation value 1 and thus its derivative approaches zero.

In Figure 1 the horizontal region of $\mathbb{D}$ occurs roughly when $\ln C_n(\epsilon) = -5$. The estimated values of ν are found by projecting that approximately horizontal line to the y axis. The value $\nu = 3.5$ is obtained. A region that might seem to correspond to a second horizontal region appears at a slightly higher value of $\ln C_n(\epsilon)$. In this example the second region is not well resolved and an ambiguity does not result. However, this is not always the case. It is possible to construct examples in which there are two horizontal regions. When this is the case, the region corresponding to the lower value of ϵ is reported as the dimension. This choice is made because the theorem from which this procedure is derived is obtained in the limit of small ϵ.

Figure 2 shows the $\ln C_n(\epsilon)$ and $\mathbb{D}$ plots for a second neural data set. This example stresses the importance of calculating C_n at several values of n, and explicitly calculating $\mathbb{D}$. For any given value of n, $\ln C_n(\epsilon)$ versus $\ln\epsilon$ in Figure 2 appears to have a well defined linear region with a slope less than the embedding dimension. However, when $\mathbb{D}$ is calculated and plotted, the horizontal regions fail to superimpose at a common value that is independent of n. Therefore, it is concluded that whatever the dimension of this time series' attractor might be, if indeed a finite dimensional attractor exists, it is greater than the largest value of n tested. This set of curves also illustrates a technical point. According to the theorem (46), the slope of the linear region should be equal to the embedding dimension whenever the embedding dimension is less than the attractor dimension. In this example this is not the case. The slope of the linear region increases with embedding dimension (therefore the procedure has not converged), but its numerical value is only about n/2. This phenomenon is not limited to neural data. It has also been seen in the analysis of laser instabilities (4) and is probably due to the finite number of data points. In operational terms, this observation is important because it indicates that a linear region with a slope less than the embedding dimension does not of itself demonstrate the presence of a finite dimensional attractor. A consistent value over a sequence of embedding dimensions must be obtained.

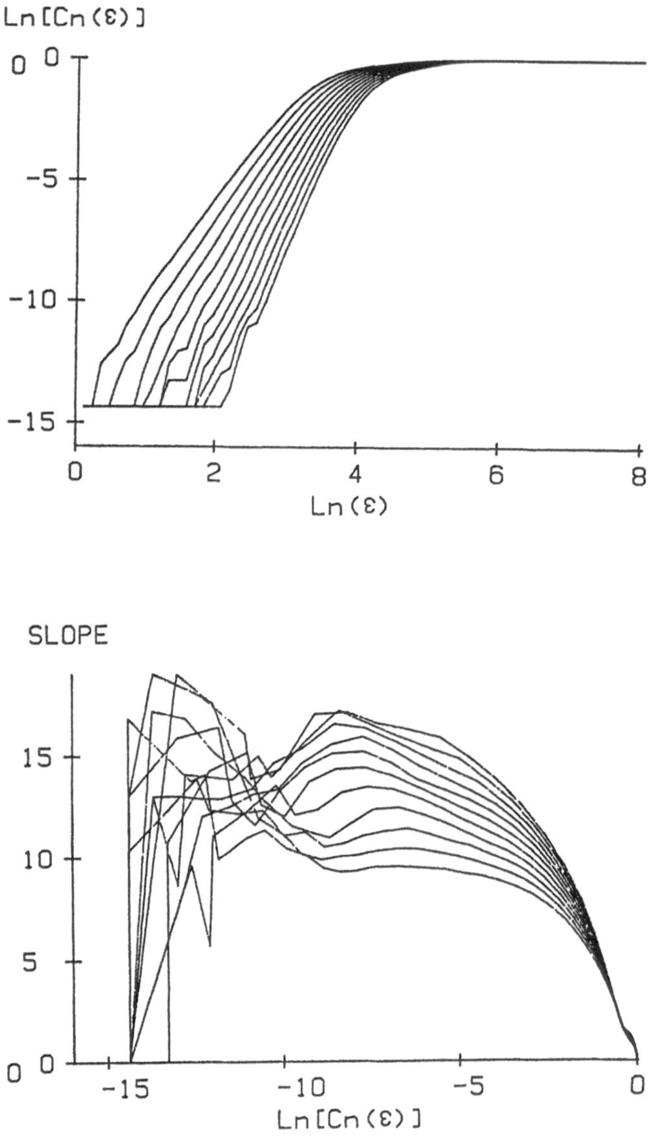

Figure 2:

Plot of $\ln C_n(\varepsilon)$ versus $\ln\varepsilon$ for embedding dimension n=20 to n=40 in steps of two. The graphs of $\mathbb{D}$ are given for n=20 to n=40 in steps of two. No plateau is found up to dimension n=40.

The importance of this procedure lies in its utility in the analysis of experimental data. The method has found application in studies of laser instabilities (1, 5), Couette-Taylor flow (14, 15), sedimentation records (98) and spontaneous activity of neurons in the motor cortex of the squirrel monkey (106).

Abraham and his colleagues (1) have investigated the data requirements of the procedure. The dimension of the logistic equation was calculated using 500, 1000 and 5000 data points and found to be equal to 0.92, 0.94 and 0.93 respectively. Similarly, the Henon attractor was calculated using 500, 1200, 4000 and 10,000 data points giving the values 1.28, 1.20, 1.24 and 1.24. As a test with experimental data, they used the output intensity of an unstable CO_2 laser. The calculated dimension was not significantly altered if 6000 data points or 500 point subsets were used. These results suggest that the procedure can converge with remarkably small data sets. The method also appears to be comparatively robust against noise in the data. Using a random number generator, they added noise to the sine function, sin(t) + 0.5r(t), where r is a random number between -1. and +1. Thus, the signal to noise ratio is only 2 to 1. Using 278 and 738 points both sets gave the dimension 1.2. Noise causes an overestimate of attractor dimension (see also Ott, Yorke and Yorke, (99)). Given the relationship between dimension and capacity, this is anticipated. Further, the calculations suggest that the effects of noise can not be reversed simply by the addition of more noisy data.

The calculations with the sine function raise an additional question about data requirements that centers on the total duration of measurements. Abraham et al (1) report "For calculations using the sinusoidal signal, the time interval between points was chosen so that the data points spanned at least 20 periods. We found that in dealing with periodic or quasiperiodic data generated from continuous maps or experiments, at least this many periods must be sampled in order to get reproducible results." Systematic investigations of the effect of duration of recording as measured by the number of cycles in the case of roughly periodic time sequences, are not reported. This merits future attention.

D. The relationship between noninteger dimension and chaotic behavior

If the attractor of a dynamical system has a noninteger dimension, it is neither a fixed point nor a periodic solution. Is an attractor with noninteger dimension a strange attractor? This depends on the definition of a strange attractor. The definition proposed by Ruelle (110, 111) includes a requirement of sensitive dependence on initial conditions in some neighborhood of the attractor. At present there is no general analytical result demonstrating that noninteger dimension implies sensitive dependence. However, empirical evidence strongly suggests that this is indeed the case. Also, it seems at least plausible at a theoretical level. A fractional dimension indicates the presence of a very complex attractor geometry. It seems probable that the flow in its vicinity would be similarly complex. As an operational procedure, we take the demonstration of a low fractional dimension and a broadband spectral component as indicative of chaotic behavior. These appear to be commonly accepted criteria.

An added complication is associated with uncertainties in the measurement of dimension. Suppose the measured value of ν is 5.2 ± 0.4. This introduces the possibility of a nonchaotic toroidal attractor composed of five possibly incommensurate frequencies. Theoretical arguments suggest that a torus of dimension greater than or equal to three may not be robust, and that small fluctuations in parameters would result in the transition to a strange attractor (26, 97, 112). However, Abraham, Gollub and Swinney (2) note that three-frequency quasiperiodicity (a three torus) has been observed in fluid experiments, which suggests that these structures may be more dynamically stable than theoretically anticipated (40, 42, 82), but even if a q-torus should be stable, the associated behavior is still complicated. Indeed, toroidal attractors form the basis of Landau's classical theory of turbulence (79, 80). Though differentiating between chaos and highly complex quasiperiodic behavior is of theoretical interest, the most important distinction to be made at this level of analysis is between deterministic behavior (as identified by finite dimension) and random, infinite dimensional behavior.

V. Spontaneous chaotic activity in neurons of the squirrel monkey

Previous reports of spontaneous chaotic neural behavior have been limited to qualitative observations of disordered activity. Recently, however, quantitative evidence for chaotic behavior in neurons of the mammalian central nervous system has been obtained (106). In that study single unit recordings were made of the interspike intervals, the time between action potentials, of spontaneously active neurons in the precentral and postcentral gyri (the areas immediately anterior and posterior to the central fissure) of the squirrel monkey. The squirrel monkey crebral cortex is a classical neurophysiological preparation that has been extensively studied (13, 101, 126). The recording area in the precentral gyrus contains part of the primary motor cortex which is responsible for the coordination of motor activity. The primary somatic sensory cortex is located in the postcentral gyrus and receives input from a chain of neurons extending from the periphery to the central nervous system. This region is responsible for integrating the afferent signals which lead ultimately to cutaneous preception. In the human a strict division is sometimes made between a primary somatic sensory cortex in the postcentral gyrus and a secondary somatic sensory cortex in the upper sylvian fissure. It has been stated that afferent inputs to the primary somatic sensory cortex derive entirely from the contralateral body while the secondary region receives bilateral input (92). This is in apparent contrast with the squirrel monkey (128) and the cat (124) where it was found that depending on the stimulus modality, units of the postcentral gyrus in these species can respond to ipsilateral as well as contralateral stimulation.

The experimental procedure used here has been described in Zimmerman and Kreisman (128). Squirrel monkeys were anesthetized by intraperitoneal injection of 65 mg/kg of alpha-chloralose dissolved in propylene glycol. A surgical level of anesthesia was maintained by supplementing doses every three to four hours during the experiment with one quarter of the initial dose. The animal was artificialy respired. A unilateral craniotomy exposed the pericentral region of the cortex. After removal of the dura, the area was bathed in warm mineral oil at 37°C. A bilateral pneumothorax was performed to reduce cortical pulsation due to respiratory movements and the animal was paralyzed by intravenous injection of gallamine triethiodide. Extracellular action potentials of a single neuron were recorded by glass micropipette electrodes with a tip diameter of 1 micrometer filled with 3M KCl. Analog tape recordings of neural discharges were subsequently played into a computer which measured the time intervals between successive action potentials. The records analyzed in this paper are of spontaneous behavior. The conventionally applied criteria used to ensure that recordings were taken from intact, healthy neurons were satisfied in all cases. Records from neurons which showed variations in their action potential amplitude or any signs of injury discharge during an observation period were discarded.

A convenient way of displaying these data is the Poincaré map, which in the simplest instance consists of plotting two consecutive intervals as an (x,y) pair. If the spike train is periodic with period = 1, each interval is equal and the plot consists of a single point. If the sequence is periodic with period = 1 and small amounts of noise are present, a small cloud would appear around a central point. Points generated by random spike records would fill the plane without apparent pattern. Presenting data in this form may make it easier to see why the dimension of spike train data is not equivalent to the dimension of digitized wave forms, for example EEG. The interspike interval recording itself is the Poincaré map of the neuron's membrane potential. It records the time of each threshold crossing. The dimension of periodic sequences of interspike intervals is zero, while the dimension of periodic waveforms is one.

An example of a Poincaré map that is neither periodic nor random is shown in Figure 3. An imperfectly defined L-shaped structure appears. L-shapes are formed by (long, short) and (short, long) pairs. This map suggests that while the spike train of Figure 3 does not display simple period = 1 rhythmicity, it does have a nonrandom structure. This anticipation is confirmed by calculation of the Renyi dimension for this data set. The results have already been presented in Figure 1, where it was found that the correlation exponent is approximately 3.5. This indicates that the sequence is chaotic rather than random.

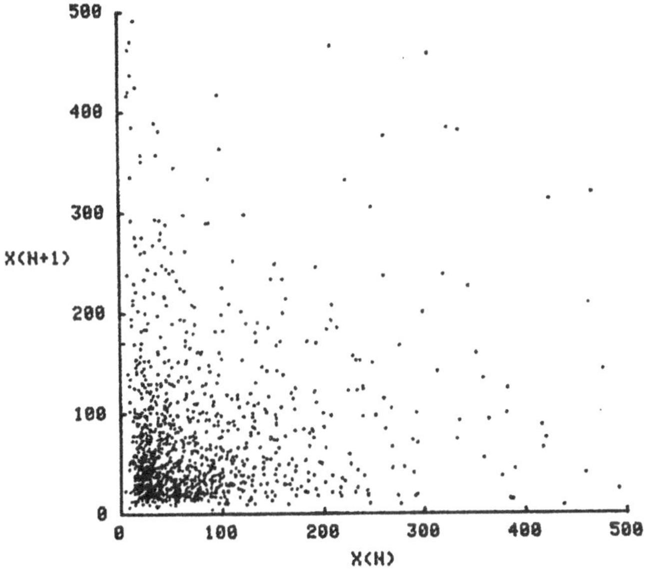

Figure 3:

Poincaré plot of the data set of Figure 1. The axes are labelled in milliseconds.

The Poincaré map of another data set is given in Figure 4. Here an extreme L-shape structure is seen. However, the calculation of C_n, shown in Figure 2, gives very different results. Calculations up to embedding dimension $n = 40$ failed to resolve a finite dimensional structure.

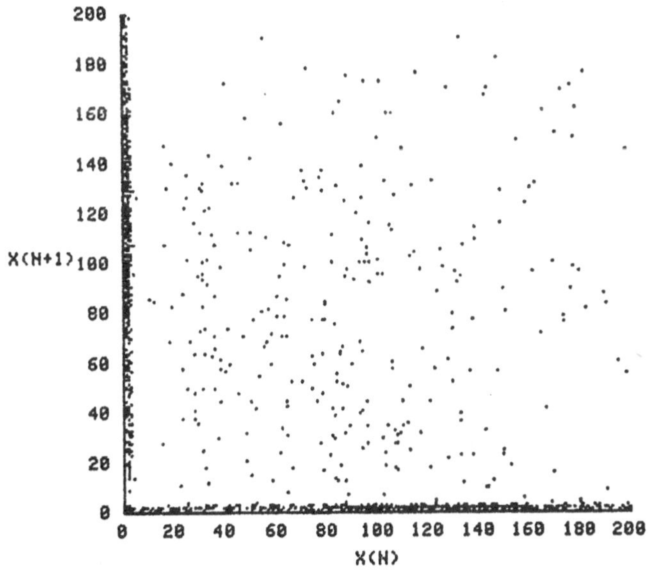

Figure 4:

Poincaré plot of the data set of Figure 2. The axes are labelled in milliseconds.

These two cases exemplify the predominant classes of neurons identified by this study. However, it would be premature to suggest that the neurons fall into two disjoint groups, those with low dimension (dimension less than 4) and those with high dimension (dimension greater than 20). The lower dimensional neurons are also slower (the modal interspike interval is on the order of 20 milliseconds) while the high dimensional neurons are much faster (the modal interval is one or two milliseconds). These intervals are measured to accuracies of one millisecond. It seems possible that for fast neurons firing repeatedly in the 1 millisecond range, the distinction between intervals of .8 and 1.3 milliseconds, for example, might be dynamically significant. This distinction would be lost in these recordings. It can be speculated that if the intervals were measured to $\pm$ .1 millisecond, a finite dimensional attractor might be resolved for faster neurons as well. This possibility can be tested numerically with simulations of chaotic neurons. Experiments with increased accuracy are also planned.

This is a preliminary study and the conclusions that can be drawn from it are limited. Nonetheless, the results do suggest that neurons in the mammalian central nervous system do not invariably display highly complex behavior. Indeed, the very low dimension of the associated attractors of the slower neurons suggest that cortical behavior can, at least in some instances, be comparatively simple.

VI. Chaotic human electroencephalographic signals

Attempts to develop procedures leading to computer assisted interpretation of EEG records have a long history (78). The specific clinical areas addressed include schizophrenia (66) and epilepsy (102, 120). These efforts have usually centered on spectral analysis and its variants. It is probably correct to observe that these programs have had limited effect on clinical practice, and one can speculate that frequency analysis does not provide a neurologically meaningful measure of brain electrical activity. Is the correlation dimension more relevant? This question can be answered only by analyzing large amounts of data from normal controls and defined patient populations. However, before this can be undertaken on a large scale, a much more fundamental question must be addressed. Are these computations feasible with electroencephalographic data?

To date, our computations with human EEG data have been limited to the resting, eyes closed alpha rhythm which is arguably one of the most regular components of the EEG. Preliminary calculations with noise-corrupted sine waves suggested that in the case of periodic or approximately periodic signals, the calculation of dimension requires sampling at least 20 points per cycle for at least 50 cycles. We therefore decided to measure 50 points per cycle for 80 cycles, giving 4000 data points. As the human alpha is roughly 10 Hz, this gives a sample interval of 2 milliseconds.

Figure 5 shows the signal and its power spectrum for points 1 to 1024. Figure 6 gives the corresponding plots of $\ln C_n(\epsilon)$ for embedding dimensions 15 to 20. The calculations for successive embedding dimensions results in a plateau corresponding to dimension of 2.4 $\pm$ .2. Calculations using the third one thousand data points gave the same value of dimension. However, in the graphs of $\mathbb{D}$ for the third one thousand data points the plateau is three times as large and corresponds to three orders of magnitude in $\ln C_n(\epsilon)$. The size of the plateau encourages confidence in the stability of the calculation. The dimension was also calculated using 20 data points. The degree of dispersion in the plateau is substantially decreased with the addition of more data. The sharply defined scaling region corresponds to a dimension of 2.6 $\pm$ .2. A cautionary note should be made. It is clear that systematic investigations will be required to determine the data needed for reliable calculations. Also, the variation in a given individual over time must be determined.

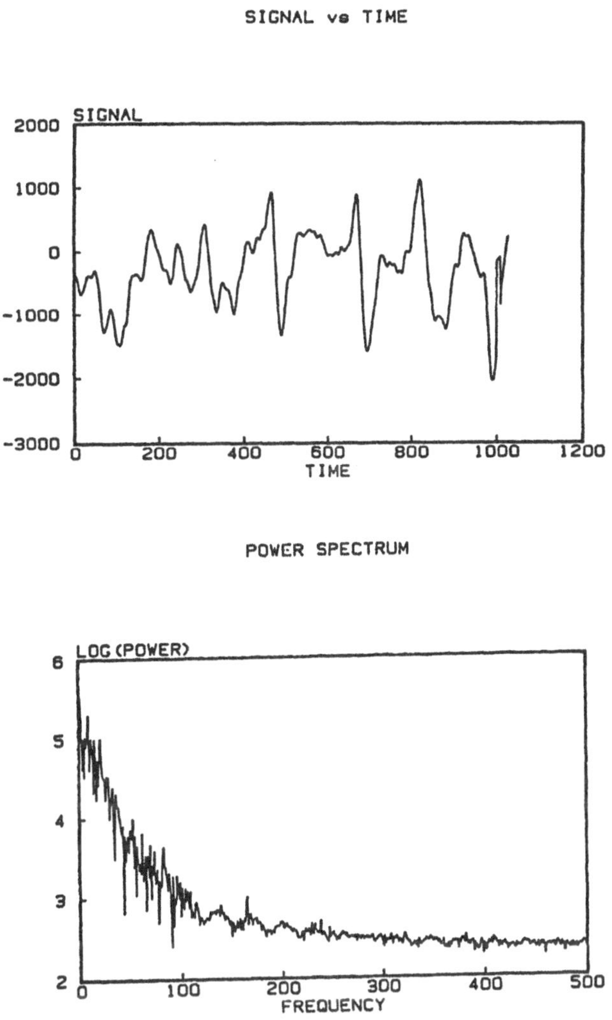

Figure 5:

A. The first 1024 data points of a human resting, eyes closed, EEG signal recorded at site O_2. The signal sampled every 2 milliseconds.
B. The corresponding power spectrum.

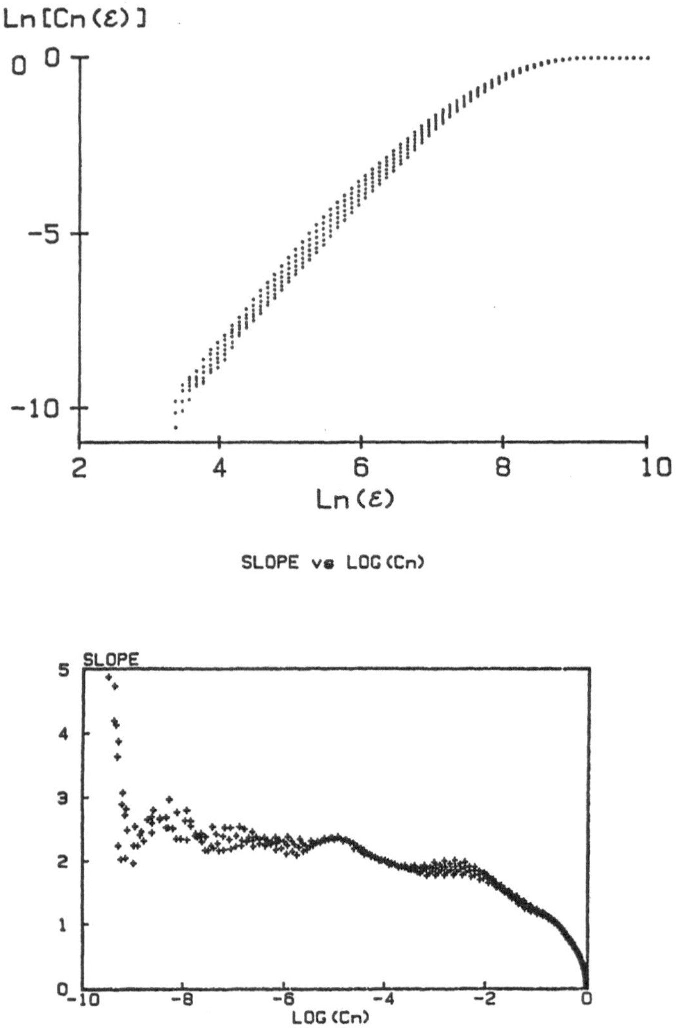

Figure 6:

Dimension calculations of the data in Figure 5 for embedding dimensions 15 to 20. The values of ϵ ran from 2 to 10 in steps of .1 and a 5-point slope calculation was used to determine ν. The value of the dimension is estimated to be 2.4 $\pm$ .2.

A pilot study was also made on the effect of cognitive activity on the alpha wave measured at electrode site O_2. The subject was instructed to count backwards from 300 in steps of 7. The signal and its power spectrum are shown in Figure 7. When compared to the spectrum in Figure 5, some broadening is discernible. Poincaré maps from both signals are shown in Figure 8 and a remarkable difference is seen. The initiation of mental activity disrupts the elegant structure of the resting Poincaré map. The calculation of the correlation dimension is shown in Figure 7. The contrast with Figure 5 is immediately recognizable. The degree of dispersion around the plateau is greatly increased and a dimension of 3.0 $\pm$ .2, to be compared with 2.4 $\pm$ .2, is obtained. The plot of $\mathbb{D}$ for the signal shown also suggests that a structure with a higher dimension, approximately 3.8, may also be present. However, the data do not permit a clear resolution of this object.

Given the limited nature of this study, any conclusions must be stated tentatively. The results do indicate that the calculations are numerically feasible and they also suggest that this analysis may be helpful in quantifying brain electrical activity.

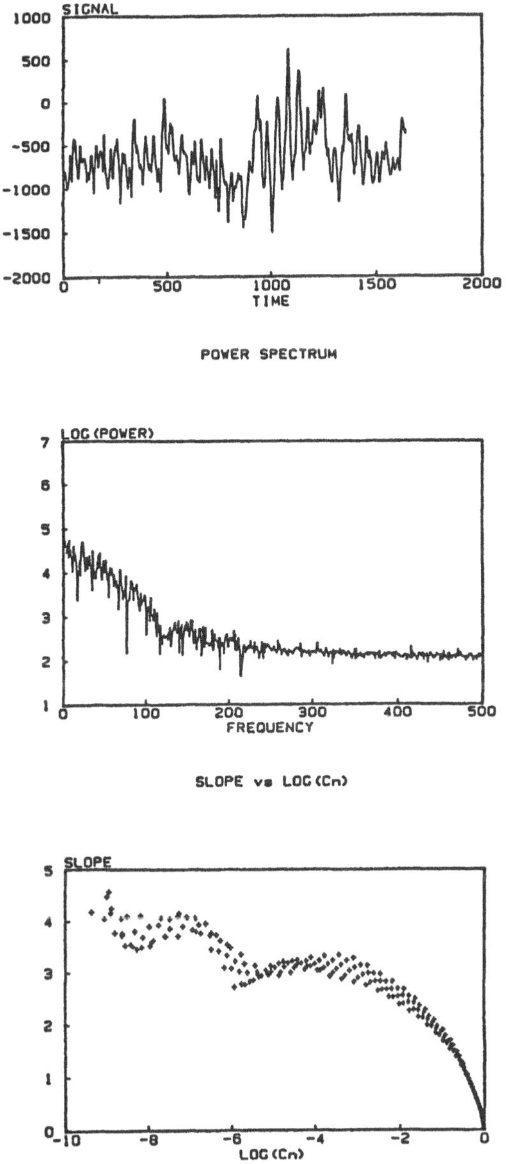

Figure 7:

A. An EEG signal recorded at site O_2. The subject's eyes are closed. The recording was made while the subject was performing mental arithmetic. The signal was sampled every 2 milliseconds.

B. The corresponding power spectrum for the first 1024 data points.

C. Calculation of the dimension using the first 1500 data points of the signal for embedding dimensions 15 to 20. The value of ϵ ran from 2 to 10 in steps of .1 The estimated value of the dimension is 3.0 $\pm$ .3.

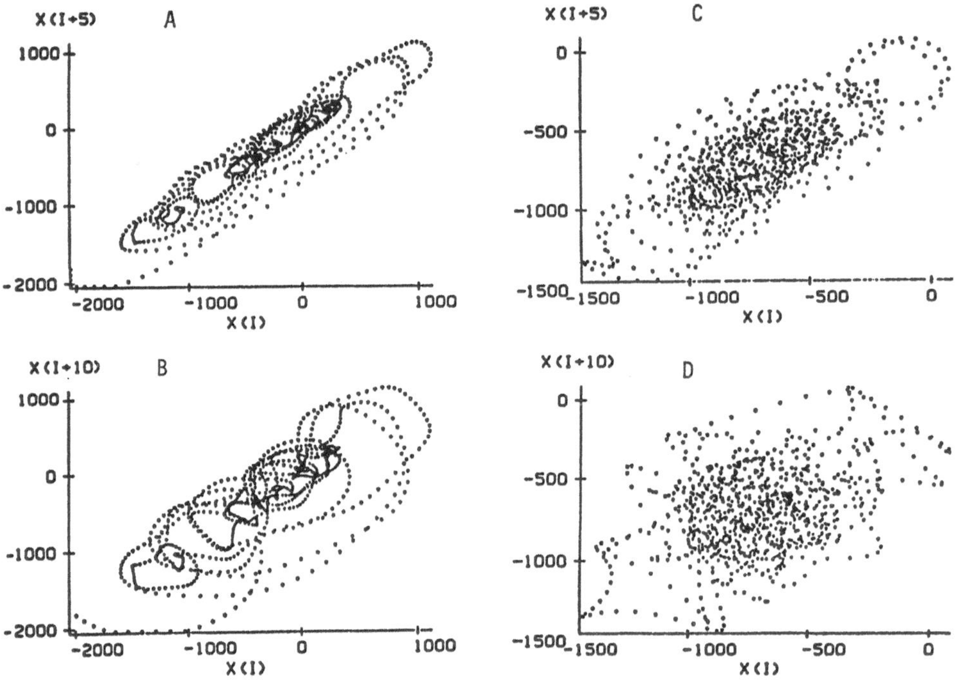

Figure 8:

Delay maps plotting the points (x (j), x(j + Delay)).

A. The first 1000 points of the signal in Figure 5 (eyes closed, resting) Delay=5.
B. The same signal as A with Delay=10.
C. The first 1000 points of the signal in Figure 7 (eyes closed, mental arithmetic) Delay=5.
D. The same signal as in C with Delay=10.

VII. Summary

Dynamical analysis began as a collection of results about abstract mathematical systems. It has evolved rapidly into a set of tools with established utility in examining physical and chemical systems. It now also seems probable that these methods will prove of value in biological research, and may come to have a role in clinical practice.

Acknowledgements

P. E. Rapp would like to acknowledge NIH Grant NS19716 from the Epilepsy Branch of the National Institute of Neurological and Communicative Disorders and Stroke. N. N. Greenbaun would like to acknowledge partial support from the Faculty and Institutional Development Program of Trenton State College. I. D. Zimmerman would like to acknowledge grant support from the National Science Foundation. T. R. Bashore would like to acknowledge NIMH Grant Number MH40627-01 and an intramural research award from the Neuroscience Development Program of the Medical College of Pennsylvania/Eastern Pennsylvania Psychiatric Institute. The assistance of R. S. Gioggia is gratefully acknowledged. The technical assistance of Andrew Goldstein, James Hornig-Rohan, Richard Latta and Maria Levitt is also gratefully acknowledged. Programming assistance of Brian Foote of the University of Illinois Cognitive Psychophysiology Laboratory is acknowledged.

1. Abraham,N.B., Albano,A.M., Das,B., deGuzman,G., Yong,S., Gioggia,R.S., Puccioni,G.P. and Tredicce,J.R. (1985). Calculating the dimension of attractors from small data sets. Phys. Lett. A. submitted.

2. Abraham,N.B., Gollub,J.P. and Swinney,H.L. (1984). Testing nonlinear dynamics. Physica. 11D, 252-264.

3. Adam,D.R., Smith,J.M., Askelrod,S., Nyberg,S., Powell,A.O. and Cohen,R.J. (1984). Fluctuations in T-wave morphology and susceptibility to ventricular fibrillation. J. Electrocard. 17, 209-218.

4. Aihara,K. and Matsumoto,G. (1984). Periodic and nonperiodic responses of a periodically forced Hodgkin-Huxley oscillator. J. theor. Biol. 109, 249-269.

5. Albano,A.M., Abounadi,J., Chyba,T.H., Searle,C.E. and Yong,S. (1985). Low-dimensional chaotic attractors for an unstable, inhomogeneously broadened, single-mode laser. J. Opt. Soc. Amer. 2B, 47-55.

6. Alexander,J.C. and Yorke,J. (1982). The fat baker's transformations. U. of Maryland. preprint.

7. an der Heiden,U., Mackey,M.C. and Walther,H.O. (1981). Complex oscillations in a simple deterministic neuronal network. Lectures. Appl. Maths. 19, 355-360.

8. Balatoni,J. and Renyi,A. (1956). On the notion of entropy. Publications Math. Inst. Hungarian Acad. Sci. 1, 9-40. (English translation: Selected Papers of A.Renyi. 1, 558-586, Akademiai, Budapest).

9. Ballard,D.H., Hinton,G.E. and Sejnowski,T.J. (1983). Parallel visual computation. Nature, Lond. 306, 21-26.

10. BenMizrachi,A. (1984). Elimination of irrelevant variables in nonlinear systems. Phys. Rev. 30A, 2708-2712.

11. BenMizrachi,A. and Procaccia,I. (1984). Universal power law for the dimension of strange attractors near the onset of chaos. Phys. Rev. Lett. 53, 1704.

12. BenMizrachi,A., Procaccia,I. and Grassberger,P. (1984). Characterization of experimental (noisey) strange attractors. Phys. Rev. 29A, 975-977.

13. Blomquist,A.J. and Lorenzini,C.A. (1965). Projection of dorsal roots and sensory nerves to cortical sensory motor regions of the squirrel monkey. J. Neurophysiol. 28, 1195-1205.

14. Brandstater,A., Swift,J., Swinney,H.L., Wolf,A., Farmer,J.D., Jen,E. and Crutchfield,J.P. (1983). Low dimensional chaos in a hydrodynamic system. Phys. Rev. Lett. 51, 1442-1445. Correction. 51, 1814.

15. Brandstater,A. and Swinney,H.L. (1984). Distinguishing low dimensional chaos from random noise in a hydrodynamic experiment. In: Fluctuations and Sensitivity in Nonequilibrium Systems. W.Horsthemke and D.K.Kondepudi, eds. pp. 166-171. Proceedings in Physics. Volume 1. Springer-Verlag, Berlin.

16. Brillouin,L. (1962). Science and Information Theory. Academic Press, NY.

17. Buxton,B.F. (1984). GEC Research Laboratory Long Range Research Laboratory Report No. 16,834A. Report on a Workshop on Statistical Physics in Engineering and Biology.

18. Carpenter,G.A. (1979). Bursting phenomena in excitable membranes. SIAM J. appl. Maths. 36, 334-372.

19. Carpenter,G.A. (1981). Normal and abnormal signal patterns in nerve cells. In: Mathematical Psychology and Psychophysiology. S.Grossberg, ed. SIAM-AMS. Proceedings. 13, 49-90. American Mathematical Society. Providence, RI.

20. Chay,T.R. (1983). Eyring rate theory in excitable membranes. Application to neuronal oscillations. J. phys. Chem. 87, 2935-2940.

21. Chay,T.R. (1984). Abnormal discharges and chaos in a neuronal model system. Biol. Cybernetics. 50, 301-311.

22. Chay,T.R. and Lee,Y.S. (1984). Impulse responses of automaticity in the purkinje fiber. Biophys. J. 45, 841-849.

23. Cohen,A. and Procaccia,I. (1985). Computing the Kolmogorov entropy from time signals of dissipative and conservative dynamical systems. Phys. Rev. A. 31A, 1872-1882.

24. Cohen,R.J. and Berger,R.D. (1983). A quantitative model for verticular response during atrial fibrillation. IEEE Trans. Biomed. BME30, 769-781.

25. Colding-Jorgensen,M. (1983). A model for the firing pattern of a paced nerve cell. J. theor. Biol. 101, 541-569.

26. Davis,P. and Ikeda,K. (1984). T3 in a model of a nonlinear optical resonator. Phys. Lett. Series A. 100A, 455-459.

27. Eckmann,J.-P. (1981). Roads to turbulence in dissipative dynamical systems. Rev. modn. Phys. 53, 643-654.

28. Ermentrout,G.B. (1984). Period doublings and possible chaos in neural models. SIAM J. appl. Maths. 44, 80-95.

29. Farmer,J.D. (1982a). Dimension, fractal measures and chaotic dynamics. In: Evolution of Order and Chaos. H.Haken, ed. pp. 228-246. Springer Verlag, Berlin.

30. Farmer,J.D. (1982b). Information dimension and the probabilistic structure of chaos. Z. Naturforsch. 37a, 1304-1325.

31. Farmer,J.D. (1982c). Chaotic attractors of an infinite dimensional dynamical system. Physica. 4D, 366-393.

32. Federer,H. (1969). Geometric Measure Theory. Springer Verlag, Berlin.

33. Frederickson,P., Kaplan,J., Yorke,E. and Yorke,J. (1982). The Liapunov dimension of strange attractors. J. diff. Eqn. 49, 185-207.

34. Froehling,H., Crutchfield,J.P., Farmer,D., Packard,N.H. and Shaw,R. (1981). On determining the dimension of chaotic flows. Physica. 3D, 605-617.

35. Gelb,A. and Vander Veld,W. (1968). Multiple Input Describing Functions and Control System Design. McGraw Hill, NY.

36. Gibson,G. and Jeffries,C. (184). Observation of period doubling and chaos in spin wave instabilities in yttrium iron garnet. Phys. Rev. Series A. 29A, 811-818.

37. Glass,L., Graves,C., Petrillo,G.A. and Mackey,M.C. (1980). Unstable dynamics of a periodically driven oscillator in the presence of noise. J. theor. Biol. 86, 455-476.

38. Glass,L., Guevara,M.R., Shrier,A. and Perez,R. (1983). Bifurcation and chaos in a periodically stimulated cardiac oscillator. Physica. 7D, 89-101.

39. Glass,L. and Mackey,M.C. (1979). Pathological conditions resulting from instabilities in physiological control systems. Ann. N.Y. Acad. Sci. 316, 214-235.

40. Gollub,J.P. and Benson,S.V. (1980). Many routes to turbulent convection. J. fluid Mech. 100, 449-470.

41. Gollub,J.P., Romer,E.G. and Socolar,J.E. (1980). Trajectory divergence for coupled relaxation oscillators: Measurements and models. J. stat. Phys. 23, 321-333.

42. Gorman,M., Reith,L.A. and Swinney,H.L. (1980). Modulation patterns, multiple frequencies and other phenomena in circular Couette flow. Ann. N.Y. Acad. Sci. 357, 10-21.

43. Grassberger,P. (1981). On the Haussdorff dimension of fractal attractors. J. stat. Phys. 26, 173-179.

44. Grassberger,P. (1983). Generalized dimensions of strange attractors. Phys. Lett. 97A, 227-230.

45. Grassberger,P. (1985). Information flow and maximum entropy measures for 1-D maps. Physica. 14D, 365-373.

46. Grassberger,P. and Procaccia,I. (1983a). Measuring the strangeness of strange attractors. Physica. 9D, 189-208.

47. Grassberger,P. and Procaccia,I. (1983b). Characterization of strange attractors. Phys. Rev. Lett. 50, 346-349.

48. Grassberger,P. and Procaccia,I. (1983c). Estimation of the Kolmogorov entropy from a chaotic signal. Phys. Rev. A. 28A, 2591-2593.

49. Grassberger,P. and Procaccia,I. (1984). Dimensions and entropies of strange attractors from a fluctuating dynamics approach. Physica. 13D, 34-54.

50. Greenside,H.S., Wolf,A., Swift,J. and Pignataro,T. (1982). Impracticality of a box counting algorithm for calculating the dimensionality of strange attractors. Phys. Rev. 25A, 3453-3456.

51. Guevara,M.R., Glass,L. and Shrier,A. (1981). Phase locking, period-doubling bifurcations and irregular dynamics in periodically stimulated cardiac cells. Science, Wash. 214, 1350-1353.

52. Guevara,M.R. and Glass,L. (1982). Phase locking, period doubling bifurcations and chaos in a mathematical model of a periodically driven oscillator: A theory for the entrainment of biological oscillators and the generation of cardiac dysrhythmias. J. math. Biol. 14, 1-24.

53. Guckenheimer,J. (1982). Noise in chaotic systems. Nature, Lond. 298, 358-361.

54. Guckenheimer,J. (1984). Dimension estimates for attractors. Contemp. Maths. 28, 357-367.

55. Guevara,M.R., Glass,L., Mackey,M.C. and Shrier,A. (1983). Chaos in neurobiology. IEEE Trans. Systems, Man and Cybernetics. SMC-13, 790-798.

56. Hausdorff, F. (1918). Dimension und ausseres Mass. Math. Annalen. 79, 157-179.

57. Hayashi,H., Ishizuka,S. and Hirakawa,K. (1983). Transition to chaos via intermittency in the Onchidium pacemaker neuron. Phys. Lett A. 98A, 474-476.

58. Hayashi,H., Ishizaka,S., Ohta,M. and Hirakawa,K. (1982). Chaotic behavior in the Onchidium giant neuron under sinusoidal stimulation. Phys. Lett. 88A, 435-438.

59. Hindmarsh,J.L. and Rose,R.M. (1984). A model of neuronal bursting using three coupled first order differential equations. Proc. R. Soc. Lond. 221B, 87-102.

60. Holden,A.V., ed. (1985). Chaos. An Introduction. Manchester University Press, Manchester, UK, in press.

61. Holden,A.V. and Ramadan,S.M. (1981). The response of a molluscan neurone to a cyclic input: Entrainment and phase locking. Biol. Cybernetics. 41, 157-163.

62. Holden,A.V., Winlow,W. and Haydon,P.G. (1982). The induction of periodic and chaotic activity in a molluscan neurone. Biol. Cybernetics. 43, 169-173.

63. Holden,A.V. and Ramadan,S.M. (1981). The response of a molluscan neuron to a cyclic input: Entrainment and phase locking. Biol. Cybern. 43, 157-163.

64. Hurewicz,W. and Wallman,H. (1941). Dimension Theory. Princeton University Press, Princeton, NJ.

65. Iooss,G., Helleman,R.H.G. and Stora,R. (1983). Chaotic Behaviour of Deterministic Systems. North-Holland Publishing, Amsterdam.

66. Itil,T.M. (1977). Qualitative and quantitative EEG findings in schizophrenia. Schizophrenia Bulletin. 3, 61-79.

67. Keener,J.P. (1981a). Chaotic cardiac dynamics. In: Mathematical Aspects of Physiology. F.C.Hoppensteadt, ed. pp. 299-325. American Mathematical Society, Providence, RI.

68. Keener,J.P. (1981b). On cardiac arrhythmias: AV conduction block. J. math. Biol. 12, 215-225.

69. Jensen,J.H., Christiansen,P.L., Scott,A.C. and Skovgaard,O. (1983). Chaos in nerve. preprint. Technical University of Denmark.

70. Kaczmarek,L.K. (1976). A model of cell firing patterns during epileptic seizures. Biol. Cybernetics. 22, 229-234.

71. Kaczmarek,L.K. and Babloyantz,A. (1977). Spatiotemporal patterns in epileptic seizures. Biol. Cybernetics. 26, 199-208.

72. Kaplan,J.C. and Yorke,J.A. (1979). Chaotic behavior of multidimensional difference equations. In: Functional Differential Equations and Approximations of Fixed Points. H.O.Peitgen and H.O.Walther, eds. Lecture Notes in Mathematics. Volume 730. Springer, Berlin.

73. King,R., Barchas,J.D. and Huberman,B.A. (1984). Chaotic behavior in dopamine neurodynamics. Proc. natn. Acad. Sci. U.S.A. 81, 1244-1247.

74. Kolata,G. (1984). Order out of chaos in computers. Science, Wash. 223, 917-919.

75. Kolmogorov,A.N. (1958). A metric invariant of transient dynamical systems and automorphisms in Lebesgue spaces. Dokl. Acad. Nauk USSR. 119, 861-864. (English summary: Math. Rev. 21, 386.)

76. Kolmogorov,A.N. (1959). Entropy per unit time as a metric invariant of automorphisms. Dokl. Akad. Nauk USSR. 124, 754-755. (English summary: Math. Rev. 21, 386.)

77. Kloeden,P., Deakin,M.A.B. and Tirkel,A.Z. (1976). A precise definition of chaos. Nature, Lond. 264, 295.

78. Ktonas,P.Y. (1983). Automated analysis of abnormal electroencephalograms. CRC Critical Reviews of Biomedical Engineering. 9, 39-97.

79. Landau,L.D. (1944). Dkl. Acad. Sci. USSR. 44, 311.

80. Landau,L.D. and Lifshitz,E.M. (1959). Fluid Mechanics. Pergamon, NY. (Section 27).

81. Libchaber,A. (1983). Experimental aspects of the period doubling scenario. Lect. Notes Phys. 179, 157-164.

82. Libchaber,A., Fauve,S. and Laroche,C. (1983). Two parameter study of the routes to chaos. Physica. 7D, 73-84.

83. Lundy,M. and Mees,A.I. (1984). Convergence of the annealing algorithm. Math. Program. in press.

84. Mackey,M.C. (1979). Periodic autoimmune hemolytic anemia: An induced dynamical disease. Bull. math. Biol. 41, 829-834.

85. Mackey,M.C. (1981a). Some models in hemopoiesis: Predictions and problems. In: Biomathematics in Cell Kinetics. M.Rotenberg, ed. pp. 23-38. Elsevier, Amsterdam.

86. Mackey,M.C. (1981b). Unravelling the connection between human hematopoietic cell proliferation and maturation. In: Regulation of Reproduction and Aging. E.V.Jensen and J.G.Vassileva-Popova, eds. Plenum Press, NY.

87. Mackey,M.C. (1985). A mitotic oscillator with a strange attractor and distribution of cell cycle times. In: Nonlinear Oscillations in Chemistry and Biology. H.G.Othmer, ed. Springer Verlag, NY.

88. Mackey,M.C. and an der Heiden,U. (1983). The dynamics of recurrent inhibition. J. math. Biol. 19, 211-225.

89. Mackey,M.C. and Dormer,P. (1981). Enigmatic hemopoiesis. In: Biomathematics and Cell Kinetics. M. Rotenberg, ed. pp. 87-103. Elsevier/North-Holland Biomedical Press, Amsterdam.

90. Mackey,M.C. and Glass,L. (1977). Oscillations and chaos in physiological control systems. Science, Wash. 197, 287-289.

91. Mandelbrot,B.B. (1983). The Fractal Geometry of Nature. Revised Edition. W.H.Freeman, San Francisco.

92. Martin,J.H. (1981). Somatic sensory system. II. Anatomical substrates for somatic sensation. In: Principles of Neural Science. E.R.Kandel and J.H.Schwartz, eds. pp. 170-183. Elsevier, North-Holland, NY.

93. Matsumoto,G., Aihara,K., Ichikawa,M. and Tasaki,A. (1983). Periodic and nonperiodic responses of membrane potential in squid giant axons under firing to sinusoidal current stimulation. J. theor. Neurobiol.. 3, 1-14.

94. Mees,A.I. and Sparrow,C.T. (181). Chaos. IEE Proc. 128D, 201-205.

95. Mori,H. (1980). Fractal dimensions of chaotic flows of autonomous dissipative systems. Prog. theor. Phys. 68, 1044-1047.

96. Nemytskii,V.V. and Stepanov,V.V. (1960). Qualitative Theory of Differential Equations. Princeton University Pres, Princeton, NJ.

97. Newhouse,S., Ruelle,D. and Takens,F. (1978). Occurrence of strange axiom A attractors near quasi-periodic flows on Tm, m $\geq$ 3. Commun. math. Phys. 64, 35-40.

98. Nicolis,C. and Nicolis,G. (1984). Is there a climatic attractor? Nature, Lond. 311, 529-532.

99. Ott,E., Yorke,E.D. and Yorke,J.A. (1985). A scaling law: How an attractor's volume depends on noise level. Physica. 16D, 62-78.

100. Packard,N.H., Crutchfield,J.P., Farmer,J.D. and Shaw,R.S. (1980). Geometry from a time series. Phys. Rev. Lett. 45, 712-716.

101. Pinneo,L.R. (1968). Brain mechanisms in the behavior of the squirrel monkey. In: The Squirrel Monkey. L.A.Rosenblum and R.W.Cooper, eds. pp. 319-346. Academic Press, NY.

102. Principe,J.C. and Smith,J.R. (1982). Microcomputer-based system for the detection and quantification of petit mal epilepsy. Comput. Biol. Med. 12, 87-95.

103. Rapp,P.E. (1975). A theoretical investigation of a large class of biochemical oscillators. Math. Biosci. 25, 165-188.

104. Rapp,P.E. (1985a). Oscillations and chaos in cellular metabolism and physiological systems. In: Chaos, An Introduction. A.V.Holden, ed. Manchester University Press, UK, in press.

105. Rapp,P.E. (1985b). Reliability in high density hierarchical devices: Possible lessons from neural systems. In: Molecular Electronic Devices. F.L.Carter, ed. Marcel Dekker, NY in press.

106. Rapp,P.E., Zimmerman,I.D., Albano,A.M., deGuzman,G.C. and Greenbaun,N.N. (1985). Dynamics of spontaneous neural activity in the simian motor cortex: The dimension of chaotic neurons. Phys. Lett. in press.

107. Renyi,A. (1959). On the dimension and entropy of probability distributions. Acta Math. Acad. Sci. Hungar. 10, 193-215. (English translation: Selected Papers of A.Renyi. 2, 320-342, Akademiai, Budapest).

108. Ritzenberg,A.L., Adam,D.R. and Cohen,R.J. (1984). Period multiplying: Evidence for nonlinear behavior in the canine heart. Nature, Lond. 307 159-161.

109. Ritzenberg,A.L., Smith,J.M., Grumbach,M.P. and Cohen,R.J. (1984b). Precursor to fibrillation in cardiac computer model. In: Computers in Cardiology. IEEE, Silver Spring, MD.

110. Ruelle,D. (1981a). Differentiable dynamical systems and the problem of turbulence. Bull. (New Series) Am. math. Soc. 5, 29-42.

111. Ruelle,D. (1981b). Small random perturbations of dynamical systems and the definition of attractors. Commun. math. Phys. 82, 137-151.

112. Ruelle,D. and Takens,F. (1971). On the nature of turbulence. Commun. math. Phys. 20, 167-192.

113. Russell,D.A., Hanson,J.D. and Ott,E. (1980). Dimension of strange attractors. Phys Rev. Lett. 45, 1175-1178.

114. Sbitnev,V.I. (1978). Transport of spikes in statistical neuron ensembles. IVa. The starting of the problem in a diffusion approximation. Biofizika. 23, 508-513. (translation: Biophysics. 23, 514-520.)

115. Sbitnev,V.I. (1979). Transport of spikes in statistical neuron ensembles. An induced epileptic focus in the model of field CA3 of the hippocampus. Biofizika. 24, 141-147. (translation: Biophysics. 24, 141-147).

116. Sbitnev,V.I. (1984). Model patterns of stochastic variations of postsynaptic activity. Biofizika. 29, 113-116. (translation: Biophysics. 29, 121-125.)

117. Schuster,H.G. (1984). Deterministic Chaos: An Introduction. VCH Publishers, Dearfield Beach, FL.

118. Sejnowski,T.J. and Hinton,G.E. (1985). Parallel stochastic search in early vision. In: Vision, Brain and Cooperative Computation. M.Arbib and A.R.Hanson, eds.

119. Shaw,R. (1981). Strange attractors, chaotic behavior and information flow. Z. Naturforsch. 36A, 80-112.

120. Siegel,A., Grady,C.L. and Mirsky,A.F. (1982). Prediction of spike-wave bursts in absence epilepsy by EEG power-spectrum signals. Epilepsia. 23, 47-60.

121. Smith,J.M. and Cohen,R.J. (1984a). Simple finite-element model accounts for wide range of cardiac dysrhythmias. Proc. natn. Acad. Sci. U.S.A. 81, 233-237.

122. Smith,J.M., Ritzenberg,A.L. and Cohen,R.J. (1984b). Finite element models of cardiac dysrhythmias. In: Proceedings 1984 Symposium on Mathematics and Computers in Biomedical Applications. NIH, Washington.

123. Takens,F. (1980). Detecting strange attractors in turbulence. In: Dynamical Systems and Turbulence. Lecture Notes in Mathematics. Volume 898. D.A.Rand and L.S.Young, eds. pp. 365-381. Springer-Verlag, NY.

124. Towe,A.L., Patton,H.D. and Kennedy,T.T. (1964). Response properties of neurons in the pericuriate cortex of the cat following electrical stimulation of the appendages. Exptl. Neurol. 10, 325-344.

125. Turner,J.S., Roux,J.C., McCormick,W.D. and Swinney,H.L. (1981). Alternating periodic and chaotic regimes in a chemical reaction: Experiment and theory. Phys. Lett. Series A. 85A, 9-12.

126. Welker,W.I., Benjamin,R.M., Miles,R.C. and Woolsey,C.N. (1957). Motor effects of stimulation of the cerebral cortex of squirrel monkey (Saimiri sciureus). J. Neurophysiol. 20, 347-364.

127. Whitney,H. (1936). Ann. Math. 37, 645.

128. Zimmerman,I.D. and Kreisman,N.R. (1970). Somatosensory cortical unit responses of long duration. Nature, Lond. 227, 1361-1363.

MATHEMATICAL METHODS

A PERIOD-DOUBLING BUBBLE IN THE DYNAMICS OF TWO COUPLED OSCILLATORS

J. C. Alexander*

Department of Mathematics and
Institute for Physical Science and Technology
University of Maryland
College Park, MD 20742 USA

A period-doubling cascade in the bifurcation diagram of two Brusselators coupled by diffusion is continued to a particular parameter regime, where it is seen numerically to be associated with other bifurcation branches, and in particular, "decascades;" we call the resulting bifurcation effect a period-doubling bubble. Moreover the dynamics of the bubble formation can be described. The emphasis in this note in on describing the phenomenon, although the (strong) possibility of describing it analytically in terms of unfolding a singularity which comes from interactions of singularities of the single oscillators is discussed, as well as a discussion of possibly similar behavior in other coupled oscillators.

1. Introduction

The purpose of this informal report is to consider numerically the behavior of two Brusselators coupled by diffusion in a certain region of parameter space (and in the spirit of a number of lectures, to advertise the numerical global bifurcation package AUTO). This is part of a program to understand the behavior of a prototypical example of coupled oscillators in a global sense. By "global" is meant the following: as adjustable parameters are varied, bifurcation branches and other qualitative dynamical phenomena continues. Conversely, it seems reasonable that behavior observed for some parameter setting can be continued via variation of parameters to more basic types of behavior. A simple case is a periodic solution continuing to a Hopf bifurcation. In a favorable case, all behavior would be "explained" this way in terms of a few basic components. Moreover, for coupled oscillators, the basic components themselves should be described in terms of interactions of basic features of the individual oscillators. The hope is that a rather complete understanding of one system will provide insights into general systems, both globally and in particular details.

* Partially supported by NSF

2. Two Brusselators

The Brusselator is the system

$$\begin{aligned}\dot{X} &= A - (B+1)X + X^2Y,\\ \dot{Y} &= BX - X^2Y.\end{aligned} \tag{2.1}$$

Here X, Y are variables and A, B are adjustable parameters. There is a single stationary point

$$X = A, \qquad Y = B/A, \tag{2.2}$$

which is stable for $B < A^2 + 1$. If B is considered a bifurcation parameter, at $B = A^2 + 1$ the system exhibits Hopf bifurcation and for $B > A^2 + 1$, there is a unique limit cycle.

The chemical model constrains A and B positive. Indeed as $A \to 0$, the Y-component of the stationary point (2.2) goes to ∞ if B is bounded away from 0. If we rescale $X' = AX$, $Y' = Y/A$, we obtain the system

$$\begin{aligned}\dot{X}' &= 1 - (B+1)X' + X'^2Y',\\ \dot{Y}' &= A^2BX' - A^2X'^2Y',\end{aligned} \tag{2.3}$$

with stationary point

$$X' = 1, \qquad Y' = B. \tag{2.4}$$

As $A \to 0$, the stationary point (2.4) becomes degenerate, with Jacobian $\left(\begin{smallmatrix}0 & 1\\ 0 & 0\end{smallmatrix}\right)$. In fact, the system is quite degenerate, since $\dot{Y}' \equiv 0$ for $A = 0$ and the standard analyses (e.g. [5, §7.3]) do not apply (the normal form [5, eq. (7.2.10)] has $b_2 = 0$). We mention this singularity, since it seems that a coupling of two of them generates the behavior we observe. However we will not proceed from this point of view, but return to it after exhibiting and discussing numerical observations.

We turn to two Brusselators coupled by linear diffusion:

$$\begin{aligned}\dot{X}_i &= A - (B+1)X_i + X_i^2Y_i - D_1(X_{i+1} - X_i),\\ \dot{Y}_i &= BX_i - X_i^2Y_i - D_2(Y_{i+1} - Y_i).\end{aligned} \qquad i = 1,2 \pmod 2 \tag{2.5}$$

where D_1, D_2 are non-negative diffusion coefficients. For moderate values of A, D_2 and small D_1, if we let B be the bifurcation parameter, a typical bifurcation diagram is as in Fig. 1. At the left, the stationary solution is stable. A supercritical secondary stationary branch bifurcates off the primary branch at $B = (1+2D_1)(A^2+2D_2)/2D_2$. The synchronous Hopf branch bifurcates at $B = A^2 + 1$. In the present case (more generally if $A^2(1+2D_1) < 2D_2(A^2 - 2D_1)$), the secondary stationary branch bifurcates first and carries the stability. However the synchronous Hopf branch throws

The Brusselator is the system of equations describing a non-existing chemical reaction [7]. However it is a simple, yet interesting dynamical system and was chosen as the prototypical example for several reasons. It is small (2 variables) and thus complications of size do not intrude. On the other hand, it exhibits the most elemental global oscillatory behavior. There are two adjustable parameters (chemical rate constants). As one is varied, a globally stable stationary equilibrium undergoes Hopf bifurcation; from that point there is a periodic limit cycle. For a complete expository discussion, see [7]. Thus the Brusselator is just complicated enough to have the basic features of an oscillator. Accordingly, any global features of the Brusselator's behavior are likely to be rather universal among oscillators. Ashkenazi and Othmer [2] consider a system modelling glycolytic oscillations, which is closely related mathematically to the Brusselator. It exhibits much the same behavior (both singly and coupled) as the Brusselator, although some of the algebra is more complicated.

Here we consider two identical Brusselators coupled by linear diffusion. This system has stationary equilibrium point, and several "obvious" features connected with this equilibrium—a synchronous Hopf bifurcation branch with both cells oscillating identically as if they were isolated, an "antisynchronous" Hopf bifurcation coming from the symmetry of interchanging cells, and a branch of secondary stationary solutions. Moreover there are secondary bifurcations coming from the interaction of these branches. The global questions are: Can all the dynamical behavior be continued to these branches and their interactions? If so, how? If not, how can such behavior be explained? Two coupled Brusselators evidently exhibit chaotic behavior [10] which of course, one Brusselator does not. We discuss a period-doubling cascade which seems to lead to chaotic behavior, and continue this to a parameter region where it can be localized. The period-doubling cascade reverses itself as the bifurcation parameter is pushed further, and we see a feature which might be called a "period-doubling bubble". We investigate the behavior via bifurcation diagrams, discussion of the flows, and graphs of the solutions against time.

It seems quite likely at least some of the features can be described analytically. In the last section, we informally discuss the possibilities. Since the Brusselator is supposed to be prototypical, other oscillators should be investigated for similar bifurcation. We explore the possibility in the last section.

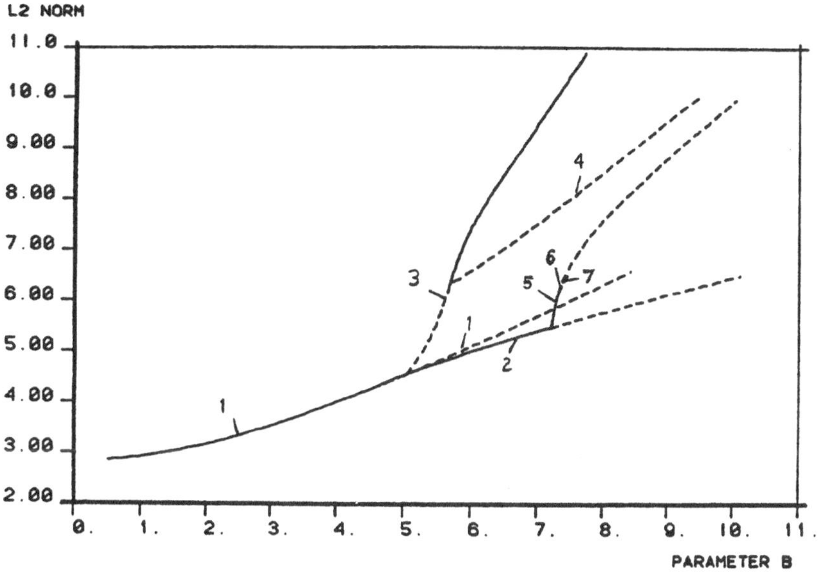

Figure 1. Bifurcation diagram for two coupled Brusselators with $A = 2.$, $D_1 = 1.$, $D_2 = 4$. The horizontal axis is B; the vertical axis is the L_2 norm of the function. Stable branches are demarked by solid lines; unstable by unsolid. The asymmetric branches each have a twin which are superimposed on the diagram. Branches: (1) the primary stationary branch, (2) secondary stationary branch (3) synchronous Hopf branch off of branch 1, (4) secondary pitchfork bifurcation off of branch 3, (5) Hopf branch off of branch 2, (6,7,...) successive period-doubling branches cascading off of branch 5.

off an unstable secondary branch (a Floquet multiplier goes through +1) and gains stability. The stable synchronous branch rapidly becomes dominant, with a large domain of attraction. Meanwhile, the stable secondary branch undergoes its own Hopf bifurcation. The resulting periodic branch undergoes period-doubling bifurcation (a Floquet multiplier goes through -1), as does the period-doubled branch, and evidently a period-doubling cascade ensues.

It is this secondary and higher-order bifurcation structure we would like to relate to more basic structure. If $\big(D_2 - (1 + 2D_1)\big)A^2 = 4D_1D_2$, there is a degenerate bifurcation at $B = A^2 + 1$ with Jacobian having eigenvalues $\pm Ai$, 0, $-2(D_1 + D_2)$, which is associated to the higher-order structure (see [5, §7.4]). However the purpose of this note is to continue the structure for large D_2.

3. The singular system

In this section, we make D_2 as large as possible, namely ∞, and study the resulting system. Pushing D_2 to ∞ makes the system singular. We change variables to

$$\begin{aligned} \xi_1 &= (X_1 + X_2)/\sqrt{2}, \qquad & \eta_1 &= (Y_1 + Y_2)/\sqrt{2}, \\ \xi_2 &= (X_1 - X_2)/\sqrt{2}, \qquad & \eta_2 &= (Y_1 - Y_2)/\sqrt{2}, \end{aligned} \tag{3.1}$$

whence (2.5) transforms to

$$\begin{aligned} \dot{\xi}_1 &= \sqrt{2}A - (B+1)\xi_1 + \tfrac{1}{2}(\xi_1^2\eta_1 + \xi_2^2\eta_1 + 2\xi_1\xi_2\eta_2), \\ \dot{\eta}_1 &= B\xi_1 - \tfrac{1}{2}(\xi_1^2\eta_1 + \xi_2^2\eta_1 + 2\xi_1\xi_2\eta_2), \\ \dot{\xi}_2 &= -(B+1)\xi_2 + \tfrac{1}{2}(\xi_1^2\eta_2 + \xi_2^2\eta_2 + 2\xi_1\xi_2\eta_1) - 2D_1\xi_2, \\ \dot{\eta}_2 &= B\xi_2 - \tfrac{1}{2}(\xi_1^2\eta_1 + \xi_2^2\eta_1 + 2\xi_1\xi_2\eta_2) - 2D_2\xi_2. \end{aligned} \tag{3.2}$$

We now set $2D_2 = \epsilon^{-1}$ and let $\epsilon \to 0$. Standard results of singular perturbation theory [9, Chap. 4] imply the order $t = 1$ behavior of (3.2) for small ϵ approximates that of the 3-dimensional system formally obtained by setting $\epsilon = 0$:

$$\begin{aligned} \dot{\xi}_1 &= \sqrt{2}A - (B+1)\xi_1 + \tfrac{1}{2}(\xi_1^2\eta + \xi_2^2\eta), \\ \dot{\eta} &= B\xi_1 - \tfrac{1}{2}(\xi_1^2\eta + \xi_2^2\eta), \\ \dot{\xi}_2 &= -(B+1)\xi_2 + \xi_1\xi_2\eta - 2D_1\xi_2. \end{aligned} \tag{3.3}$$

where with respect to (3.2), $\eta = \eta_1$ and $\eta_2 \equiv 0$, (i.e. $Y_1 \equiv Y_2$). Increasing D_1 tends to damp out effects, so below we set $D_1 = 0$.

This system has a stationary point at

$$\xi_1 = \xi_2 = \sqrt{2}A, \qquad \eta = \sqrt{2}B/A \tag{3.4}$$

and the bifurcation diagram of (3.3) is virtually the same as Fig. 1. The secondary stationary branch bifurcates from the primary branch (3.3) at $B = 1$. The primary (synchronous) Hopf bifurcation occurs at $B = A^2 + 1$. There is no antisynchronous Hopf branch because $Y_1 \equiv Y_2$. The secondary stationary branch undergoes Hopf bifurcation when [1]

$$A^2 = \frac{(B-1)(B+1)^3}{2B(B^2+B-1)}. \tag{3.5}$$

This defines B as a monotone function of A (that is, as A increases, the secondary Hopf bifurcation point moves to the right), and $B \to 1$ as $A \to 0$. Analytic expressions for the higher-order bifurcation points are not available. For smaller A, the bifurcations all (numerically) occur for smaller B. Moreover, since the B-value of the Hopf bifurcation off the secondary branch given by (3.5) is less than $A^2 + 1$, the B-value of the Hopf bifurcation off the primary branch [1], one could hope that the period-doubling cascades could be moved to a region of the bifurcation diagram where the synchronous periodic branch is unstable (or doesn't exist), thereby making it easier to numerically study the cascades. One would then be able to investigate such phenomena for coupled oscillators in a 3-dimensional system, rather than a 4-dimensional system. Accordingly we set $A = .3$ and generate the bifurcation diagram of Fig. 2.

The secondary branch from the synchronous Hopf branch now comes off almost vertically and is seen to be the other end of the Hopf branch off the secondary stationary branch. The period-doubling branches bifurcate and then return to their base branches. It is this picture we call "period-doubling bubble".

4. The flow

In a later section we consider the possibility the structure of Fig. 2 is not special to the Brusselator. In light of such a possibility, it is worth while to understand the dynamics of the system in the region of the bifurcation. In this section we consider the vector field and flow of (3.3).

Fix A, B and set $D_1 = 0$. Note that the coordinate planes in (ξ_1, ξ_2, η)-space are boundaries for an orbit originating in the positive orthant. Indeed, the system is invariant under $\xi_2 \to -\xi_2$), so the ξ_2-plane is invariant. In this plane $X_1 = X_2$, and the flow in this plane is that of a single Brusselator. For an orbit starting with (ξ_1, ξ_2, η) positive, let the orbit first hit one of the planes $\xi_1 = 0$ or $\eta = 0$ at time t. If at this time $\xi_1 = 0$, $\eta \geq 0$, then $\dot{\xi}_1 > 0$. If $\eta = 0$, $\xi_1 > 0$, then $\dot{\eta} > 0$. In neither case does t exist.

The primary stationary point p_0 has coordinates $\xi_1 = \sqrt{2}A$, $\xi_2 = 0$, $\eta = \sqrt{2}B/A$. In the plane $\xi_2 = 0$ where the field is that of a single Brusselator, the flow is well-understood. For $B < A^2 + 1$, the orbits spiral in to the stationary point p_0. For $B > A^2 + 1$, there is a single limit cycle; the orbits exterior to the cycle spiral into it, those interior to the cycle spiral out to it. As B is varied beyond $A^2 + 1$, the cycle rapidly assumes its characteristic triangular relaxation cycle shape (see [8]). The differing characteristic times on the different sections of the triangle give rise to the characteristic graph of ξ_1 (or X) against time—a sharp peak followed by a long recovery period.

The hyperbolic cylinder $\xi_2 > 0$, $\xi_1\eta = B+1$ is an isocline for ξ_2. If $\xi_1\eta > B+1$, then $\dot{\xi}_2 > 0$ and conversely. If $B > 1$ there is a second stationary point p_1 on the cylinder with coordinates

$$(\xi_1^{(1)}, \xi_2^{(1)}, \eta^{(1)}) = (\sqrt{2}A, \sqrt{2}A\frac{B-1}{B+1}, \frac{1}{\sqrt{2}}\frac{B+1}{A}).$$

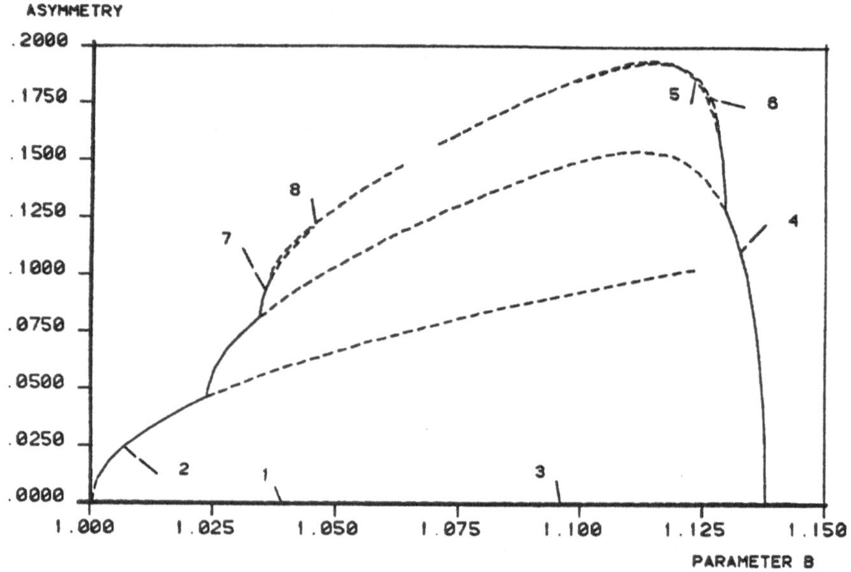

Figure 2. Bifurcation diagram of two coupled Brusselators with $A = .3$, $D_1 = 0$, $D_2 = \infty$. Horizontal axis is B; vertical axis is deviation from symmetry ($\max|\xi_2|$, which is different from Fig. 1). Thus the primary stationary branch and the synchronous Hopf branch both lie along the horizontal axis. Stable branches are demarked by solid lines, unstable by unsolid. Branches: (1) primary stationary branch, (2) secondary stationary branch which bifurcates from branch 1 at $B = 1.0000$ and continues monotonically to $B = \infty$, (3) synchronous Hopf branch which bifurcates from branch 3 at $B = 1.0900$ and continues monotonically to $B = \infty$, (4) Hopf branch which bifurcates from secondary stationary branch at $B = 1.0238$ and which at its other end bifurcates from branch 3 in a pitchfork bifurcation at $B = 1.1373$, (5,6) successive right-hand ends of period-doubling branches bifurcating respectively at $B = 1.1290$ and $B = 1.1276$, (7,8) corresponding left-hand ends bifurcating respectively at $B = 1.0346$ and $B = 1.0366$. The computer detects another period-doubling bifurcation off of branch 8 at $B = 1.0370-$. The gaps between the left- and right-hand ends are artifacts of not continuing the computer calculations. The asymmetric branches each have a twin which are superimposed on the diagram. If the Feigenbaum renormalization theory [5, §6.8.2], [3] is valid for this example, the computed Feigenbaum ratio $\delta^{-1} = .207$ for the left bifurcations agrees quite well with the universal $\delta^{-1} = .214$ [3] and implies the left cascade limits at about $B = 1.037$. The computed $\delta^{-1} = .162$ for the right cascade, but this is based on only the first three bifurcations. It implies the right cascade limits at about $B = 1.127$.

For $1 < B < A^2 + 1$, this point is attracting. Consider the ξ_1 coordinate pointing to the right, the ξ_2 coordinate vertical, and the η coordinate pointing back out of the (ξ_1, ξ_2) plane. A typical orbit spirals into p_1, approaching it in a plane tilted upwards to the right. In particular, the one-dimensional unstable manifold of p_0 behaves this way. It would seem that a dynamical connection from p_0 on p_1 could be shown by the Conley index; it is a matter of constructing the right isolating block.

At Hopf bifurcation, a limit cycle forms around p_1. For small A (and presumably all A), the bifurcation is supercritical [1], so the limit cycle is stable. At bifurcation, the tangent plane of the center manifold is given by the equation

$$4H(1+H^2)(\xi_1-\xi_1^{(1)})-(1+4H^2+H^4)(1-H^4)(\xi_2-\xi_2^{(1)})-2H(1-H^2)^2(\eta-\eta^{(1)})=0,$$

where $H^2 = (B-1)/(B+1)$. This plane intersects the plane $\xi_2 = \xi_2^{(1)}$ in the line

$$4H(1+H^2)(\xi_1-\xi_1^{(1)}) - 2H(1-H^2)^2(\eta-\eta^{(1)}) = 0, \qquad \xi_2 = \xi_2^{(1)}.$$

The tangent plane is above the plane $\xi_2 = \xi_2^{(1)}$ if this quantity is positive and conversely. If $A = .3$, the angle between the planes is about $22.5°$. The branch of this cycle in the bifurcation diagram is of course the Hopf branch off the secondary stationary branch. For reasonably small A, this limit cycle also becomes rather triangular in shape. On the first leg of the triangle, in the region $\xi_1\eta < B+1$, both ξ_1 and ξ_2 decrease sharply. On the second leg η increases and the orbit point moves into the region $\xi_1\eta < B+1$. This motion is relatively slow. On the third leg, ξ_1 and ξ_2 increase sharply and η decreases sharply. The graph of ξ_1 or ξ_2 against time has a spiked appearance, much like that of X for a single Brusselator. The third and first legs of the triangle respectively form the rise and descent of the spike. It is not so apparent from (3.3), but X_2 is relatively constant. The cycle does not move far from the plane $\xi_1 - \xi_2 = \sqrt{2}A$. Consider the effect on the vector field (3.3) of an increase of B. At any point (ξ_1, ξ_2, η), the components $\dot{\xi}_1$, $\dot{\xi}_2$ are decreased, so that for example, the first leg of the triangle drops further down towards the plane $\xi_2 = 0$. The component η is increased (changing B does not affect $\dot{\xi}_1 + \dot{\xi}_2$), so the orbit moves closer to p_0 on the second leg. Thus as B increases further, the cycle, on its first leg, drops closer and closer to the plane $\xi_2 = 0$ and hence on its second leg, behaves more and more like an orbit of a single Brusselator. In particular, it would like to spiral around p_0. However it is geometrically impossible for the second leg of the orbit to continue to a spiral. At some B, the orbit would have to intersect the unstable manifold of p_0, which is impossible.

Rather, as B increases, at some value, this cycle loses stability and bifurcates, doubling its period, into an orbit consisting at first of two virtually identical triangular halves. One of these halves lifts up, as B increases, and retracts and "slides down" around the unstable manifold of p_0 to form a loop around the unstable manifold near the plane $\xi_2 = 0$. The picture is somewhat complicated by the fact that the unstable manifold of p_0 limits on the cycle. The resultant cycle looks like the triangular cycle described above, except it spirals once around the unstable manifold of p_0 near the plane $\xi_2 = 0$ before entering on the third leg. The bifurcation branch of this orbit is the first period-doubling branch of the bifurcation diagram.

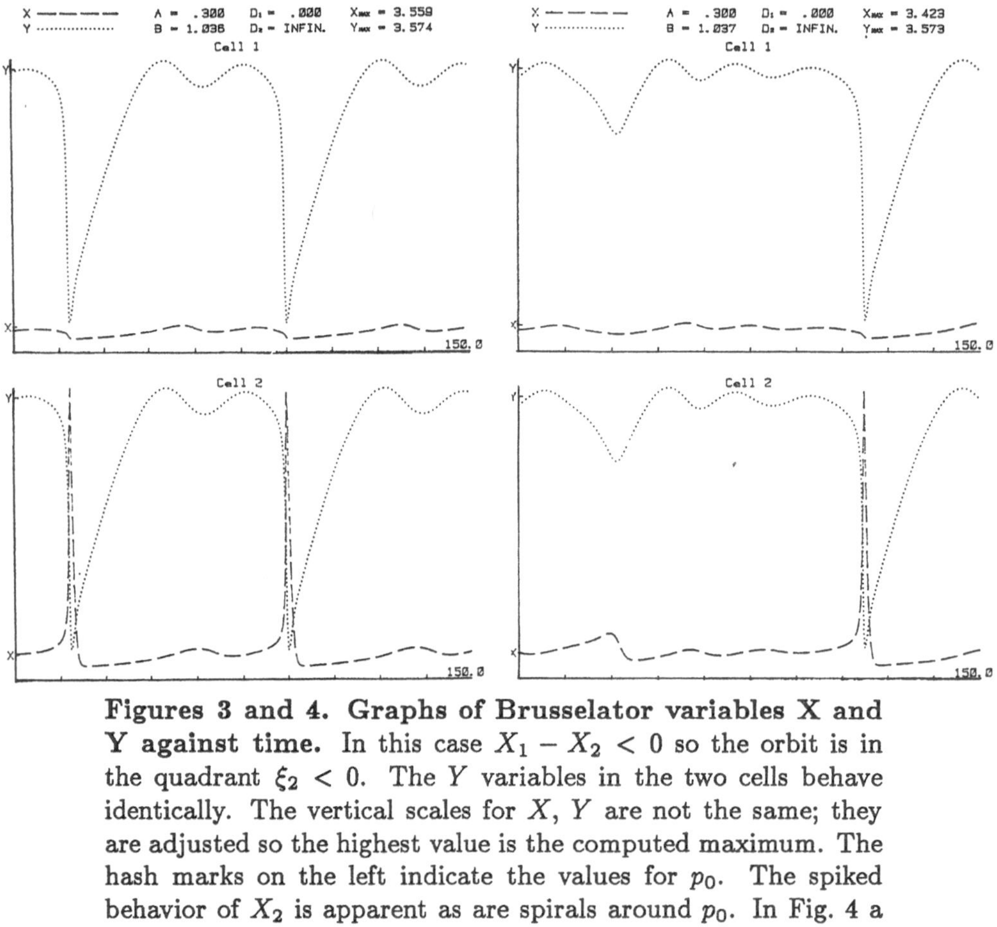

Figures 3 and 4. Graphs of Brusselator variables X and Y against time. In this case $X_1 - X_2 < 0$ so the orbit is in the quadrant $\xi_2 < 0$. The Y variables in the two cells behave identically. The vertical scales for X, Y are not the same; they are adjusted so the highest value is the computed maximum. The hash marks on the left indicate the values for p_0. The spiked behavior of X_2 is apparent as are spirals around p_0. In Fig. 4 a spike has just about retracted to an oscillation.

As B increases further, the process repeats to create a cycle that spirals around the unstable manifold of p_0 twice, then four times, and so on. As B increases past the the limit of the period-doubling cascade, other numbers of spirals appear, or the orbit is not periodic and the number of spirals behaves chaotically. In effect, the orbit drops near to the plane $\xi_2 = 0$ and spirals in towards p_0. It never reaches the plane $\xi_2 = 0$ however, and during its spiralling comes so close to the unstable manifold of p_0 that it shoots up and away from p_0 to the right. The spirals are evident as damped oscillations in graphs of the variables against time. They can be seen in Fig. 3 which plots X_1, X_2, $Y_1 = Y_2$ (the program actually integrates (3.3) and makes a change of variables to plot). The near constancy of one of the X variables (in this case X_1) is evident, as well as the spike in the other X variable. During a transitions just beyond a period doubling, as one piece of the period-doubling orbit slides down around the unstable manifold of p_0, the spike on that piece should retract and turn into an oscillation. This phenomenon is presented in Fig. 4. Alternately, a chaotic orbit may have a variety of spike heights. The bifurcations from the right-hand branches of Figure 2 can be described in much the same way. We leave the details to the interested reader.

A note about the production of Figures 1–4: Computations were done on a UNIVAC 1100/82. Figures 1 and 2 were generated by the global bifurcation package AUTO of E. Doedel. The results, in so far as they can be checked, are accurate to 5 digits. For example, the Hopf bifurcation points on the primary and secondary stationary branches, which can be computed analytically, are accurate to that many digits. Branch 4 in Figure 2 was actually computed starting from its right end point. The computed left end point matched the computed Hopf bifurcation point on branch 2 to 6 digits. Figures 3 and 4 were generated by directly integrating system (3.3) by the stiff integrator DSODE of A. Hindmarsh. Computations were done with a local control of 13 digits. None the less, as is usual for chaotic dynamics, quantitative details of the dynamics, such as the number of spirals, are quite sensitive to numerical effects.

5. Analytic considerations

The emphasis in the report has been on a description of observed behavior. In this section we discuss possible analytic explanations, as well as other oscillators. The discussion here becomes more speculative, since the analytic details have not been carried through.

A dynamicist will recognize the strong similarity between the flow described in the last section and Šilnikov type behavior [5,§6.5]. Šilnikov's theorem concerns the flow near a homoclinic orbit which begins and ends (α and ω limit points) on a spiral saddle stationary point. The point p_0 is indeed a spiral saddle for $2D_1 < B-1 < A^2$ (with eigenvalues $\frac{1}{2}\left((A^2 - B + 1) \pm i\sqrt{(3A^2 - B + 1)(A^2 + B - 1)}\right)$, $B - 1 - 2D_1$; also an auxiliary condition on the sizes of the eigenvalues is satisfied). However there is no homoclinic orbit. The stable manifold of p_0 is the plane $\xi_2 = 0$. If the symmetry is broken between the two cells (say by having two parameters A_1, A_2), homoclinic orbit should appear for correct tuning of A_1, A_2, B.

In line with the philosophy espoused in the opening paragraphs, the best description of the behavior would be in terms of an unfolding of a singularity which comes from some features of the individual oscillators. Šilnikov behavior would be an adjunct of the unfolding. Accordingly we look for the nearest singularity. As A is decreased, the Hopf bifurcation points of Figure 2 move to the left; both collapse to $B = 1$ as $A \to 0$ (for $D_1 = 0$). The η component of p_0 moves off to ∞ as $A \to 0$. A rescaling, as in the case of one Brusselator, leads to a degenerate singularity with Jacobian having an algebraically triple zero of geometric multiplicity 2. As in the case of one Brusselator, the system is too degenerate—the normal form is not generic. If we set $D_1 \neq 0$ however, the resultant singularity is well-known. In this case, the degenerate singularity occurs when $A^2 = 2D_1$, $B = A^2 + 1$ and the eigenvalues of the Jacobian are 0, $\pm iA$. It seems likely much of the observed behavior (perhaps not the bubble effect) can be described in terms of unfolding this singularity (see [5], esp. Fig. 7.4.11(d) and the discussion in §7.4 and §7.5 and [6]).

We next discuss the origin of this singularity. Tracing through the analysis, we find this singularity in the coupled system arises from the degenerate singularity of the single (rescaled) oscillator with Jacobian $\left(\begin{smallmatrix}0 & 1\\ 0 & 0\end{smallmatrix}\right)$ by setting $D_2 = \infty$ to remove the degeneracy of the normal form. Thus we have the possibility that other oscillators with such a singularity will have behavior similar to that we have described when coupled. For examples, we briefly consider some other well-known oscillators.

Consider first the oscillator of Ashkenazi and Othmer [2] used to model self-catalytic glycolytic oscillations. The system is

$$\begin{aligned} \dot{X} &= \delta - \kappa X - XY^2, \\ \dot{Y} &= \kappa X + XY^2 - Y. \end{aligned} \tag{5.1}$$

Let $\hat{\kappa} = \delta^2/(\delta^2 + \kappa)$. Then the single stationary point of (5.1) is

$$X = \hat{\kappa}/\delta, \qquad Y = \delta, \tag{5.2}$$

and the Jacobian of the system at that point is

$$\begin{pmatrix} \hat{\kappa}\delta^2 & -2\hat{\kappa} \\ \hat{\kappa}\delta^2 & 2\hat{\kappa} - 1 \end{pmatrix}. \tag{5.3}$$

The Jacobian equals $\left(\begin{smallmatrix}0 & -1\\ 0 & 0\end{smallmatrix}\right)$ when $\delta = 0$, $\hat{\kappa} = \frac{1}{2}$. However the X component of the stationary point (5.2) goes to ∞. A rescaling $X' = \delta X$, $Y' = Y/\delta$ creates a degenerate singularity, just as in the case of the Brusselator. Coupling two such systems (as was done in [2]) and setting the X-diffusion to ∞ creates a 3-dimensional system with a singularity with Jacobian having one zero and two pure imaginary eigenvalues. We conjecture this system exhibits a period-doubling bubble.

Finally we consider the Oregonator which models the Belousov-Zhabotinskii reaction [4] (see also [8, Chap. 13]). Tyson [11, eq. 5] has changed variables and put the Oregonator in the form

$$\begin{aligned} \epsilon\dot{x} &= -\alpha x - \beta y - qx^2 - xy, \\ \dot{y} &= -\gamma x - \delta y + fz - xy, \\ p\dot{z} &= x - z, \end{aligned} \tag{5.5}$$

for parameters α, β, γ, δ, ϵ, f, p (the primary stationary point is $x = y = z = 0$). If for example, we set $p^{-1} = 0$ and also $\alpha\gamma = \beta\delta$, the single Oregonator has an algebraically double eigenvalue 0 of geometric multiplicity one and a nongeneric normal form—just what we were looking for (the other eigenvalue is $-\alpha/\epsilon - \delta$). If our general hypothesis is correct, two coupled Oregonators exhibit a period-doubling bubble which can be traced to an appropriate singularity of the system with z diffusion coefficient equal ∞.

Thus such singularities in single oscillators seem rather common. If the process of combining such singularities and setting diffusion constants equal ∞ leads to period-doubling bubbles, such bubbles should be also rather common and lead to global period-doubling behavior. If higher-order terms in the single oscillator affect the details, such a sweeping statement is too bold; singularities formed this way will have other structure. They are still worth investigating since they will affect the global bifurcation structure of the coupled system.

References

[1] J.C. Alexander, "Spontaneous oscillations in two 2-component cells coupled by diffusion," *J. Math. Biol.*, submitted, 1984.

[2] M. Ashkenazi & H.G. Othmer, "Spatial patterns in coupled bio-chemical oscillators," *J. Math. Biol.* **5**(1978), 305–350.

[3] M.J. Feigenbaum, "Universal behavior in nonlinear systems," *Los Alamos Sci.* **1**(1980), 4-29.

[4] R.J. Field & R.M. Noyes, "Oscillations in chemical systems, IV. Limit cycle behavior in a model of a real chemical reaction," *J. Chem. Phys.* **60**(1974), 1877–1884.

[5] J. Guckenheimer & P. Holmes, *Nonlinear Oscillations, Dynamical Systems and Bifurcation of Vector Fields*, Appl. Math. Sci. **#42**, Springer-Verlag, 1983.

[6] P. Holmes, "Unfolding a degenerate nonlinear oscillator: a codimension two bifurcation" in *Nonlinear Dynamics*, R.H.G. Helleman, ed., Ann. N.Y. Acad. Sci. **#357**(1980), 473–488.

[7] R. Lefever & I. Prigogine, "Symmetry-breaking instabilities in dissipative systems II," *J. Chem. Phys.* **48**(1968), 1695–1700.

[8] G. Nicolis & I. Prigogine, *Self-Organization in Nonequilibrium Systems*, Wiley, 1977.

[9] R. O'Malley, *Introduction to Singular Perturbations*, Academic Press, 1974.

[10] I. Schreiber & M. Marek, "Strange attractors in coupled reaction-diffusion cells," *Physica* **5D**(1982), 258–272.

[11] J.J. Tyson, "Analytic representation of oscillations, excitability, and travelling waves in a realistic model of the Belousov-Zhabotinskii reaction," *J. Chem. Phys.* **66**(1977), 905–915.

BISTABLE BEHAVIOR IN COUPLED OSCILLATORS

D.G. Aronson
School of Mathematics
University of Minnesota
Minneapolis, MN 55455 USA

E.J. Doedel
Department of Computer Science
Concordia University
Montreal, Quebec H3G IM8 CANADA

H.G. Othmer
Department of Mathematics
University of Utah
Salt Lake City, UT 84112 USA

1. Introduction. We consider a very simple model of two identical nonlinear oscillators, each with an asymptotically stable limit cycle, coupled together by a linear diffusion path. The system depends on two parameters: the natural frequency of the individual oscillators and the intensity of the coupling. Our main result is that the coupled system exhibits bistable behavior for an open set of parameter values which includes moderate values of the parameters rather than just very large or very small values.

The individual oscillators which we deal with are described by the equations

$$\dot{\underset{\sim}{x}} = F(\underset{\sim}{x};\alpha,\beta) \ , \tag{1}$$

where

$$\underset{\sim}{x} = \begin{pmatrix} x \\ y \end{pmatrix} \quad \text{and} \quad F(\underset{\sim}{x};\alpha,\beta) = \begin{pmatrix} \alpha x + \beta y - x^2(x^2+y^2) \\ -\beta x + \alpha y - y^2(x^2+y^2) \end{pmatrix} .$$

As is well known, the system (1) has a unique limit cycle

$$\underset{\sim}{x} = \begin{pmatrix} \sqrt{\alpha}\ \cos \beta t \\ -\sqrt{\alpha}\ \sin \beta t \end{pmatrix}$$

which is asymptotically stable and attracts all of $\mathbb{R}^2 \setminus \{(0,0)\}$. Roughly speaking, (1) is the truncated normal form of an oscillator with weak angular dependence on amplitude. For our purpose in this paper the value of $\alpha \neq 0$ is irrelevant. We therefore normalize it to be 1 and write $F(\underset{\sim}{x};\beta)$ instead of $F(\underset{\sim}{x};1,\beta)$.

To form the coupled system we index the individual oscillators and write

$$\begin{aligned} \dot{\underset{\sim}{x}}_1 &= F(\underset{\sim}{x}_1;\beta) + D(\underset{\sim}{x}_2 - \underset{\sim}{x}_1) \\ \dot{\underset{\sim}{x}}_2 &= F(\underset{\sim}{x}_2;\beta) - D(\underset{\sim}{x}_2 - \underset{\sim}{x}_1) \ , \end{aligned} \tag{2}$$

where

$$\underset{\sim}{x}_j = \begin{pmatrix} x_j \\ y_j \end{pmatrix}$$

and D is a 2×2 constant matrix. In this paper, we shall consider only the special choice of diffusion matrix

$$D = \delta \begin{pmatrix} 1 & 1 \\ 1 & 1 \end{pmatrix},$$

where δ is a nonnegative parameter which measures the intensity of the coupling. With this choice of D, we shall write the system (2) as

$$\dot{\underset{\sim}{X}} = \mathcal{F}(\underset{\sim}{X};\delta,\beta) , \tag{3}$$

where

$$\underset{\sim}{X} = \begin{pmatrix} \underset{\sim}{x}_1 \\ \underset{\sim}{x}_2 \end{pmatrix} .$$

We shall exploit the rather extravagant symmetries which we have built into the system to extract a great deal of analytic information about the behavior of solutions. Combining this with numerical information gained from using AUTO, we have been able to construct a comprehensive picture of the surprisingly rich dynamics of our system. Here we describe only one aspect of that picture. Much more information and details of the proofs can be found in [ADO].

2. Invariant Tori and Manifolds. For $\delta = 0$ the system (3) is uncoupled and, for each $\beta_0 \in \mathbb{R}^+$, there is an invariant parallel flow two-torus $T_{0\beta_0}$ foliated by the limit cycles of the two oscillators at arbitrary phase. This torus is an attracting normally hyperbolic manifold and it follows from standard perturbation theory for invariant manifolds [F,HPS] that it persists as an invariant attracting normally hyperbolic two-torus $T_{\delta\beta}$ for (δ,β) in a neighborhood of $(0,\beta_0)$. The torus $T_{\delta\beta}$ no longer consists of orbits at arbitrary phase. Instead, as can be shown by an averaging argument [ADO], it is composed of a stable in-phase orbit ω_0, an unstable π - radians - out - of - phase orbit ω_π, and the unstable manifold associated with ω_π. As we shall see below, the tori $T_{\delta\beta}$ do not persist for arbitrarily large δ. At present we do not understand the mechanisms involved in their break up, but they are being studied.

Because of the various symmetries in the system (3), it is easy to verify that the two dimensional manifolds

$$\Phi \equiv \{(x_1,y_1,x_2,y_2) \in \mathbb{R}^4 : x_1 + x_2 = 0, y_1 + y_2 = 0\}$$

and

$$\Pi \equiv \{(x_1,y_1,x_2,y_2) \in \mathbb{R}^4 : x_1 - x_2 = 0,\ y_1 - y_2 = 0\}$$

are invariant. Clearly $\omega_0 \subset \Phi$ and $\omega_\pi \subset \Pi$. It is natural to introduce new coordinates which reflect these invariances. Let

$$u_1 = (x_1+x_2)/2 ,\ v_1 = (y_1+y_2)/2 ,\ u_2 = (x_1-x_2)/2 \quad \text{and} \quad v_2 = (y_1-y_2)/2 .$$

In these coordinates

$$\mathbb{O} = \{(u,v,0,0) \in \mathbb{R}^4 : (u,v) \in \mathbb{R}^2\} ,$$
$$\Pi = \{(0,0,u,v) \in \mathbb{R}^4 : (u,v) \in \mathbb{R}^2\} ,$$

and (3) becomes

$$\dot{\underset{\sim}{U}} = \mathscr{X}(\underset{\sim}{U};\beta,\delta) , \tag{4}$$

where

$$\underset{\sim}{U} = \begin{pmatrix} u_1 \\ v_1 \\ u_2 \\ v_2 \end{pmatrix} \text{ and } \mathscr{X}(\underset{\sim}{U};\beta,\delta) = \begin{pmatrix} u_1+\beta v_1-u_1^3-3u_1u_2^2-u_1(v_1^2+v_2^2)-2u_2v_1v_2 \\ -\beta u+v_1-v_1^3-3v_1v_2^2-v_1(u_1^2+u_2^2)-2u_1u_2v_2 \\ (1-2\delta)u_2+(\beta-2\delta)v_2-u_2^3-3u_1^2u_2-u_2(v_1^2+v_2^2)-2u_1v_1v_2 \\ -(\beta+2\delta)u_2+(1-2\delta)v_2-v_2^3-3v_1^2v_2-v_2(u_1^2+u_2^2)-2u_1u_2v_1 \end{pmatrix} .$$

On both $\mathbb{O}$ and Π, the system (4) reduces to a two dimensional system. In particular, on $\mathbb{O}$ the coupling is irrelevant and (4) reduces to the equations for an uncoupled oscillator. Thus the limit cycle ω_0 is given by

$$\omega_0(t) = (\cos\beta t , -\sin\beta t , 0 , 0) .$$

On Π the two dimensional system in polar coordinates (ρ,θ) is

$$\begin{aligned} \dot{\theta} &= -\beta - 2\delta\cos 2\theta \\ \dot{\rho} &= (1-2\delta-2\delta\sin 2\theta)\rho - \rho^3 . \end{aligned} \tag{5}$$

This system can be integrated explicitly for $0 \leq \delta \leq \min(1/2,\beta/2)$ to give the limit cycle ω_π in the form

$$\omega_\pi(t) = (0,0,\rho(t)\cos\theta(t) , \rho(t)\sin\theta(t)) ,$$

where

$$\rho^2(t) = (1-2\delta)(1-k^2)/\{1+k\cos 2(\theta(t)-\varphi)\} \tag{6}$$

with

$$k = 2\delta/\{(1-2\delta)^2+\beta^2\}^{1/2} \text{ and } 2\varphi = \arctan(1-2\delta)/\beta ,$$

and

$$\theta(t) = -\ell\pi - \arctan\left(\frac{\beta+2\delta}{\beta-2\delta}\right)^{1/2} \tan(\beta^2-4\delta^2)^{1/2}t$$

for $\ell \in \mathbb{Z}$ and

$$t \in ((2\ell-1)\pi/2(\beta^2-4\delta^2)^{1/2} , (2\ell+1)\pi/2(\beta^2-4\delta^2)^{1/2}] .$$

The period of ω_π is

$$T_{\delta\beta} = 4\pi/(\beta^2-4\delta^2)^{1/2} .$$

It is clear from (6) that $\omega_\pi|\Pi$ is an ellipse in Π. For $\beta > 1$ this ellipse shrinks to the origin as $\delta \uparrow 1/2$. Thus, for $\beta > 1$, ω_π disappears in a Hopf bifurcation at the origin when $\delta = 1/2$. On the other hand, for $\beta > 1$ this Hopf bifurcation does not occur. Instead, when $\delta = \beta/2$ there is an infinite period bifurcation in which the ellipse $\omega_\pi|\Pi$ develops rest points (saddle-nodes) at $\theta = \pm\ \pi/2$. When this occurs, $\omega_\pi|\Pi$ is no longer a limit cycle, but is an invariant circle consisting of the saddle-nodes and heteroclinic orbits joining them (Figure 1(b)). For $\delta > \beta/2$ the saddle-nodes split into pairs of nondegenerate rest points and $\omega_\pi|\Pi$ remains an invariant circle (Figure 1(c)). We shall not go into this any further here and the interested reader should consult [ADO].

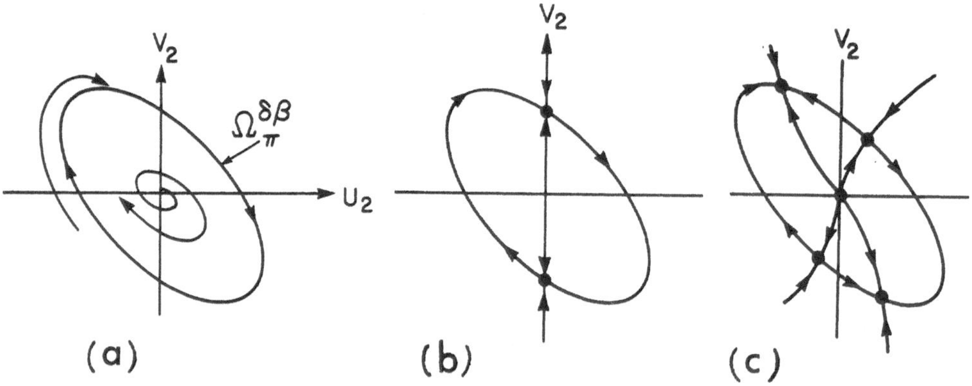

Figure 1. The ellipse $\omega_\pi|\Pi$ for $\beta > 1$. (a) $\delta < \beta/2$, (b) $\delta = \beta/2$, and (c) $\delta > \beta/2$ but not too large.

It is not difficult to show that $\omega_0|\Phi$ is asymptotically stable in Φ and that $\omega_\pi|\Pi$ is asymptotically stable in Π. We now turn to the consideration of the stability of these orbits in the full space $\mathbb{R}^4$.

3. Stability. According to the well known Floquet theory [H] the stability of a T-periodic orbit $\omega = \omega(t)$ of (4) is determined by the Floquet multipliers, that is, by the eigenvalues of $X(T;\delta,\beta)$ where $X(t;\delta,\beta)$ is the fundamental matrix solution of the variational system

$$\dot{\underset{\sim}{V}} = \mathcal{X}'(\omega(t);\delta,\beta)V \tag{7}$$

with $\underset{\sim}{V}(0) = I$. Since $\dot{\omega}(t)$ is a solution of the variational system, one of the Floquet multipliers is always 1. The Floquet stability theorem states that $\omega(t)$ is asymptotically stable if all but one of the multipliers lies inside the unit circle in $\mathbb{C}$ and unstable if any multiplier lies strictly outside the unit circle.

To facilitate the discussion of stability we introduce the Floquet signature $\sigma(\omega)$ of an orbit $\omega = \omega(t)$. Let $\lambda_j (j = 1,2,3,4)$ denote the Floquet multipliers corresponding to $\omega(t)$. Then $\sigma(\omega) \equiv (\alpha_1,\alpha_2,\alpha_3,\alpha_4)$ where

$$\alpha_j = \begin{cases} + & \text{if } |\lambda_j| > 1 \\ 0 & \text{if } |\lambda_j| = 1 \\ - & \text{if } |\lambda_j| < 1 \end{cases} .$$

For $\omega = \omega_0$ or ω_π the variational system breaks up into two independent subsystems of the form

$$\dot{\xi} = K(\omega(t))\xi \tag{8a}$$

$$\dot{\zeta} = \{K(\omega(t)) - 2D\}\zeta . \tag{8b}$$

We adopt the convention that the first two entries in the Floquet signature refer to the multipliers associated with the subsystem (8a) while the last two refer to the multipliers associated with (8b).

Since $\omega_0|\Phi$ is asymptotically stable in Φ and $\dot{\omega}_0$ lies in Φ we find that $\sigma(\omega_0) = (0,-,\alpha_3,\alpha_4)$ for all $(\delta,\beta) \in \mathbb{R}^+ \times \mathbb{R}^+$. Using straightforward estimates one can show, in addition, that $\alpha_3 = \alpha_4 = -$ for all $(\delta,\beta) \in \mathbb{R}^+ \times \mathbb{R}^+$. Thus, <u>the in-phase orbit ω_0 is asymptotically stable in $\mathbb{R}^4$ for all $(\delta,\beta) \in \mathbb{R}^+ \times \mathbb{R}^+$</u>.

In view of the asymptotic stability of $\omega_\pi|\Pi$ in Π and the fact that $\dot{\omega}_\pi$ lies in Π, we have $\sigma(\omega_\pi) = (\alpha_1,\alpha_2,0,-)$. The multipliers λ_1 and λ_2 are the eigenvalues of $Y(T;\delta,\beta)$, where $Y(\cdot;\delta,\beta)$ is the fundamental matrix solution of (8a) with

$$K(\omega_\pi(t)) \equiv K_\pi(t) \equiv \begin{pmatrix} 1-2\rho^2(t)-\rho^2(t)\cos 2\theta(t) & \beta-\rho^2(t)\sin 2\theta(t) \\ -\beta-\rho^2(t)\sin 2\theta(t) & 1-2\rho^2(t)+\rho^2(t)\cos 2\theta(t) \end{pmatrix} .$$

The least period of $K_\pi(t)$ is $\hat{T}_{\delta\beta} = T_{\delta\beta}/2$. Let

$$\hat{Y}_{\delta\beta} \equiv Y(\hat{T}_{\delta\beta};\delta,\beta) .$$

Then $Y(T_{\delta\beta},\delta,\beta) = (\hat{Y}_{\delta\beta})^2$, and the stability of ω_π depends on the eigenvalues μ_1 and μ_2 of $\hat{Y}_{\delta\beta}$.

In general, it is very difficult (indeed, usually impossible) to compute the μ's explicitly or to deduce anything about them individually from the algebraic properties of $K_\pi(t)$. However, there is a simple formula for the product $\mu_1\mu_2$ in terms of the integral of $\mathrm{tr}K_\pi$, [H]. In the present case this integral can be evaluated explicitly [ADO] and we obtain

$$\log \mu_1\mu_2 = 2\pi(4\delta - 1)/(\beta^2 - 4\delta^2)^{1/2} . \tag{9}$$

Let

$$\mathfrak{D} \equiv \{(\delta,\beta) \in \mathbb{R}^2 : 0 \leq \delta \min(\beta/2, 1/2),\ \beta \in \mathbb{R}^+\} .$$

It follows from (9) that $\mu_1\mu_2 > 1$ everywhere in

$$\mathfrak{R} \equiv \mathfrak{D} \cap \{(\delta,\beta) \in \mathbb{R}^2 : \delta > 1/4\} .$$

Thus $\lambda_1\lambda_2 = (\mu_1\mu_2)^2 > 1$ and it follows that $\sigma(\omega_\pi) = (+,-,0,-)$ or $(+,+,0,-)$ for

$(\delta,\beta) \in \mathcal{R}$. In either event we conclude that ω_π <u>is unstable for all</u> $(\delta,\beta) \in \mathcal{R}$. Let

$$\mathcal{L} \equiv \mathcal{D} \cap \{(\delta,\beta) \in \mathbb{R}^2 : 0 \leq \delta < 1/4\} .$$

Clearly, $(\delta,\beta) \in \mathcal{L}$ does not imply the stability of ω_π . However, ω_π is asymptotically stable if $(\delta,\beta) \in \mathcal{L}$ and the corresponding μ's are complex. Thus we now turn our attention to showing that the subset of $\mathcal{L}$ on which the μ's are complex is not empty.

To study the μ's it is most convenient to write the system (8a) in polar form

$$\dot{\Psi} = -2\beta + 2S(\Theta)\sin(\Psi - \Theta) \tag{10a}$$

$$\frac{\dot{R}}{R} = 1 - 2S(\Theta) - S(\Theta)\cos(\Psi - \Theta) , \tag{10b}$$

where $\Theta = 2\theta$, $S(\Theta) = \rho^2(\Theta/2)$ and $\theta = \theta(t)$. This system is not autonomous, but if we append the equation

$$\dot{\Theta} = -2\beta - 4\delta\cos\Theta \tag{10c}$$

for Θ , the resulting system is autonomous. Equations (10a) and (10c) are independent of R so that, once they have been solved, R can be obtained by quadrature.

Since $\dot{\Theta} < 0$ for $\beta < 2\delta$, we can take Θ as the independent variable and write

$$\frac{d\Psi}{d\Theta} = \frac{\beta - \delta(\Theta)\sin(\Psi - \Theta)}{\beta + 2\delta\cos\Theta} . \tag{11}$$

The right hand side of (11) is 2π -periodic in (Θ,Ψ) . Thus (11) defines a flow on a two-torus $\mathcal{T}$. Let $\Psi(\Theta;\Psi_0,\delta,\beta)$ denote the solution of (11) with $\Psi(\pi) = \Psi_0$. Then, for $(\delta,\beta) \in \mathcal{D}$, the maps

$$C_{\delta\beta} : \Psi_0 \mapsto \Psi(-\pi; \Psi_0,\delta,\beta)$$

form a two parameter family of order preserving diffeomorphisms of the circle $\Theta = \pi$ into itself. We associate with the map $C_{\delta\beta}$ the rotation number defined by

$$r(\delta,\beta) \equiv \lim_{N\to\infty} \frac{\Psi(-(2N-1)\pi; \Psi_0,\delta\beta) - \Psi_0}{2\pi N} .$$

Since there are no critical points for the (Θ,Ψ) - flow when $(\delta,\beta) \in \mathcal{D}$, $r(\delta,\beta)$ exists, is independent of Ψ_0 , and is continuous in $\mathcal{D}$. Moreover, $r(\delta,\beta)$ is an integer if and only if $C_{\delta\beta}$ has a fixed point [H].

The rotation number is related to the μ's by the following observation which is proved in [ADO]. <u>The matrix</u> $\hat{Y}_{\delta\beta}$ <u>has a real non-zero eigenvalue if and only if</u> $C_{\delta\beta}$ <u>has a fixed point, that is, if and only if</u> $r(\delta,\beta)$ <u>is an integer</u>. For integers $n \geq 1$ let

$$H_n \equiv \{(\delta,\beta) \in \mathcal{D} : r(\delta,\beta) = n\} .$$

The subset of $\mathcal{L}$ on which the μ's are complex is

$$\mathcal{L} \setminus \bigcup_{n=1}^{\infty} H_n \quad ,$$

and we shall show that this set is not empty.

On the Hopf bifurcation line $\delta = 1/2$ for $\beta > 1$ we have $S(\Theta) \equiv 0$ so that equation (11) reduces to

$$\frac{d\Psi}{d\Theta} = \frac{\beta}{\beta + \cos\Theta} \ .$$

Thus $C_{\delta\beta}$ is conjugate to a rigid rotation. Set $\epsilon = 1 - 2\delta$ and let τ be defined by

$$\frac{d\tau}{d\Theta} = -\frac{\beta}{\beta + \cos\Theta} \quad , \quad \tau(\pi) = 0 \ .$$

Then, for $\beta > 1$, $z(\tau) \equiv \Psi(\Theta(\tau))$ satisfies an equation of the form

$$\frac{dz}{d\tau} = -1 + \epsilon h(\tau/p, z, p, \epsilon) \ ,$$

where $p = \beta/(\beta^2 - 1)^{1/2}$ and h is a smooth 2π-periodic function of τ/p and z as long as p is bounded and ϵ is sufficiently small. By a theorem of Bushard [B] we conclude that for each integer $n \geq 2$ the region H_n is a cusp-like region with continuous boundary emanating from the point $h_n \equiv (1/2, \beta_n)$ where $\beta_n = n/(n^2 - 1)^{1/2}$. Moreover, generically each of the H_n is a wedge with non-zero angular opening at h_n (Figure 2). We shall refer to the H_n as resonance horns.

For integers n and m with $n \neq m$ we have $H_n \cap H_m = \emptyset$. In addition, because of the continuity of $r(\delta,\beta)$, none of the horns H_n can terminate in $\mathcal{D}$. Using various large and small parameter techniques one can show that $r(\delta,\beta) < 2$ for sufficiently small δ or large β . Thus the horns H_n for $n \geq 2$ cannot terminate on the parts of $\partial\mathcal{D}$ where $\delta = 0$ or $\beta = \infty$. On the other hand, $r(\delta,\beta) \to \infty$ as $(\delta,\beta) \to (\beta_0/2, \beta_0)$ for $\beta_0 \in (1/2,1)$, so that none of the H_n can terminate on the infinite period bifurcation line $\delta = \beta/2$ for $\beta \in (1/2,1)$. We conclude that for $n \geq 2$ the H_n must terminate on the segment

$$\ell \equiv \{(\beta/2, \beta) \in \mathbb{R}^2 : \beta \in (0,1/2)\}$$

of the infinite period bifurcation line $\delta = \beta/2$.

When $\delta = \beta/2$, equation (11) becomes

$$\frac{d\Psi}{d\Theta} = \frac{1 - \{S(\Theta)/\beta\}\sin(\Psi - \Theta)}{1 + \cos\Theta} \tag{12}$$

which is singular for $\Theta = \pm\pi$. For $\beta \in (0,1/2)$, that is, for $(\delta,\beta) \in \ell$, the section $\Theta = \pi$ of $\mathcal{T}$ is an invariant circle connecting saddle-nodes at $(\Theta,\Psi) = (\pi, \Psi_\beta^{\pm})$, where

$$\Psi_\beta^- = \pi + \arcsin\frac{\beta}{S(\pi)} \quad \text{and} \quad \Psi_\beta^+ = 2\pi - \arcsin\frac{\beta}{S(\pi)} \ .$$

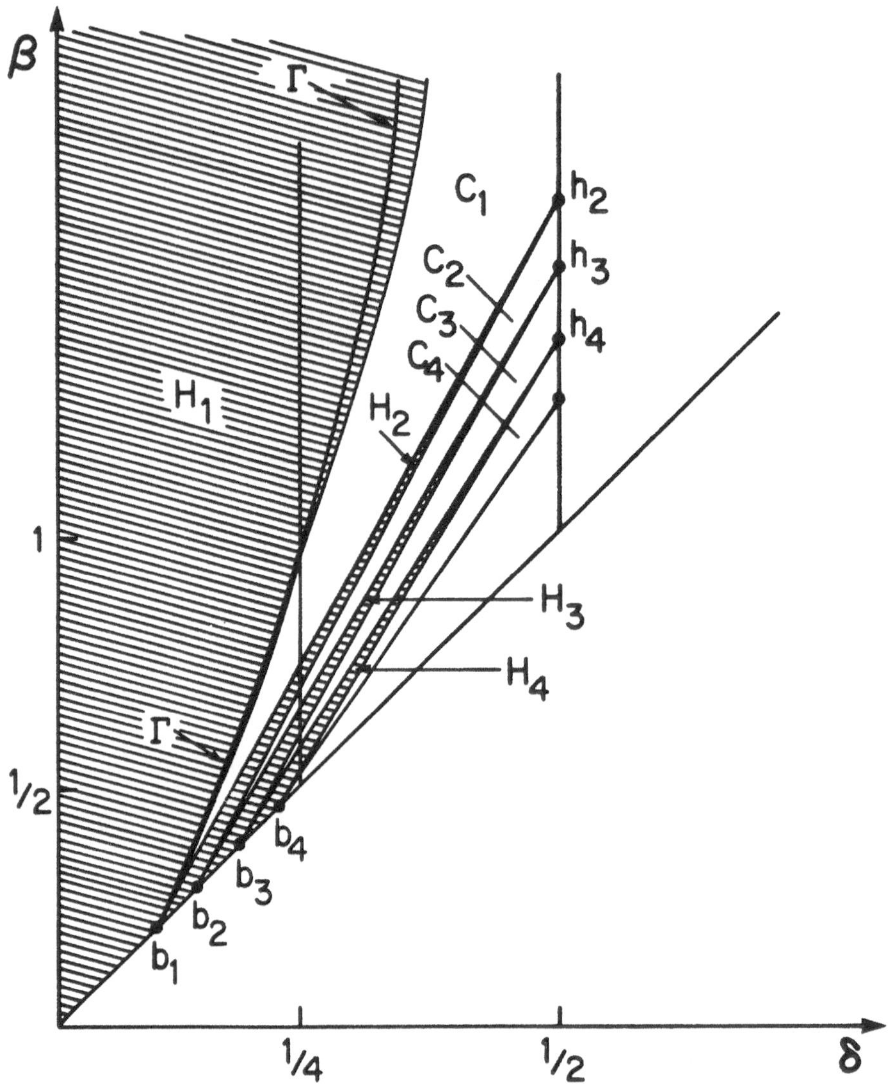

Figure 2. The first four resonance horns H_p (shaded) and the corresponding regions where the eigenvalues of $\hat{Y}_{\delta\beta}$ are complex.

There is a unique orbit $\Psi = \Psi^+(\cdot,\beta)$ which leaves (π,Ψ_β^+) for decreasing Θ and a unique orbit $\Psi = \Psi^-(\cdot,\beta)$ which leaves $(-\pi,\Psi_\beta^-)$ for increasing Θ. Let $\mathcal{E}$ denote the subset of β-values in $(0,1/2)$ for which

$$\Psi^+(\Theta;\beta) \to \Psi_\beta^- \pmod{2\pi} \quad \text{as} \quad \Theta \downarrow -\pi ,$$

that is, $\mathcal{E}$ is the set of $\beta \in (0,1/2)$ for which $\Psi^+(\cdot;\beta)$ and $\Psi^-(\cdot;\beta)$ coincide modulo 2π. For $\beta \in (0,1,2)\setminus\mathcal{E}$ we have

$$\Psi^+(\Theta;\beta) \to \Psi_\beta^+ \pmod{2\pi} \quad \text{as} \quad \Theta \downarrow -\pi .$$

In [ADO] we show that the exceptional set $\mathcal{E}$ is discrete so that we can write $\mathcal{E} = \{b_j\}$, where

$$0 < b_1 < b_2 < \ldots < 1/2$$

and $b_j \uparrow 1/2$ as $j \to \infty$. Let $b_0 \equiv 0$. For $\beta \in A_j \equiv (b_{j-1},b_j)$ there is a well defined rotation number r_j which simply counts the number of times $\Psi^+(\cdot,\beta)$ winds around $\mathfrak{T}$. Clearly r_j is an integer. From the arguments mentioned above which show that $r < 2$ near $\partial\mathfrak{D}$, one can estimate b_1 and show that $r_1 = 1$. Let $P = (\beta/2,\beta)$ with $\beta \in (b_{j-1},b_j)$. There is a neighborhood $N(P)$ of P in $\mathfrak{D}$ such that for all $(\delta,\beta) \in N(P)$ the map $C_{\delta\beta}$ has two fixed points and $r(\delta,\beta) = r_j$. Moreover, $\sigma(\omega_\pi) = (+,-,0,-)$ throughout $N(P)$, [ADO].

So far we know that, for $n \geq 2$, each of the resonance horns H_n intersects $\mathcal{L}$ on the segments A_j for which $r_j = n$. Our numerical evidence strongly suggests that each of these resonance horns H_n intersects $\mathcal{L}$ on the unique segment A_n and that $r_n = n$. To simplify the exposition we shall assume that this is true. If it is not true, our conclusion that ω_π is asymptotically stable on a nonempty open subset of $\mathcal{L}$ is still valid, but the description of the stability region is much more complicated.

Figure 2 shows schematically the resonance horns H_1,H_2,H_3 and H_4. The "horn" H_1 is a region which intersects $\mathcal{L}$ on the segment $(0,b_1)$ and is asymptotic to $\delta = 1/2$ as $\beta \to \infty$. In general, boundaries of the horns are as smooth as the vector field. For (δ,β) in the open region C_p between the horns H_p and H_{p+1}, the rotation number $r(\delta,\beta) \in (p,p+1)$ so that the eigenvalues of $\hat{Y}_{\delta\beta}$ are complex. Thus ω_π is asymptotically stable for

$$(\delta,\beta) \in \bigcup_{p=1}^{\infty} C_p \cap \mathcal{L} .$$

If $(\delta,\beta) \in (\partial C_p) \cap \mathcal{L}$ for any $p \geq 1$, the eigenvalues of $\hat{Y}_{\delta\beta}$ are both at $\pm\eta$ for some $\eta \in (0,1)$. If the parameter point now passes into the interior of either H_p or H_{p+1} the eigenvalues move away from each other along the real axis, but as long as (δ,β) is sufficiently close to $(\partial C_p) \cap \mathcal{L}$ they will remain inside the unit circle. Thus ω_π is asymptotically stable in a full neighborhood S_p of each $C_p \cap \mathcal{L}$. Typical S_p's are shown schematically in Figure 3. The boundaries of the S_p are bifurcation curves. The bifurcation curve which is the right hand

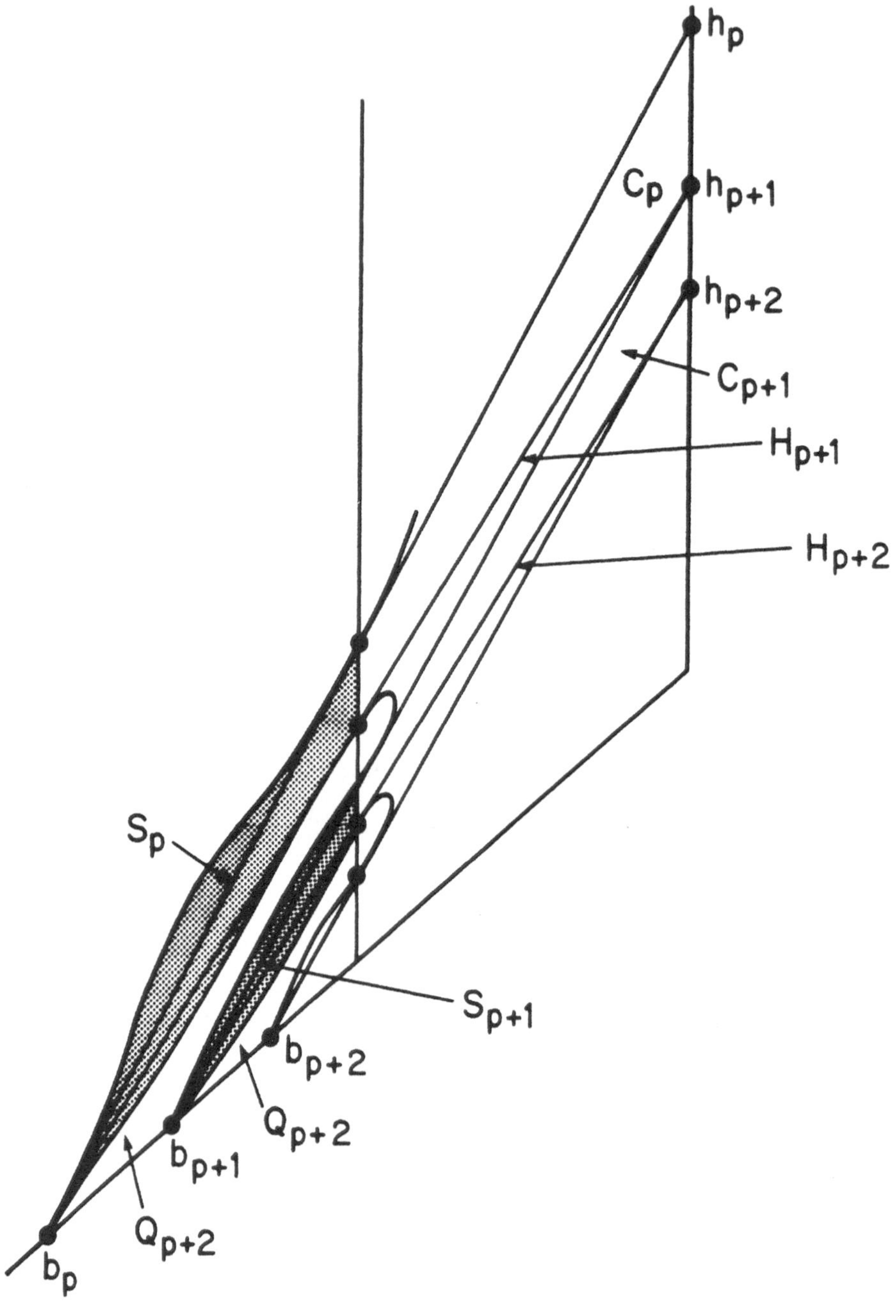

Figure 3. Typical stability regions S_p and the corresponding bifurcation curves shown schematically.

boundary of H_1 is labeled Γ in Figure 2. The Floquet signature $\sigma(\omega_\pi) = (-,-,0,-)$ in $\bigcup_{p=1}^{\infty} S_p$ and $\sigma(\omega_\pi) = (+,-,0,-)$ in the regions labeled Q_p in Figure 3. Elsewhere in $\mathcal{S}$, $\sigma(\omega_\pi) = (+,+,0,-)$. Figure 4 shows the results of some computations of the bifurcation curves

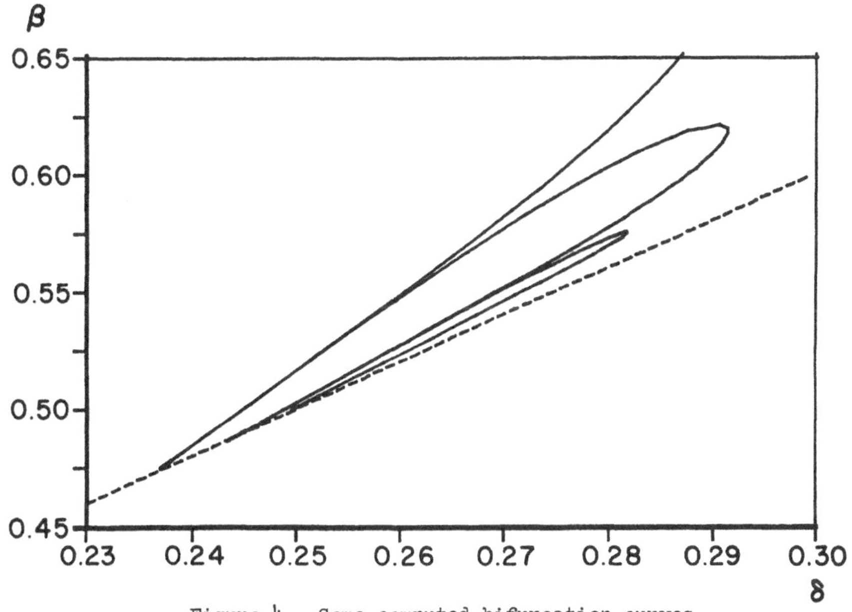

Figure 4. Some computed bifurcation curves.

REFERENCES

[ADO] D.G. Aronson, E.J. Doedel, and H.G. Othmer, An analytical and numerical study of the bifurcations in a system of linearly coupled oscillators, preprint, 1985.

[B] L.B. Bushard, Periodic solutions and locking-in on the periodic surface, Int. J. Non-linear Mech., 8 (1973), 129-141.

[F] N. Fenichel, Persistence and smoothness of invariant manifolds for flows, Ind. Univ. Math. J., 21 (1971), 193-226.

[H] J.K. Hale, Ordinary Differential Equations, Krieger 1980.

[HPS] M. Hirsch, C. Pugh, and M. Shub, Invariant Manifolds, Lecture Notes in Math. 583, Springer-Verlag 1977.

CONTINUATION OF ARNOLD TONGUES IN MATHEMATICAL MODELS OF PERIODICALLY FORCED BIOLOGICAL OSCILLATORS

Leon Glass
Department of Physiology
McGill University
Montreal, Quebec, Canada

Jacques Bélair
Département de Mathématiques et de Statistique
Université de Montréal
Montréal, Québec, Canada

1. Introduction

Periodic stimulation of spontaneously oscillating physiological rhythms has powerful effects on the intrinsic rhythm. As the frequency and amplitude of the periodic stimulus are varied, a large number of different coupling patterns are set up between the stimulus and the spontaneous oscillator. In one class of rhythms, there are a fixed number, N, of cycles of the stimulus for each M cycles of the spontaneous rhythm, and the spontaneous oscillation occurs at fixed phase (or phases) of the periodic stimulus. We call such rhythms N:M phase locking. In addition to phase-locked rhythms, it is also possible to observe irregular or aperiodic rhythms in which fixed phase relationships and regular repeating cyclic patterns are not observed. The following generalizations are applicable to a large number of experiments on periodic forcing of biological oscillators (Guevara et al., 1981; Glass et al., 1984; Petrillo and Glass, 1984).

i) The stable zones of phase locking which are most commonly observed correspond to low order ratios between the number of cycles of the forcing stimulus and the intrinsic rhythm (i.e., 2:1, 3:2, 1:1, 2:3, 1:2). Although other N:M ratios with larger values of N and M can also be observed, these occupy smaller areas in the frequency-amplitude parameter space and they are consequently easily overlooked or obscured by noise.

ii) The stable rhythms are organized in the frequency-amplitude plane in an orderly fashion (see below).

iii) At very low stimulation amplitudes it is difficult to maintain stable phase locking.

iv) If the regions of frequency-amplitude parameter space between stable phase-locking zones are studied, then it is generally possible to find stimulation parameters which give rise to irregular dynamics.

Despite the gross similarities, differences between the organization of phase locking zones are also found. For example, experimental studies on the entrainment of periodically stimulated cells from chick heart revealed 2:2 rhythms, (Guevara et al., 1981; Glass et al., 1984) whereas no 2:2 rhythms were observed in the periodic forcing of respiratory rhythm in cats (Petrillo and Glass; 1984). Does such a discrepancy reveal intrinsic differences between the two situations, or is it merely a reflection of experimental difficulties which make detailed experimental comparisons impossible?

A problem of some interest is to characterize the dynamics in the two-dimensional frequency-amplitude parameter space. We have been particularly intrigued about whether or not one can establish a topological equivalence between the global bifurcations observed in different periodically forced biological oscillators. The present communication is meant as a summary of theoretical results relating to global organization of bifurcations in periodically forced nonlinear oscillators modelling biological systems. We conclude that although topological equivalence is apparently found at low stimulus strengths, as the stimulus strength increases there are a large number of different topological organizations for the periodic and aperiodic dynamics.

2. Mathematical Background

Mathematical analysis of the effects of periodic forcing on biological oscillations utilize techniques developed in the study of bifurcations of maps of the circle into itself,

$$\phi_{i+1} = g(\phi_i, b, \tau) \qquad [1]$$

where $g: S^1 \to S^1$, where b and τ correspond to the strength and the period of the periodic forcing. The variable ϕ_i can refer either to the phase of a marker event of the stimulus with respect to the driven oscillator or the phase of a marker event of the driven oscillator with respect to the stimulus. Equation [1] is a one-dimensional equation and hence must be considered as an approximation to the dynamics in any realistic biological system. The topological degree of g measures the number of times g goes around the unit circle as ϕ goes around the unit circle once. In practical situations in which g has been measured it either has been discontinuous (topological degree is not defined) or continuous with either topological degree 1 or 0 (Winfree, 1980; Glass and Winfree, 1984).

In the limit of weak forcing, g is a 1:1 invertible map of degree 1. Analysis of the bifurcations of invertible circle maps was undertaken by Poincaré (1928) and subsequently by many others (Arnold, 1965; Herman, 1977). The original motivation for the analysis of invertible circle maps was the study of flows defined by differential equations without singular points defined on the torus. Poincaré (1928) constructed a cross section to the flow, and then considered the map which gives the successive points at which a single trajectory cuts the cross section after one complete circuit around the torus. Existence and uniqueness of the solutions of the differential equation guarantee that the map, Eq. [1], is well-defined, invertible and of topological degree 1. Poincaré invented the concept of the rotation number which gives the average increment in ϕ as the number of iterations of the map becomes infinite. Since trajectories cannot cross, the rotation number for circle maps generated by flows on the torus is independent of the initial condition. If there is a fixed point of the circle map, then the rotation number will be rational. Otherwise, the rotation number is irrational (quasiperiodic dynamics). For quasiperiodic dynamics the orbit in general densely covers the unit circle. For noninvertible circle maps the original definition of the rotation number must be extended (see Guevara and Glass, 1982; Bélair and Glass, 1985).

A prototypical example of circle maps originally studied by Arnold (1965) is

$$\phi_{i+1} = \phi_i + \tau + b\sin(2\pi\phi_i) \qquad [2]$$

For $b<1/(2\pi)$, this is an invertible circle map. The organization of the rotation numbers in (b,τ) parameter space was described by Arnold (1965). For b constant, ρ is a continuous nondecreasing function of τ. For each rational number M/N with $b<1/(2\pi)$, there exists, in general, a cusp-shaped region in which ρ = M/N for all initial conditions. These regions are called Arnold tongues or Arnold horns. Herman (1977) has shown that the region in (b,τ) space which does not lie in the union of the Arnold tongues (i.e., the set of parameter values for which there is quasiperiodic dynamics) has positive measure.

In simple models which have been proposed for the entrainment of biological oscillators, the dynamics can be represented by 1:1 invertible circle maps at low stimulation strengths. As stimulation amplitude increases the models are no longer represented by 1:1 invertible circle maps. In noninvertible circle maps the rotation number is no longer unique and the definition of the Arnold tongue is problematical.

Boyland (1984) defined the Arnold tongue T_r to be the set of parameter values (b,τ) for Eq. [2] for which there is a periodic orbit with rotation number r. This definition is not adequate for the class of problems we consider and we suggest an alternative. For parameter values for which the map, Eq. [1], is invertible any given point ϕ^* will be a periodic point of period N and rotation number M/N along some line $\tau = \tau(b)$. For values of (b,τ) for which the map is invertible the Arnold tongue $T_{M/N}$ is the union of all such continuation lines with rotation number M/N. This is equivalent to the Boyland definition (Bélair and Glass, 1985). The continuation of $T_{M/N}$ to values of (b,τ) for which the map is noninvertible is now described. Each continuation line associated with $T_{M/N}$ in the invertible region has a unique extension to the noninvertible region along which ϕ^* is a periodic point of period N (but not necessarily rotation number M/N). The Arnold tongue is then the union of all such continuation lines for parameter values at which ϕ^* is a stable periodic orbit of period N.

3. Periodic Forcing of Integrate-and-Fire Models

One class of models which are frequently employed in biology are called integrate-and-fire models. Suppose that there is an oscillatory threshold , $\theta(t)$

$$\theta(t) = 1 + k \sin(2\pi t) \qquad [3]$$

and linearly rising and falling activities as depicted in Fig. 1.

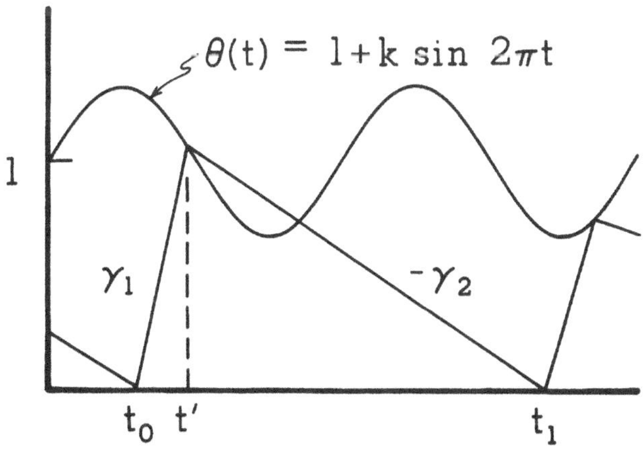

Fig. 1. The integrate-and-fire model

There are three parameters, the slopes of the rising and falling phases of the activity (denoted γ_1 and $-\gamma_2$, respectively) and the modulation of the sine wave, k. In the limit of no modulation k=0, the intrinsic frequency of the oscillator is given

$$f_0 = \gamma_1\gamma_2/(\gamma_1 + \gamma_2) \quad [4]$$

This model has also been proposed for the entrainment of the circadian rhythm (Winfree, 1980). The goal is to understand in detail the bifurcations and the dynamics in this model as a function of the parameters for all initial conditions. The charm of this problem is that it is exceedingly simple to state but a complete solution has not yet been obtained.

Assume that t_0 is known (Fig. 1). Then t', the time the activity first reaches threshold can be found by solving the equation

$$\gamma_1(t'-t_0) = 1 + k\sin(2\pi t') \quad [5a]$$

This is a transcendental equation in t' and admits of no analytical solution. Once t' is computed using Eq. [5a], then t_1 can be computed from the relation

$$\gamma_2(t_1-t') = 1 + k\sin(2\pi t') \quad [5b]$$

In Eq. [5b] t_1 can be computed explicitly as a function of t'. Using Eqs. [5a,b] the successive starting times of each cycle, $t_1, t_2, \ldots$ can be numerically computed once the initial condition and the parameters are specified. Clearly, if we define $\phi_i = t_i$ (mod 1) then the process is defined by a circle map, Eq. [1]. For $k>\gamma_1/(2\pi)$ the circle map will be discontinuous, and for $k>\gamma_2/(2\pi)$ the circle map will be non-monotonic (and hence non-invertible). Three special cases will be discussed.

In the first special case, we consider the limit in which $\gamma_2 \to \infty$ (Fig. 2). In this case, there is a slow rise to the threshold and a vertical fall. Early studies analyzed this model in the context of periodically forced relaxation oscillations (Builder and Roberts, 1939) (LG thanks T.T. Allen for pointing out this reference to us). Unaware of this early study, the model was proposed in the context of entrainment of biological oscillators (Glass and Makey, 1979; Glass et al., 1980). For $k<f_0/2\pi$ the dynamics are described by a continuous invertible map of the unit circle, and the discussion above concerning Arnold tongues

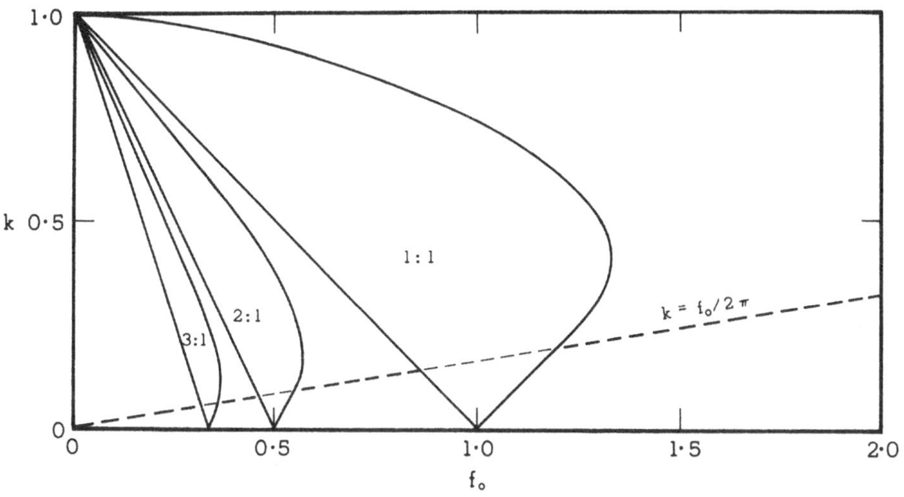

Fig. 2. Phase locking zones for the integrate-and-fire model with $\gamma_2 = \infty$, $f_o = \gamma_1$. See also Builder and Roberts (1939).

is applicable. For $k>f_o/2\pi$ the dynamics are given by a discontinuous, non-invertible monotonic function. The analysis of this problem has been given by Keener (1980) with a number of applications in Keener (1981), Keener et al. (1981). In this case, it is known that there will still be N:M phase locking for all integers N and M relatively prime. There are also parameter values which give rise to aperiodic dynamics. There are however two differences between the properties of the aperiodic dynamics in the case in which the dynamics are described by the discontinuous piecewise monotonic maps, and the invertible maps. First, the successive iterates no longer form a dense orbit on the unit circle, but rather a Cantor set. Second, the probability that one will choose a set of parameter values associated with the aperiodic dynamics is now zero (this means the Lebesgue measure associated with the aperiodic dynamics is zero).

Another special case is the situation in which $\gamma_1 = \gamma_2$ (Fig. 3). To the best of our knowledge, this case has not been considered previously. We give a brief summary of our results to date, and will hopefully present a more detailed analysis at some future time. For $k<f_o/\pi$ the map is invertible. The phase locking ratios N:2 where N is an odd integer, are present only for $f_o = 2/N$ (i.e., the Arnold tongue in this case does not extend over an interval of f_o with k fixed). In this degenerate situation, all points on the unit circle are periodic and neutrally stable. In the non-invertible case, $k>f_o/\pi$, the map is both discontinuous and non-monotonic. The boundaries of the N:1 zones overlap giving

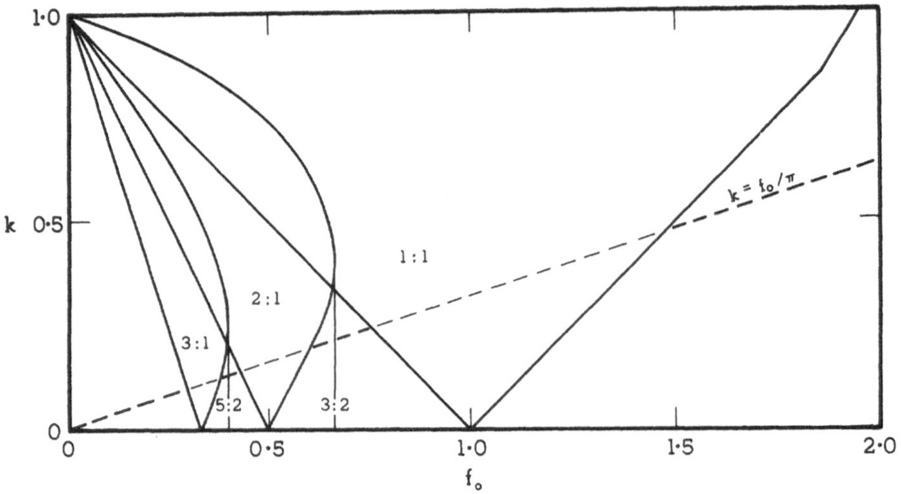

Fig. 3. Phase locking zones for the integrate-and-fire model with $\gamma_1 = \gamma_2$, $f_o = \gamma_1/2$.

rise to bistability. The N:2 zones (N odd) vanish at a finite value of k which can be explicitly computed. This vanishing is a result of the periodic points being lost into the discontinuity of the circle map as k increases. It can be analytically shown that no period doubling bifurcations are present in the N:1 zones. We note certain qualitative similarities of the bifurcations here and those found in the periodically forced van der Pol equation (Flaherty and Hoppensteadt, 1978). A connection between periodically forced van der Pol oscillators and integrate-and-fire models is given in Petrillo and Glass (1984).

Now consider a third special case in which $\gamma_1 \to \infty$ (Fig. 4). It is not hard to see (Perez and Glass, 1982) that we obtain Eq. [2] where $b = k/\gamma_2$ and $\tau = 1/\gamma_2$. For $k > f_o/2\pi$ the map is still continuous but it is no longer monotonic and hence it is non-invertible. Recently, there has been interest in the transition from invertibility to non-invertibility, at the value $b = 1/(2\pi)$ (Feigenbaum et al., 1982; Ostlund et al., 1983). It is also of interest to examine the dynamics in which the map is non-invertible, and several recent papers deal with this question (Perez and Glass, 1982; Glass and Perez, 1982; Fraser and Kapral, 1984; Boyland, 1984; Bélair and Glass, 1985).

Each of the Arnold tongues present for $k < f_o/2\pi$ extends to the region $k > f_o/2\pi$. Each tongue, however, splits into two branches. Once again,

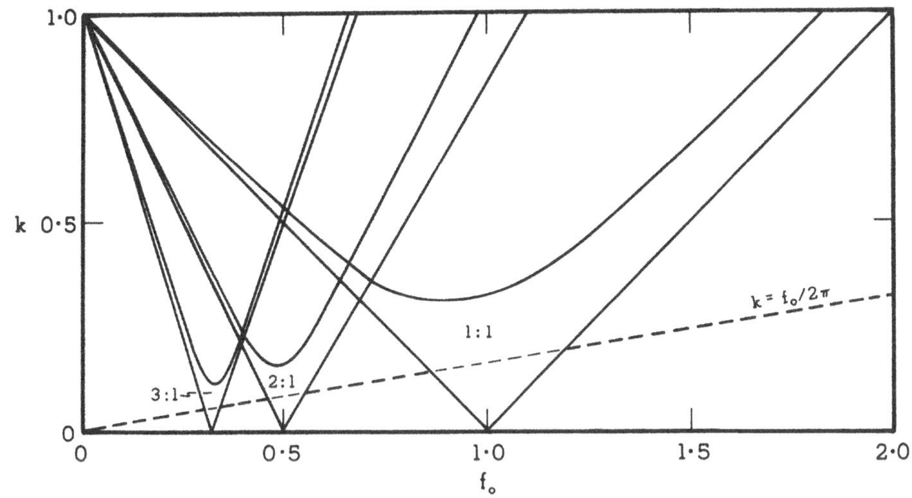

Fig. 4. Phase locking zones for the integrate-and-fire model with $\gamma_1 = \infty$, $f_o = \gamma_2$. Corresponds to Eq. [2] with $b = k/\gamma_2$, $\tau = 1/\gamma_2$.

the Arnold tongues can cross leading to a situation in which two different rotation numbers can be found for the same values of the parameters. The other main feature is that there are complex sequences of bifurcations which are present in the Y-shaped regions formed by each Arnold tongue. The structure of the bifurcations in each of the Arnold tongues is discussed in detail elsewhere (Glass and Perez, 1982; Boyland, 1984; Ostlund, et al., 1983; Fraser and Kapral, 1984; Bélair and Glass 1985).

4. Entrainment of Limit Cycle Models

Many biological rhythms are best represented mathematically as limit cycle oscillations in differential equations. We now consider the effects of δ function pulses delivered to a simple mathematical model of a biological oscillator in which there is a rapid return to the limit cycle following a perturbation. To analyze this problem we use a technique originally presented in studies on the periodic forcing of neural oscillators (Perkel et al., 1964).

The main theoretical idea can be appreciated from a consideration of the effects of periodic stimulation of the simple model for limit cycle oscillations originally proposed by Winfree (1975). The equations are conveniently written in a polar coordinate system, and are

$$dr/dt = ar(1-r)$$
$$d\phi/dt = 2\pi \qquad [6]$$

where a is a positive constant. There is a limit cycle at r = 1 which is globally attracting in the limit $t \to \infty$ for all initial conditions except for the origin. Assume that there is a periodic stimulus consisting of a δ function impulse which is represented by a horizontal translation by an amount b. Call ϕ_i the phase immediately preceding the ith stimulus. Then in the limit $a \to \infty$ the phase immediately preceding the (i+1)st stimulus is

$$\phi_{i+1} = g(\phi_i, b) + \tau \qquad [7a]$$

where

$$\cos(2\pi g) = \frac{b + \cos 2\pi\phi}{(1+2b\cos 2\pi\phi + b^2)^{1/2}} \qquad [7b]$$

where τ is the time interval between periodic stimuli measured relative to the intrinsic cycle length of the limit cycle oscillator. In this system Eq. [7] represents a 1:1 invertible circle map of degree 1 for $0<b<1$, and a degree 0 map for $b>1$. This system displays period doubling bifurcation and chaotic dynamics (Guevara and Glass, 1982; Hoppensteadt and Keener, 1982). A recent study has analyzed in some detail the bifurcations for $b>1$ (Keener and Glass, 1984). Here we give partial results related to the geometry of the Arnold tongues.

Except for the Arnold tongues associated with the 1:M periodic orbits the Arnold tongues do not extend to arbitrarily high values of the parameter b. The Arnold tongues apparently have a "mushroom" configuration. Consider two Arnold tongues associated with N:M and N':M' phase locking. Then the Arnold tongue associated with the mediant, N+N':M+M' will send branches extending to the intersections of its "mother" zones and the line b=1. A schematic figure, distorted purposely to show the geometry is given in Fig. 5. The points of intersection of the Arnold tongues with the line b=1 represent points of accumulation at which an infinite number of stable periodic orbits converge, a subset of which has been described in the above construction. The rotation number in this problem can change inside the Arnold tongue as defined here (Guevara and Glass, 1982; Keener and Glass, 1984).

5. Discussion

The effects of periodic stimulation on biological oscillators is not simply of mathematical interest. Practical applications of great

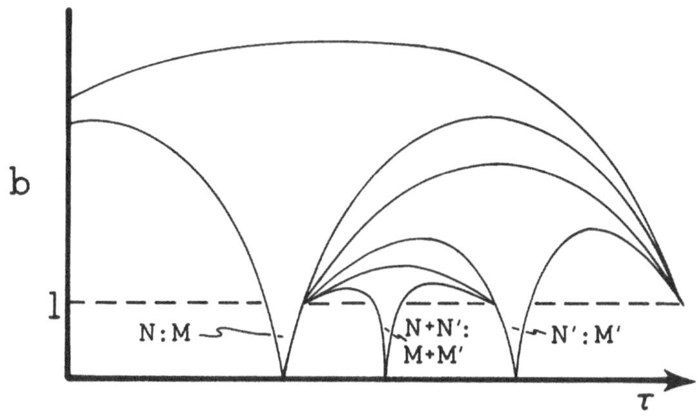

Fig. 5. Schematic organization of phase locking zones for limit cycle model, Eq. [7]. The b-axis and phase locking zones are distorted to show the geometry.

significance in the health sciences include the stimulation of cardiac tissue by electronic pacemakers, the entrainment of the respiratory rhythm by mechanical ventilators and the entrainment of the circadian rhythm by the 24-hour light dark cycles. In some operating modes, pacemakers and mechanical ventilators can also be influenced by the patient's intrinsic rhythm. The mathematical richness of these systems, and the attendant possibilities for complex bifurcations under parametric changes are not well appreciated by health professionals. Of course, in real biological systems in which there is noise, the very fine details of the bifurcations which are theoretically predicted will be impossible to observe. Nevertheless, it seems well worthwhile to pursue experimental studies on the effects of periodic forcing in diverse systems. It should be of some interest to try to unravel the bifurcations which are observed in systems of practical importance, particularly at levels of large stimulation intensity.

The mathematical questions arising from the study of the periodically forced biological oscillators are still poorly understood. Even the bifurcations in the comparatively simple one-dimensional systems considered here are not understood in all detail. In systems in which there is not instantaneous relaxation to the limit cycles, it will be necessary to study bifurcations in maps of at least dimension 2. Some of the gross features of the bifurcations in two-dimensional maps may mimic the bifurcations in one-dimensional maps (for example, see Aronson, et al., 1982) but there may also be differences in the bifurcations which at this point are not at all understood. In particular, the relationship, if any, between the bifurcations in the one-dimensional circle maps considered here and periodically forced nonlinear oscillators such as the van der Pol equation and the Brusselator remain to be clarified.

In conclusion, experimental studies on the periodic forcing of biological oscillators yields beautiful data reflecting the rich bifurcations present in the dynamics. Determining the details of these bifurcations in diverse systems is a challenge both for experimentalists and theorists.

Acknowledgements

We have greatly benefited from our collaborations with M.C. Mackey, M.R. Guevara, and J.P. Keener on the work discussed here. This research has been partially supported by a grant from the Natural Sciences and Engineering Research Council. JB thanks NSERC and FCAC for Fellowship support. LG was a resident in the Institute for Nonlinear Science, University of California, San Diego during 1984-85.

References

ARNOLD, V.I. (1965). Translations A.M.S. Series 2, 46, 213.
ARONSON, D.G., CHORY, M.A., HALL, G.R., and MCGEHEE, R.P. (1982). Commun. Math. Phys. 83, 303.
BÉLAIR, J. and GLASS, L. (1985). Physica D, In Press.
BOYLAND, P. (1984). Preprint.
BUILDER, G. and ROBERTS, N.F. (1939). A.W.A. Technical Review 4, 165.
FEIGENBAUM, M.J., KADANOFF, L.P. and SHENKER, S. (1982). Physica 5D, 370.
FLAHERTY, J. and HOPPENSTEADT, F. (1978). Stud. Appl. Math. 58, 5.
FRASER, S. and KAPRAL, R. (1984). Phys. Rev. A30, 1017.
GLASS, L., GUEVARA, M.R., BÉLAIR, J. and SHRIER, A. (1984). Phys. Rev. A29, 1348.
GLASS, L. and MACKEY, M.C. (1979). J. Math. Biol. 7, 339.
GLASS, L. and PEREZ, R. (1982). Phys. Rev. Lett. 48, 1772.
GLASS, L. and WINFREE, A.T., (1984). Am. J. Physiol. 246 (Regulatory Integrative Comp. Physiol. 15):15, R251.
GUEVARA, M.R. and GLASS, L. (1982). J. Math. Biol. 14, 1.
GUEVARA, M.R., GLASS, L. and SHRIER, A. (1981). Science 214, 1350.
HERMAN, M.R. (1977). In: Lecture Notes in Mathematics, Vol. 597, Geometry and Topology, pp. 271-293. Berlin:Springer-Verlag.
HOPPENSTEADT, F.C. and KEENER, J.P. (1982). J. Math. Biol. 15, 339.
KEENER, J.P. (1980). Trans. AMS 26, 589.
KEENER, J.P. (1981). J. Math. Biol. 12, 215.
KEENER, J.P., HOPPENSTEADT, F.C., and RINZEL, J. (1981). SIAM J. Appl. Math. 41, 503.
KEENER, J.P. and GLASS, L. (1984). J. Math. Biol. 21, 175.
OSTLUND, S., RAND, D., SETHNA, J. and SIGGIA, E. (1983). Physica 8D, 303.
PEREZ, R. and GLASS, L. (1982). Phys. Lett. 90A, 441.
PERKEL, D.H., SCHULMAN, J.H., BULLOCK, T.H., MOORE, G.P. and SEGUNDO, J.P. (1964). Science 145, 61.
PETRILLO, G.A. and GLASS, L. (1984). Am. J. Physiol. 246 (Regulatory

Integrative Comp. Physiol. 15):R311.
POINCARÉ, H. (1928). Oeuvres I, Paris: Gauthier-Villar.
WINFREE, A.T. (1975). Physics Today 28, 34.
WINFREE, A.T. (1980). The Geometry of Biological Time, New York: Springer.

AVERAGING AND SYNCHRONIZATION OF WEAKLY COUPLED SYSTEMS

Humberto Carrillo Calvet[1]
Facultad de Ciencias
Universidad Nacional Autónoma de México,
D.F. 04510, México

Abstract

The averaging method for ordinary differential equations is briefly reviewed and new results on the uniform validity of its approximations over infinite intervals of time are given. As an application a synchronization theorem for weakly coupled systems is proved.

1. The Averaging Method

In this perturbation method one considers the equation

(E) $\qquad \dot{x} = \varepsilon f(t,x,\varepsilon)$

where $f\colon \mathbb{R} \times \mathbb{R}^n \times [0,\infty) \to \mathbb{R}^n$ is continuous, has a continuous partial derivative with respect to x and is almost periodic in t, uniformly with respect to (x,ε) in compact sets. Under these conditions $f(t,x,\varepsilon)$ has a mean value:

$$f_0(x) := \lim_{T\to\infty} \frac{1}{T} \int_0^T f(t,x,0)\, dt.$$

The averaging method consists in approximating the solutions of equation (E) by the solutions of the *averaged equation*,

(AE) $\qquad \dot{\bar{x}} = \varepsilon f_0(\bar{x}),$

for small values of the parameter ε.

This procedure simplifies the original system (E) in, at least, reducing its dimension by one. The fact that it is widely used in non-linear oscillation theory is going to be corroborated by its frequent appearance in the present collection of papers.

To give an intuitive justification of the method we will use a

1 Partially supported by Consejo Nacional de Ciencia y Tecnología, México.

new time scale ($\tau = \varepsilon t$) in equation (E) and (AE) to have:

$$(E') \qquad \frac{dx}{d\tau} = f(\tau/\varepsilon, x, \varepsilon)$$

$$(AE') \qquad \frac{d\bar{x}}{d\tau} = f_0(\bar{x})$$

Now it is obvious that equation (E') involves two time scales: y varies according to the time scale in which τ is measured, but the system is being forced on a faster scale of time τ/ε. For small values of ε, $f(\tau/\varepsilon, x, \varepsilon)$ oscillates rapidly as τ changes. Therefore, it is reasonable to consider that in each moment τ, the solution y will feel only the average effect of the vector field $f(\tau/\varepsilon, x, \varepsilon)$ at the point $x(\tau)$, which is given by the mean value $f_0(x\ (\tau))$.

2. Foundations of the Averaging Method

The following result is a generalization of Bogolyubov's theorem on the validity of the averaging method over finite time intervals of the form $[t_0, \frac{1}{\varepsilon}]$.

Theorem 1. (Ref. [1], [2])

Let us denote by $x_\varepsilon(t)$ a solution of equation (E) for the value ε of the parameter. Assume that $\phi(t)$ is a solution of the equation $\dot{x} = f_0(x)$. (Under these conditions, for each $\varepsilon > 0$, the function $\bar{x}_\varepsilon(t) = \phi(\varepsilon t)$ is a solution of the averaged equation (AE)). Then, given a tolerance $\eta > 0$ and T as large as we please, there are ε_0 and δ, positive, such that $(t_0, \varepsilon)\ [0,\infty)x(0,\varepsilon_0)$ and $|x_\varepsilon(t_0)-\phi(\varepsilon t_0)| < \delta$ implies that

$$|x_\varepsilon(t)-\phi(\varepsilon t)| < \eta \qquad \text{for all } t\in[t_0, T/\varepsilon].$$

This theorem justifies the averaging approximation during intervals of time that can be as long as we wish, with the condition that ε is taken small enough. However, it does not follow that for two solutions $x_\varepsilon(t)$ and $\bar{x}_\varepsilon(t)$ that satisfy the same initial condition we will have that

$$\lim_{\varepsilon\to 0} |x_\varepsilon(t)-\bar{x}_\varepsilon(t)| = 0$$

uniformly with respect to t $(0,\infty)$. In fact, this is not true and it can happen that, without mattering how small we choose the value of ε, in the long run, the two solutions x_ε and $\bar{x}_\varepsilon$ find themselves very far apart from each other [2]. This constitutes a strong limitation of the method. However, it has been proved ([4], [5], [6], [7]) that the averaging approximation does not fail for strongly stable solutions. After giving some definitions, we will state a very general result on the uniform validity of the averaging method over infinite intervals of time.

Let be $\phi: \mathbb{R} \times (0,\infty) \to \mathbb{R}^n$ and let us denote by x_ε a solution of equation (E). We will say that $\phi(t,\varepsilon)$ has a *stable neighborhood* (*s.n.*) as $\varepsilon \to 0$ for the equation (E), if for any $\eta > 0$ there are positive numbers $\varepsilon_0(\eta)$ and $\delta(\eta)$ such that $(t_0,\varepsilon) \in [0,\infty) \times (0,\varepsilon_0)$ and $|x_\varepsilon(t_0) - \phi(t_0,\varepsilon)| < \delta$ implies that $|x_\varepsilon(t) - \phi(t,\varepsilon)| < \eta$ for all $t > t_0$.

Similarly if for each $\varepsilon > 0$, γ_ε is a set in $\mathbb{R}^n$, we will say that γ_ε has an *orbitally stable neighborhood* (*o.s.n.*) as $\varepsilon \to 0$ for the equation (E) if for any $\eta > 0$ there are positive numbers $\varepsilon_0(\eta)$ and $\delta(\eta)$ such that $(t_0,\varepsilon) \in [0,\infty) \times (0,\varepsilon_0)$ and $d(x_\varepsilon(t_0),\gamma) < \delta$ implies that $d(x_\varepsilon(t),\gamma) < \eta$ for all $t \geq t_0$. Here $d(x,\gamma)$ represents the distance between the point x and the set γ.

Theorem 2. (Ref. [3]).

Let ϕ be a solution of $\dot{x} = f_0(x)$ and $x_\varepsilon(t;t_0,x_0)$ the solution of equation (E) that satisfies the initial condition $x_\varepsilon(t_0;t_0,x_0) = x_0$.

(a) If ϕ is uniformly asymptotically stable and bounded then the solution of the averaged equation, $\bar{x}_\varepsilon(t) = \phi(\varepsilon t)$, has a *s.n.* as $\varepsilon \to 0$ for equation (E). Also, if x_0 is a point in the domain of attraction of ϕ, then given $\eta > 0$ there is $\varepsilon_0(\eta) > 0$ such that for any $0 < \varepsilon < \varepsilon_0$ there exists $T(\varepsilon) > 0$ such that $|x_\varepsilon(t;t_0,x_0) - \phi(\varepsilon t)| < \eta$ for all $t \geq T$.

(b) If ϕ is orbitally uniformly asymptotically stable and bounded with orbit γ then γ has an *o.s.n.* as $\varepsilon \to 0$ for the equation (E). If x_0 is a point in the domain of attraction of γ, then given $\eta > 0$ there is $\varepsilon_0(\eta) > 0$ such that for any $0 < \varepsilon < \varepsilon_0$ there exists $T(\varepsilon) > 0$ such that $d(x_\varepsilon(t;t_0,x_0), \gamma) < \eta$ for all $t \geq T$.

Remarks.

(i) Neither (a) nor (b) require exponential stability of the solution ϕ, and then the theorem also applies in the nonlinear stability case.

(ii) In an autonomous equation asymptotically stable (*a.s.*) static solutions are uniformly *a.s.* and orbitally *a.s.* periodic solutions are orbitally uniformly *a.s.*. Then, these conditions are enough to guarantee the assertions (a) and (b) respectively.

(iii) The statement (a) is also true for any solution stable under persistent disturbances and bounded. Furthermore, if we extend in a natural way the notion of stability under persistent disturbances for sets in $\mathbb{R}^n$ then we can obtain a similar result to that of (b) for this more general situation [3].

Example. Consider the van der Pol equation with rapidly oscillating, forcing and coefficients.

(1) $$\ddot{x} + \mu(t/\varepsilon)(x^2-1)\dot{x} + \omega(t/\varepsilon)\, x = g(t/\varepsilon,\ x,\ \dot{x}).$$

Let us assume that the functions μ and ω are almost periodic with positive mean values $\bar{\mu}$ and $\bar{\omega}$ respectively. Let us also assume that $g(t,x,y)$ is almost periodic in t uniformly with respect to (x,y) in compact sets, and has mean value zero. Reescaling the time ($\tau = t/\varepsilon$) and writing equation (1) as a system, we have

(2) $$\frac{dx}{d} = \varepsilon y$$
$$\frac{dy}{d\tau} = -\varepsilon\mu(\tau)(x^2 -1)\, y - \varepsilon\omega(\tau)x + g(\tau,x,y).$$

The corresponding averaged system

(3) $$\frac{d\bar{x}}{d\tau} = \varepsilon\bar{y}$$
$$\frac{d\bar{y}}{d\tau} = -\ \varepsilon\bar{\mu}(\bar{x}^2-1)\bar{y} - \varepsilon\bar{\omega}\ \bar{x}$$

has a limit cycle. By Theorem 2 the set γ has an orbitally stable neighborhood, this means that the solutions of (2) that start close to γ remain close for all future time. Also, in its domain of attraction, γ is "attracting" the solutions of (E) in the sense of assertion (b).

The limit cycle of van der Pol equation is an exponential attractor. This is because its Poincaré map is linearly asymptotically stable. However, the theorem can also be applied to limit cycles that do not satisfy this property [2].

3. Synchronization.

Let us consider the equation

(4) $$\dot{\theta} = f(t,\theta)$$

where $t\in\mathbb{R}$, $\theta = (\theta_1,\dots,\theta_n)\in\mathbb{R}^n$ and f is 2π-periodic in each component of the vector θ. Phase-only equations like (4) appear in models of n coupled ring devices [8].

Let be $\omega = (\omega_1,\dots,\omega_n)$ with each ω_i a natural number. We say that the system (4) has the *rational synchronization* property with *rotation vector* ω, if for each t_0 $\mathbb{R}$ there is an open set $\Gamma\subset\mathbb{R}^n$ such that, any solution $\theta(t)$ of (4) with $\theta(t_0)\in\Gamma$ satisfy the relations:

(5) $$\lim_{t\to\infty} \theta_1:\theta_2:\dots:\theta_n = \omega_1:\omega_2:\dots:\omega_n$$

That is,

$$\lim_{t\to\infty} \frac{\theta_i}{\theta_j} = \frac{\omega_i}{\omega_j} \qquad \text{for } i,\ j = 1,\dots,\ n.$$

Considering that equation (1) represents a one parameter family of vector fields on an n-dimensional torus, the condition (5) means that the orbit of $\theta(t)$ winds, asymptotically, ω_i times around the i-axis of this torus for each ω_j windings around the j-axis. To have rational synchronization with rotation vector $\omega = (1,\dots,1)$ means that the frequencies are asymptotically the same and it is called just *synchronization*.

4. Weakly coupled systems.

Here we will study a system in amplitude (x) and phase (θ) variables of the form:

(6) $$\begin{aligned} \dot{x} &= \varepsilon G(x,\theta,\varepsilon) \\ \dot{\theta} &= \omega + \varepsilon F(x,\theta,\varepsilon) \end{aligned}$$

where $G: \mathbb{R}^m \times \mathbb{R}^n \times [0,\infty) \to \mathbb{R}^m$ and $F: \mathbb{R}^m \times \mathbb{R}^n \times [0,\infty)$ $\mathbb{R}^n$ are continuos, 2π-periodic in each component of the vector θ, of class C^1 with respect to (x,θ) for each ε fixed an F is bounded as a function of x for each (θ,ε) fixed.

Let $\mathbb{Z}$ denote the set of integer numbers. We will suppose that $\{\omega,\Omega_2,\dots,\Omega_n\}\subset\mathbb{Z}^n$ is an orthogonal basis of $\mathbb{R}^n$ and A is the nxn matrix whose rows are $\omega,\Omega_2,\dots,\Omega_n$. We also adopt the following notation:

$$G_0(x,u) = \frac{1}{\omega\cdot\omega} \lim_{T\to\infty} \frac{1}{T} \int_0^T G(x,A^{-1}\binom{v}{u},\ 0)dv$$

$$F_0(x,u) = \frac{1}{\omega\cdot\omega} (F_0^2(x,u),\dots,F_0^n(x,u))$$

where $u \in \mathbb{R}^{n-1}$, $v \in \mathbb{R}$, the dot is the scalar product of $\mathbb{R}^n$, and

$$F_0^i(x,u) = \lim_{T\to\infty} \frac{1}{T} \int_0^T \Omega_i \cdot F(x,A^{-1}\binom{v}{u},0)dv$$

for $i = 2,\dots,n$.

The following theorem gives conditions under which the angular variables of the system (6) synchronizes rationally to ω for small values of the parameter ε; it generalizes a result by Hoppensteadt and Kenner (Ref. [9]).

Theorem 3.

If the auxiliary system

$$\text{(7)} \qquad \begin{aligned} \frac{dx}{dv} &= G_0(x,u) \\ \frac{du}{dv} &= F_0(x,u) \end{aligned}$$

has a bounded and orbitally uniformly asymptotically stable solution, then there exists and $\varepsilon_0 > 0$ such that, for $0 < \varepsilon < \varepsilon_0$, the phase variables $(\theta_1,\dots,\theta_n)$ of the system (6) sinchronize rationally with rotation vector ω.

The proof of this theorem will give us an example of an argument that requires the validity of the averaging method over the unbounded interval of time $[t_0,\infty)$.

Proof. In the new variables $v = \omega \cdot \theta$ and $u_i = \Omega_i \cdot \theta$ for $i=2,\dots,n$, equation (6) becomes:

$$\text{(8)} \qquad \begin{aligned} \dot{x} &= \varepsilon G(x,A^{-1}\binom{v}{u},\varepsilon) \\ \dot{u}_i &= \varepsilon\Omega_i \cdot F(x,A^{-1}\binom{v}{u},\varepsilon) \qquad \text{(for } i = 2,\dots,n) \\ \dot{v} &= \omega\cdot\omega + \varepsilon\omega\cdot F(x,A^{-1}\binom{v}{u},\varepsilon). \end{aligned}$$

Since $F(x,A^{-1}\binom{v}{u},\varepsilon)$ is bounded, there exists $\varepsilon_1 > 0$ such that for $0 < \varepsilon < \varepsilon_1$ the component $v(t)$ of any solution of equations (8) tends monotonically to infinite.

Eliminating time in equations (8) we obtain

$$\frac{dx}{dv} = \frac{\varepsilon G(x, A^{-1}\binom{v}{u}, \varepsilon)}{\omega\cdot\omega + \varepsilon\omega\cdot F(x, A^{-1}\binom{v}{u}, \varepsilon)}$$

(9)

$$\frac{du_i}{dv} = \frac{\varepsilon\Omega_i \cdot F(x, A^{-1}\binom{v}{u}, \varepsilon)}{\omega\cdot\omega + \varepsilon\omega\cdot F(x, A^{-1}\binom{v}{u}, \varepsilon)} \qquad (\text{for } 1 = 2,\dots,n)$$

whose corresponding averaged equation is the auxiliary system (7). As this averaged system has an orbitally uniformly *a.s.* solution, Theorem 2 (b) implies the existence of $\Delta \subset \mathbb{R}^{m+n-1}$ and $\varepsilon_0 < \varepsilon_1$ such that, for $0 < \varepsilon < \varepsilon_0$, it happens that: if (x,u) is a solution of (9) and $(x(v_0), u(v_0)) \in \Delta$ with $v_0 > 0$, then (x,u) is bounded for $v > v_0$. Fix $v_0 > 0$ and let Γ be the set of (y,z,v) in $\mathbb{R}^m \times \mathbb{R}^{n-1} \times \mathbb{R}$ such that $v > v_0$ and $(y,z) = (x(v),u(v))$ with (x,u) a solution of equation (9) that satisfies $(x(v_0), u(v_0)) \in \Delta$. This set, Γ, is open and for

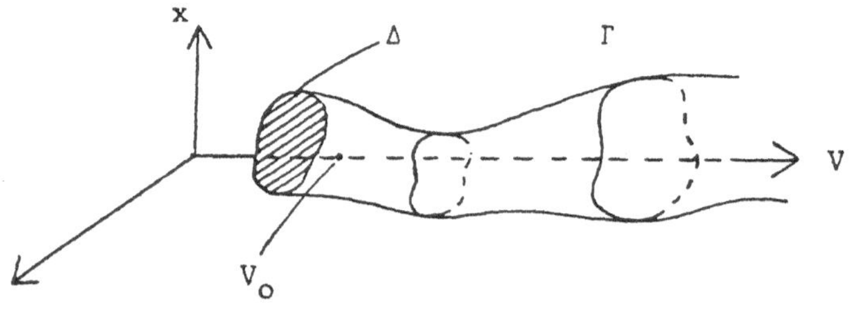

$\varepsilon < \varepsilon_0$, if (x,θ) is a solution of (6) with $(x(t_0),\theta(t_0)) \in \Gamma$, then, for $k = 2,\dots,n$, the function $u_k(t) = \Omega_K \cdot \theta(t)$ is bounded for $t > t_0$. Hence,

$$\lim_{t\to\infty} \frac{\Omega_K \cdot \theta(t)}{\omega \cdot (t)} = 0$$

From this it follows that

$$\lim_{t\to\infty} \frac{\theta(t)}{\omega\cdot\theta(t)} = \lim_{t\to\infty} \left[\frac{\omega}{\omega\cdot\omega} + \sum_{k=2}^{n} \frac{\Omega_k\cdot\theta(t)}{\omega\cdot\theta(t)} \frac{\Omega_k}{\Omega_k\cdot\Omega_k}\right] = \frac{\omega}{\omega\cdot\omega}$$

Therefore, for $\varepsilon < \varepsilon_0$, if $(x(t_0), \theta(t_0)) \in \Gamma$, then

$$\lim_{t\to\infty} \frac{\theta_i(t)}{\theta_j(t)} = \lim_{t\to\infty} \frac{\theta_i(t)}{\omega\cdot\theta(t)} \frac{\omega\cdot\theta(t)}{\theta_j(t)} = \frac{\omega_i}{\omega_j} \qquad (\text{for } i,j = 1,\dots,n).$$

This proves Theorem 3.

References

[1]. N.N. Bogolyubov and J.A. Mitropolsky. Asymptotic methods in the theory of nonlinear oscillations. (Chapter 6). Gordon and Breach Science Publishers, N.Y., 1961.

[2]. H. Carrillo Calvet. Perturbation results on the long run behavior of nonlinear dynamical systems. To appear in: Differential Equations. Proceedings of the Lefschetz Conference,1984. Contemporary Mathematics, AMS.

[3]. H. Carrillo Calvet. Stability under persistent disturbances and averaging. In preparation.

[4]. V.M. Volosov. Averaging in systems of differential equations. Russ. Math. Surveys, Vol. 17, No. 6, 1962.

[5]. V.M. Volosov. Averaging on an unbounded interval. Soviet Mathematics. Translation of Doklady, Vol. III, part II, 1962.

[6]. C. Banfi. Sull' approssimazione di processi non stationari in meccanica non lineare. Boll. Un. Mat. Italiana 22, 442-450.

[7]. V. Eckhaus. New approach to the asymptotic theory of nonlinear oscillations and wave propagation. J. Math. Anal. Appl. 49, 575-611.

[8]. A.T. Winfree. The geometry of biological time. Springer-Verlag, Berlin, 1980.

[9]. F.C. Hoppensteadt and J. Keener. Phase locking of biological clocks. J. Math. Biology 15 (1982), 339-349.

Variational Principles for Periodic Solutions of Autonomous Ordinary Differential Equations.

Giles Auchmuty
Department of Mathematics
University of Houston
University Park, Houston, TX 77004.

1. INTRODUCTION.

In many chemical and biological oscillators, one would like to find approximations to the observed periodic solutions. Usually this is done by trying to show that the initial value problem for the differential equation has a stable limit cycle and trying to compute the resulting solution. When the equation is an autonomous ordinary differential equation in more than two variables, this is often very difficult.

In this paper, we shall use convex analysis to derive some variational principles for these periodic solutions. The advantages of having a variational formulation of a problem are well-known. There are a number of standard analytical methods for obtaining approximate solutions of variational problems and the numerical simulation of such problems has been highly developed. The methods are global and are independent of the number of variables in the equation. Consequently variational characterizations of periodic solutions are sometimes easier to analyze than the initial value problem for the equation, especially if there are three or more unknowns in the equation.

The next section is devoted to formulating the problem and then, in Section 3, we describe the variational principles and find criteria which guarantee that minima exist. The equations obeyed by these minima are derived. In Section 4, we construct the functionals for some particular equations arising in chemical kinetics. When terms from convex analysis and the calculus of variations are used in this paper without explicit definition, they should be taken as in Ekeland and Temam (3). The author appreciates the support of this research provided by the Welch Foundation.

2. PERIODIC SOLUTIONS OF O.D.E.'s.

Consider the problem of finding periodic solutions of an autonomous ordinary differential equation

$$\frac{dv(t)}{dt} = f(v(t)) \tag{2.1}$$

Here $v: \mathbb{R} \to \mathbb{R}^n$ is the unknown vector-valued function and $f: \mathbb{R}^n \to \mathbb{R}^n$ is assumed to be locally Lipshitz on $\mathbb{R}^n$.

Let $p(t)$ be a periodic solution of period T of (2.1). Define $\tau = {}^t/_T$ and $u(\tau) = p(\tau T)$. Then u is periodic of period 1 in τ and obeys

$$\omega \frac{du(\tau)}{d\tau} = f(u(\tau)) \tag{2.2}$$

and

$$u(0) = u(1) \tag{2.3}$$

Here $\omega = T^{-1}$ is a frequency. Thus the problem of finding periodic solutions of (2.1) is equivalent to finding a pair (u,ω) obeying (2.2) and (2.3).

When there is a solution (u,ω) of (2.2) - (2.3) for some value ω_1 of ω then there will also be solutions (u_m,ω_m) for $\omega_m = \frac{\omega_1}{m}$ and m any positive integer. This occurs because when p is a solution of (2.1) of period T, then it is also a solution of period mT for any positive integer m. Thus we whall concentrate attention on the largest values of ω for which (2.2) - (2.3) has a solution.

Also if u is a solution of (2.2) - (2.3) and one defines

$$w(t) = \begin{cases} u(t+\theta) & \text{for } 0 \leq t \leq 1-\theta \\ u(t+\theta-1) & \text{for } 1-\theta \leq t \leq 1 \end{cases}$$

then w will also be a solution of (2.2) - (2.3). Here $0 < \theta < 1$ is arbitrary. Thus if there is one non-trivial solution of (2.2) - (2.3) there will be a one-parameter family of such solutions.

3. VARIATIONAL PRINCIPLES FOR PERIODIC SOLUTIONS.

Rewrite equation (2.2) in the form

$$\omega \frac{du}{dt} = g(u) + h(u) \tag{3.1}$$

Henceforth we shall assume,

(A1): g and h are both Lipshitz continuous on any compact subset of $\mathbb{R}^n$,

(A2): g is the derivative of a convex function $G: \mathbb{R}^n \to \mathbb{R}$.

Very often one assumes (A3): $g(u) = Au$ with A being a real symmetric and positive definite $n \times n$ matrix. In this case

$$G(u) = \tfrac{1}{2}(Au,u) \tag{3.2}$$

is the corresponding convex function where $(u,v) = \sum_{i=1}^{n} u_i v_i$ is the usual inner product on $\mathbb{R}^n$.

Choose $1<p<\infty$ and let $X = W^{1,p}([0,1];\mathbb{R}^n)$ be the Sobolev space of all functions $v:[0,1] \to \mathbb{R}^n$ whose first distributional derivatives $\dot{v}_i(t)$, all lie in the real Lebesgue space $L^p(0,1)$. Then X is a Banach space under the norm

$$||v||_{1,p} = \left[\sum_{i=1}^{n}\int_0^1 [|v_i(t)|^p + |\dot{v}_i(t)|^p]\, dt\right]^{1/p}$$

Let K be the closed subspace of X of all functions which obey

$$\int_0^1 \dot{v}_i(t)dt = 0 \qquad \text{for } 1 \le i \le n. \tag{3.3}$$

This implies (2.3). Here, and henceforth, we use $\dot{v}_i(t)$ for $\frac{dv_i}{dt}(t)$.

We shall also use the L^p-norms

$$||v||_p = [\sum_{i=1}^{n}\int_0^1 |v_i(t)|^p dt]^{1/p}$$

Define $\phi: X \to (-\infty,\infty]$ by

$$\phi(u) = \begin{cases} \int_0^1 G(u(t))dt & \text{if } u \in K \\ \infty & \text{otherwise} \end{cases} \tag{3.4}$$

Let X^* be the dual space of X with respect to the pairing

$$\langle u,v\rangle = \sum_{j=1}^{n}\int_0^1 u_j(t)v_j(t)dt \tag{3.5}$$

and let $\phi^*: X^* \to (-\infty,\infty]$ be the dual (or polar) functional defined by

$$\phi^*(v) = \sup_{u\in X} [\langle u,v\rangle - \phi(u)] \tag{3.6}$$

The variational principle (P) for solutions of (3.1) - (2.3) is to minimize the functional $\mathcal{F}$ on $X \times (0,\infty)$ where

$$\mathcal{F}(u,\omega) = \phi(u) + \phi^*(\omega\dot{u} - h(u)) + \langle u,h(u)\rangle \tag{3.7}$$

The next theorem shows that this functional is always bounded below and gives criteria for a minimizer to be a periodic solution of (3.1).

<u>Theorem 1.</u> For any $(u,\omega)\in K \times [0,\omega)$ one has $\mathcal{F}(u,\omega) \ge 0$. (3.8)

Moreover $\mathcal{F}(\hat{u},\hat{\omega}) = 0$ iff $(\hat{u},\hat{\omega})$ is a solution of (3.1) and (2.3).

<u>Proof.</u> From the definition of ϕ^*, one has

$$\phi(u) + \phi^*(v) \ge \langle u,v\rangle \tag{3.9}$$

for all $u\in X$, $v\in X^*$. When $u\in X$ then $\dot{u}\in L^p([0,1];\mathbb{R}^n)$ so $\dot{u}\in X^*$. Also u is continuous and bounded on [0,1] so h(u) is continuous and bounded and hence is in X^*.

Put $v = \omega\dot{u} - h(u)$ in (3.9) and note that $\langle \dot{u},u \rangle = 0$ for all u in K. Thus

$$\phi(u) + \phi^*(\omega u - h(u)) \geq -\langle u,h(u)\rangle$$

for all u in K. This implies (3.8).

From a basic result in convex analysis, equality holds in (3.9) iff $u \in K$ and $v \in \partial\phi(u)$, where $\partial\phi(u)$ is the subdifferential of ϕ at u. In this problem, $\partial\phi(u) = \{g(u)\}$ so equality holds in (3.8) iff $u \in K$ and $\omega\dot{u}-h(u) = g(u)$ which is (2.3) and (3.1).

Comments. 1. This result is similar to a result obtained by Brezis and Ekeland (1) where they derived a variational principle for the heat equation.

2. As noted in section 2, if there is a non-constant periodic solution of (3.1) then there is a one parameter family of them. Hence there will be a corresponding family of minima for $\mathcal{F}$.

When G is given by (3.2), and p=2 in the definition of X and K, then one can show that

$$\phi^*(v) = \tfrac{1}{2}\int_0^1 (A^{-1}v(t),v(t))dt$$

if $v \in L^2((0,1));\mathbb{R}^n)$. Thus

$$\mathcal{F}(u,\omega) = \tfrac{1}{2}\langle Au,u\rangle + \tfrac{1}{2}\langle A^{-1}(\omega\dot{u}-h(u)),\omega\dot{u}-h(u)\rangle + \langle u,h(u)\rangle \tag{3.10}$$

$$= \omega^2\mathcal{F}_1(u) - \omega\mathcal{F}_2(u) + \mathcal{F}_3(u)$$

where

$$\mathcal{F}_1(u) = \tfrac{1}{2}\langle A^{-1}\dot{u},\dot{u}\rangle$$

$$\mathcal{F}_2(u) = \langle A^{-1}\dot{u},h(u)\rangle$$

and

$$\mathcal{F}_3(u) = \tfrac{1}{2}\langle Au,u\rangle + \tfrac{1}{2}\langle A^{-1}h(u),h(u)\rangle + \langle u,h(u)\rangle$$

$$= \tfrac{1}{2}\langle Au + h(u),\ u + A^{-1}h(u)\rangle.$$

In this case, one has the following result.

Lemma 2. Suppose g, h obey (A1) - (A3) and p=2 in the definition of X, K. Then for each $\omega \geq 0$ functional $\mathcal{F}(.,\omega)$ defined by (3.10) is continuous and weakly lower semicontinuous (wlsc) on K.

Proof. Since A is real symmetric and positive definite so is A^{-1}. Thus $\mathcal{F}_1(u)$ is convex, continuous and there exists m>0 such that

$$\mathcal{F}_1(u) \geq m\|\dot{u}\|_2^2 \tag{3.11}$$

Consequently, $\mathcal{F}_1$ is also w.l.s.c. on K.

Suppose $\{u_n: n \geq 1\}$ is a sequence in K which converges weakly to a function u in K. Then the sequence is bounded in K and from the imbedding theorems u_n converges strongly to u in $Y = C([0,1];\mathbb{R}^n)$. Y is the space of continuous $\mathbb{R}^n$-valued functions on [0,1] with the sup norm. Since h is Lipshitz on compact subsets of $\mathbb{R}^n$ one has $h(u_n)$ converges to $h(u)$ in Y.

Thus $\mathcal{F}_2(u) - \mathcal{F}_2(u_n) = \langle A^{-1}(\dot{u}-\dot{u}_n), h(u)\rangle + \langle A^{-1}\dot{u}_n, h(u) - h(\dot{u}_n)\rangle$.

When $n \to \infty$ the first term on the right converges to zero as $\dot{u}_n$ converges weakly to $\dot{u}$ in $L^2([0,1];\mathbb{R}^n)$. The second term goes to zero as $||A^{-1}\dot{u}_n||_2$ is bounded and $h(u_n)$ converges to $h(u)$ in L^2. Thus $\mathcal{F}_2$ is weakly continuous. In a similar manner $\mathcal{F}_3$ is weakly continuous. Hence $\mathcal{F}_2$ and $\mathcal{F}_3$ are also strongly continuous and the lemma is proven.

Theorem 3. Assume the conditions of lemma 2 and that there exists $R>0$ s.t. $||u||_2>R$ implies $|\mathcal{F}_2(u)| \leq C||\dot{u}||_2||u||_2$ (3.12)

and $\mathcal{F}_3(u) \geq \phi(||u||_2)$. (3.13)

Here C is a constant and $\lim_{s\to\infty} \frac{\phi(s)}{s^2} = \infty$ Then for each $\omega>0$, $\mathcal{F}(u,\omega)$ attains its minimum value on K.

Proof. Let $\alpha(\omega) = \inf_{u \in K} \mathcal{F}(u,\omega)$.

One has $\alpha(\omega) \geq 0$, and $\mathcal{F}$ is $\omega.\ell.s.c.$ on K. To prove that $\mathcal{F}(.,\omega)$ attains its infimum on K, it suffices to prove that $\mathcal{F}(.,\omega)$ is coercive From (3.11) - (3.13), one sees that if $||u||_2 \geq R$, then

$$\mathcal{F}(u,\omega) \geq \omega^2 m\, ||\dot{u}||_2^2 - C||\dot{u}_2||\;||u||_2 + \phi(||u||_2).$$
$$\geq \omega^2 m(||\dot{u}||_2 - C'||u||_2)^2 + \phi(||u||_2) - d\;||u||_2^2$$

If $||u||_{1,2}^2 \to \infty$ then both $||\dot{u}||_2 \to \infty$ and $||u||_2 \to \infty$, so from the assumptions on ϕ one has $\mathcal{F}(u) \to \infty$. Thus $\mathcal{F}$ is coercive and it attains its minimum value on K.

Note that if $\hat{u}$ minimizes $\mathcal{F}(.,\omega)$ on K, then $\hat{u}$ will be a periodic solution of (3.1) iff $\alpha(\omega) = 0$, from theorem 1.

Theorem 4. Suppose g, h obey (A1) - (A3) and p=2 in the definition of X. Assume h is continuously differentiable on $\mathbb{R}^n$ and let $D(u) = Dh(u)$ be its Jacobian matrix. If $\hat{u}$ minimizes $\mathcal{F}(u,\omega)$ on K, then $\hat{u}$ obeys (2.3) and

$$\omega^2\ddot{u}(t)-\omega(D(u)-AD(u)^TA^{-1})\dot{u}(t)-A(I+D(u)^TA^{-1})(Au+h(u))=0 \tag{3.14}$$

where the superscript T denotes a transpose.

Proof. When $\mathcal{F}(.,\omega)$ is Gateaux differentiable on K, then if $\hat{u}$ minimizes $\mathcal{F}(.,\omega)$ on K one has that

$$\langle D_u\mathcal{F}(\hat{u},\omega),\ w\rangle = 0$$

for all $w \in K$; or equivalently $D\mathcal{F}(\hat{u},\omega) = 0$ in K^*. Here $D_u\mathcal{F}(u,\omega)$ is the derivative of $\mathcal{F}$ with respect to u. To obtain (3.14) it must be computed. One can easily show that

$$\langle D_u\mathcal{F}_1(u),\ w\rangle = \langle A^{-1}\dot{u},\dot{w}\rangle$$
$$= -\langle A^{-1}\ddot{u},w\rangle \qquad \text{since } u,w \in K.$$

Similarly $\langle D\mathcal{F}_2(u),w\rangle = \langle A^{-1}\dot{w},h(u)\rangle + \langle A^{-1}\dot{u},\ D(u)w\rangle$

$$= \langle D(u)^TA^{-1}\dot{u} - A^{-1}D(u)\dot{u},w\rangle$$

and $\langle D_3\mathcal{F}(u),w\rangle = \langle Au\ w\rangle + \langle h(u),w\rangle + \langle u,D(u)w\rangle + \langle A^{-1}h(u),D(u)w\rangle$

$$= \langle Au+h(u),w\rangle + \langle A^{-1}(Au+h(u)),D(u)w\rangle$$

$$= \langle (I+D(u)^TA^{-1})(Au+h(u)),w\rangle.$$

Thus $D_u\mathcal{F}(u) = -\omega^2A^{-1}\ddot{u} + \omega(D(u)^TA^{-1} - A^{-1}D(u))\dot{u}$

$$+ (I+D(u)^TA^{-1})(Au+h(u)).$$

Thus the extremality condition (3.14) holds since A is non-singular.

Corollary. Suppose the conditions of theorem 4 hold and $(\hat{u},\hat{\omega})$ is a local minimum of $\mathcal{F}$ on $K\times[0,\infty)$. Then $(\hat{u},\hat{\omega})$ obeys (3.14) and

$$\langle A^{-1}\dot{u},\omega\dot{u}-h(u)\rangle = 0. \tag{3.15}$$

Proof. Since $(\hat{u},\hat{\omega})$ is a local minimum on $X\times[0,\infty)$ we must have $\frac{\partial\mathcal{F}}{\partial\omega}(\hat{u},\hat{\omega}) = 0$ and thus (3.15) holds at $(\hat{u},\hat{\omega})$.

It is worth noting that if (u,ω) obeys (3.1) then they obey (3.14) but, of course, the converse isn't necessarily true.

4. Examples.

(i) Consider the problem of finding periodic solutions of

$$\frac{dv}{dt} = Bv \tag{4.1}$$

where B is an $n\times n$ skew-symmetric real matrix. From linear spectral theory one knows that such equations have periodic solutions whose frequencies are related to the purely imaginary eigenvalues of B.

An analog of (3.1) is

$$\omega \frac{du}{dt} = \varepsilon I u + (B-\varepsilon I)u$$

where $\varepsilon>0$. Thus $A=\varepsilon I$ in (3.2) and one finds

$$\mathcal{F}(u,\omega) = \frac{\omega^2}{2\varepsilon}<u,u> - \frac{\omega}{\varepsilon}<\dot{u},(B-\varepsilon I)u> + \tfrac{1}{2}\varepsilon<Bu,Bu>$$

$$= \varepsilon^{-1}[\frac{\omega^2}{2}<\dot{u},\dot{u}> - \omega<\dot{u},Bu> + \tfrac{1}{2}<Bu,Bu>] \qquad (4.2)$$

as $<\dot{u},u> = 0$, for $u \varepsilon K$.

The variational problem of finding periodic solutions of (4.1) thus is reduced to the least squares problem of finding when (4.2) is zero.

(ii) Now consider the problem of finding periodic solutions of the model chemical reaction known as the Brusselator. (see Nicolis & Prigogine (4) Chapter 7).

Let $v_1(t) = X(t)/A$, $v_2(t) = AY(t)/B$, then the equations may be written

$$\frac{d}{dt}\begin{pmatrix} v_1 \\ v_2 \end{pmatrix} = \begin{pmatrix} 1 - (B+1)v_1 + Bv_1^2 v_2 \\ A^2(v_1 - v_1^2 v_2) \end{pmatrix} \qquad (4.3)$$

It is known (see Chin and Zhen (2)) that when $B>1+A^2$ this system has a unique periodic solution in the positive quadrant.

An analog of (3.1) in this case is

$$\omega \frac{d}{dt}\begin{pmatrix} u_1 \\ u_2 \end{pmatrix} = \begin{pmatrix} 1 & 0 \\ 0 & 1 \end{pmatrix}\begin{pmatrix} u_1 \\ u_2 \end{pmatrix} + \begin{pmatrix} 1-(B+2)u_1 + Bu_1^2 u_2 \\ A^2 u_1(1-u_1 u_2) - u_2 \end{pmatrix} .$$

Here $A=I$ in (3.2) and the last vector is $h(u)$.

With this splitting, one finds that

$$\mathcal{F}(u,\omega) = \tfrac{1}{2}<\omega\dot{u}-u-h(u),\omega\dot{u}-u-h(u)>$$

is again the least squares functional for this nonlinear problem.

(iii) This method need not lead to the standard least squares-functional for (3.1). To see this consider the Field-Noyes model for the Belousov-Zhabotinski reaction. It may be written

$$\frac{d}{dt}\begin{pmatrix} u \\ v \\ w \end{pmatrix} = \begin{pmatrix} s & s & 0 \\ 0 & -s^{-1} & s^{-1}f \\ 0 & 0 & -a \end{pmatrix}\begin{pmatrix} u \\ v \\ w \end{pmatrix} - \begin{pmatrix} su(v+qu) \\ s^{-1}uv \\ 0 \end{pmatrix}$$

where s,a,f and q are positive parameters, see Tyson (5) Chapter III. In this chapter he also shows that this system has periodic solutions under certain conditions.

An analog of (3.1) for this system is

$$\omega\frac{d}{dt}\begin{pmatrix} u_1 \\ u_2 \\ u_3 \end{pmatrix} = \begin{pmatrix} s & 0 & 0 \\ 0 & s^{-1} & 0 \\ 0 & 0 & 1 \end{pmatrix}\begin{pmatrix} u_1 \\ u_2 \\ u_3 \end{pmatrix} + \begin{pmatrix} su_2(1-u_1)-squ_1^2 \\ s^{-1}u_3f-s^{-1}u_2(u_1+2) \\ au_1-(a+1)u_3 \end{pmatrix}$$

$$= Au + h(u)$$

This time $A^{-1} = \mathrm{diag}\,(s^{-1},s,1)$, and the functional $\mathcal{F}$ given by (3.10) becomes

$$\mathcal{F}(u,\omega) = \frac{\omega^2}{2}<A^{-1}\dot{u},\dot{u}> - \omega<A^{-1}\dot{u},h(u)>$$

$$+ \tfrac{1}{2}<Au+h(u),\ A^{-1}(Au+h(u))>.$$

with A, A^{-1} and h as above. This functional has at most quartic expressions in the u_i's in the integrand.

For each of these problems we have converted the original question of finding periodic solutions of an autonomous o.d.e. into an optimization problem of minimizing a functional on an appropriate space of functions. To actually compute these periodic oscillations one has to solve these variational problems, either analytically or numerically. This will be described elsewhere.

REFERENCES

1. Brezis, H. and Ekeland, I., "Un principle variational associé à certaines équations paraboliques. Le cas independant du temps." C. R. Acad. Sc. Paris 282 (1976) pp. A-971-974.

2. Chin, Y-S and Zhen, X.,"Qualitative Investigation of the Differential Equation of Brusselator in Biochemistry," Kexue Tongbao 25 (1980) pp. 273-276.

3. Ekeland, I and Temam, R., Convex Analysis and Variational Problems North-Holland, Amsterdam, 1976.

4. Nicolis, G. and Prigogine, I., Self-Organization in Nonequilibrium Systems, (1977), J. Wiley.

5. Tyson, J. J. The Belousov-Zhabotinskii Reaction, Lecture Notes in Biomathematics V. 10, (1976), Springer-Verlag.

A NUMERICAL ANALYSIS OF WAVE PHENOMENA IN A REACTION DIFFUSION MODEL

E. J. Doedel
Applied Mathematics 217-50
California Institute of Technology
Pasadena, California 91125 USA

J. P. Kernevez
Département de Génie Informatique
Université de Technologie de Compiègne
60206 Compiègne, France

Abstract. A numerical analysis of wave phenomena in a reaction diffusion model is given. These take place in the neighborhood of a singularity in the underlying ordinary differential equations governing the spatially uniform solutions. Many of the states, whether traveling waves, stationary waves (patterns) or uniform states, coexist and are asymptotically stable. The number of states and their complexity increases as a size parameter becomes larger.

1. The S A System. We consider a system of differential equations that models an enzyme-catalyzed reaction. There are two substrates in this model, viz. an *activator* A and an *inhibitor* (at high concentration) S. The enzyme is imbedded in a membrane where the reactions take place, and an unstirred layer separates the membrane from a well-stirred solution where S and A are held at a constant level. A model of such a system is given by (cf. [4])

$$\begin{aligned} s_t &= s_{xx} - \mu[\rho R(s,a) - (s_0 - s)], \\ a_t &= \beta a_{xx} - \mu[\rho R(s,a) - \alpha(a_0 - a)], \end{aligned} \tag{1.1}$$

where s and a denote the concentrations of two chemical species S and A inside the membrane, and s_0 and a_0 are the (constant) concentrations of S and A in the outside reservoir. The reaction rate is given by

$$R(s,a) = \frac{a}{\kappa_1 + a} \quad \frac{s}{1 + s + \kappa_2 s^2}.$$

There are several known examples of such *activator-inhibitor* or $S - A$ systems. For example these equations constitute a reasonably realistic model of the reaction

$$Oxaloacetate \quad + \quad NADH \quad \rightarrow \quad Malate \quad + \quad NAD,$$

catalyzed by *Malate Deshydrogenase.* In this case appropriate values for the constants κ_1 and κ_2 in the reaction rate are $\kappa_1 = 3.4$ and $\kappa_2 = 0.023$. The value of the constant α depends on the type of membrane: for cellophane $\alpha = 0.2$ and for gelatine $\alpha = 0.5$. Below we shall take $\alpha = 0.2$. Throughout we also set $s_0 = 145$, $\mu = 3$, $\rho = 210$, and $\beta = 5$. We treat a_0 as a free parameter.

For brevity we shall also write system (1.1) in the form

$$u_t = Du_{xx} + f(u, \lambda), \tag{1.2}$$

where $u \equiv (s, a)$, $D \equiv \binom{1\,0}{0\,\beta}$, and where λ is the free parameter (here a_0).

2. Bifurcation Behavior of the ODE. The ordinary differential equations governing the spatially uniform states, i.e., the equations (1.1) without diffusion terms s_{xx} and βa_{xx}, are (after scaling)

$$\begin{aligned} \frac{ds}{dt} &= (s_0 - s) - \rho R(s, a), \\ \frac{da}{dt} &= \alpha(a_0 - a) - \rho R(s, a), \end{aligned} \tag{2.1}$$

while for the general system (1.2) the associated ODE is

$$\frac{du}{dt} = f(u, \lambda). \tag{2.2}$$

Given our choices of κ_1, κ_2, s_0, ρ, and α above, there is one free parameters, viz. a_0. The corresponding one-parameter bifurcation diagram is shown in Figure 1.

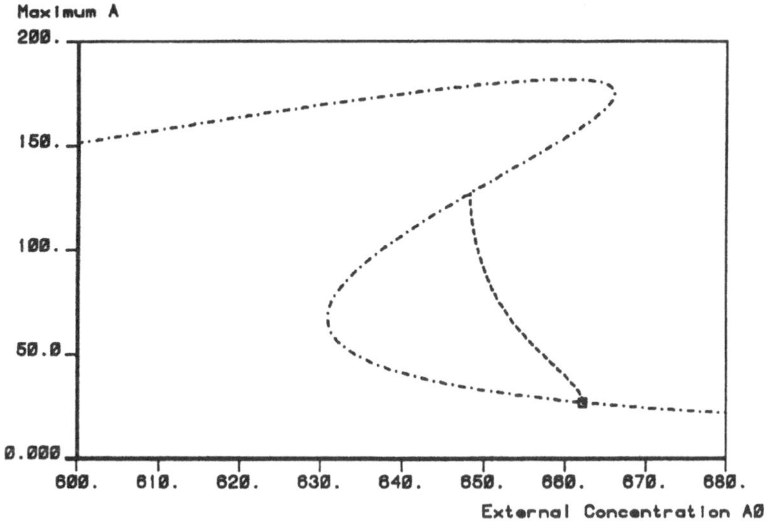

Figure 1 Bifurcation Diagram of the ODE (2.1).

The S-shaped curve (·—·—·) denotes steady states of (2.1) (i.e., the spatially uniform states of (1.1)), the open square (□) is a Hopf bifurcation, and the emanating branch of periodic solutions is indicated by a dashed line (– – –). The left termination point of the branch of periodic solutions is not a Hopf bifurcation, rather the branch terminates in an orbit of infinite period containing one saddle point (a homoclinic orbit). In the diagram the location of this saddle point on the branch of steady states and the point representing the homoclinic orbit actually coincide.

Important for our objective of finding wave solutions to (1.1) are the coexistence of different spatially uniform states and the presence of oscillatory solutions to (2.1), as seen in Figure 1. The homoclinic orbit is also of significance when solitary waves on an infinite spatial interval are sought (For the present model the structure of these is analyzed in Section 8 of [1]).

If a second parameter is varied (e.g., s_0) then the two limit points in Figure 1 can be made to coincide in a cusp singularity. Similarly the Hopf bifurcation can be merged with one of the limit points, giving the singularity analyzed in [2]. In a two parameter diagram the curve of Hopf bifurcation points and a curve of homoclinic orbits would emanate from this latter singularity. For the equations (2.1) such a two-parameter analysis is given in Section 7 of [1].

One may further merge the two singularities referred to above in a yet higher order singular point by varying a third parameter. But we have not determined the minimal conditions that ensure an unfolding containing the structure that we shall describe below.

3. Traveling Waves on an Infinite Interval. First we investigate the existence of traveling waves to (1.1) in general, i.e., on an infinite interval. If we look for traveling wave solutions to (1.1) of the form $s(x,t) = s(x-ct)$, $a(x,t) = a(x-ct)$, then (1.1) reduces to two coupled second order ordinary differential equations, viz.,

$$s'' + cs' - \mu[\rho R(s,a) - (s_0 - s)] = 0,$$
$$a'' + \frac{c}{\beta}a' - \frac{\mu}{\beta}[\rho R(s,a) - \alpha(a_0 - a)] = 0,$$

which we can rewrite in equivalent first order form

$$\begin{aligned} s' &= q, \\ a' &= r, \\ q' &= -cq + \mu[\rho R(s,a) - (s_0 - s)], \\ r' &= -\frac{c}{\beta}r + \frac{\mu}{\beta}[\rho R(s,a) - \alpha(a_0 - a)]. \end{aligned} \tag{3.1}$$

More generally, for (1.2) we have the *reduced system*

$$\begin{aligned} u' &= v, \\ v' &= -D^{-1}[cv + f(u,\lambda)]. \end{aligned} \tag{3.2}$$

The fixed points of (3.1) satisfy $q = r = 0$, and

$$\begin{aligned} \rho R(s,a) - (s_0 - s) &= 0, \\ \rho R(s,a) - \alpha(a_0 - a) &= 0, \end{aligned} \tag{3.3}$$

while for the general case (3.2) the fixed points satisfy $v = 0$ and

$$f(u,\lambda) = 0, \tag{3.4}$$

and these are independent of the wave speed c. Furthermore, these fixed points are precisely those of (2.1) and (2.2) respectively, i.e., of the system without the diffusion terms. Thus the fixed point diagram of (3.1) (with a_0 as parameter) coincides with the fixed point branch of (2.1) given in Figure 1. But the Jacobian of (3.1) evaluated along the fixed point branch depends on c, and hence the Hopf bifurcation in Figure 1 need not be present for (3.1).

However, one can show that the bifurcation diagram for (3.1) will approach that of Figure 1 when the wave speed c becomes large. In particular the Hopf bifurcation in Figure 1 guarantees the existence of traveling waves of (3.1) and the homoclinic orbit guarantees the existence of a solitary wave, if c is sufficiently large.

As we show now, the system (3.1) also gives rise to Hopf bifurcation for relatively small c, so that correspondingly the parabolic system (1.1) allows low speed traveling waves. A particular bifurcation diagram for (3.1) is shown in Figure 2a. Here c is fixed at $c = 0.05$. As noted above, the fixed point ($\equiv$ spatially uniform) structure is identical to that of Figure 1. The Hopf bifurcation takes place at a somewhat different location and the emanating branch of wave solutions behaves differently also. However, as in Figure 1, the branch terminates in a homoclinic orbit, where the wave length tends to infinity (We use the terms ***wave solution*** and ***wave length*** rather than ***periodic solution*** and ***period*** respectively, in order to emphasize that we deal with periodicity in the independent variable $z \equiv x - ct$).

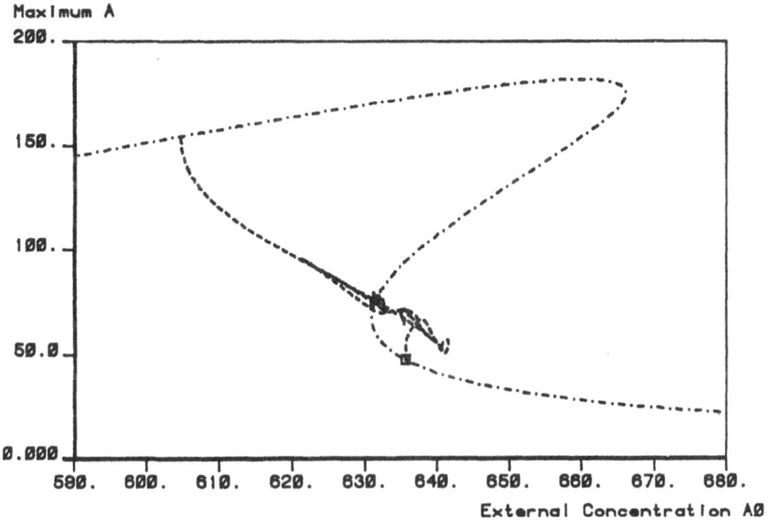

Figure 2a Bifurcation Diagram for Equation (2.2) with $c = 0.05$.

From Figure 2a we conclude that for a_0 between approximately 605 and 640 there are traveling waves with wave speed equal to $c = 0.05$, and for a particular value of a_0 near $a_0 = 605$ there exist a solitary wave having that speed. We cannot as yet draw any conclusions on the stability of these waves, since there is no direct relation between stability of solutions of the ODE (3.1) and the corresponding wave solutions to (1.1).

Actually, the bifurcation behavior in Figure 2a for $c = 0.05$ is considerably more complicated than that in Figure 1 for the ODE. Some of this behavior is shown enlarged in Figure 2b.

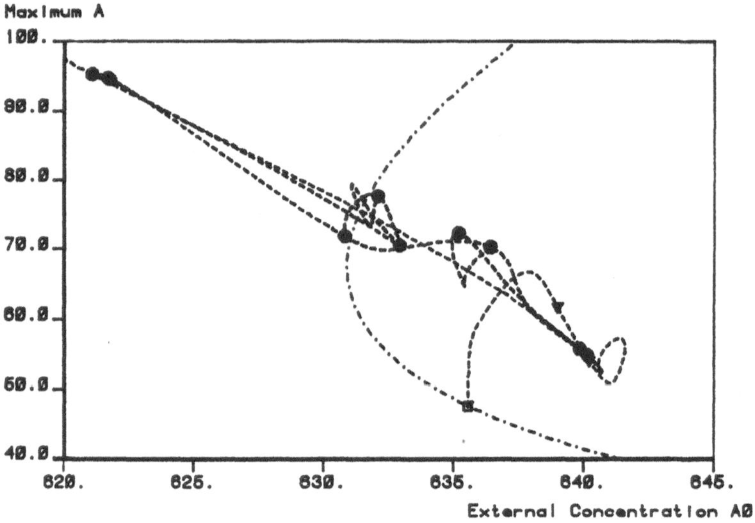

Figure 2b Blow up of Figure 2a.

The •'s denote "wave length doubling" bifurcations, the ▿ represent a torus bifurcation, and, as before, the □ is a Hopf bifurcation. Several complete branches originating at wave length doubling bifurcations were traced out numerically. These branches contain limit points and further wave length doubling bifurcations. In fact, there appear to be *nested sequences* of such bifurcations, resulting in very complicated spatially periodic behavior.

4. Traveling Waves on a Ring. We now turn to our primary objective: the construction of wave solutions for varying a_0 in a fixed geometry. Specifically, *we fix the value of the wave length* from here on: $L = 22$. From the numerical results schematically represented in Figure 2a,b we can construct traveling wave solutions having this wave length in a systematic way: Along each of the branches of traveling waves we monitor the wave length L and we accurately determine those having $L = 22$. If necessary we put two or more waves from Figure 2a,b "in series" in order to obtain the desired wave length. We found five distinct waves using this procedure. For larger L the number of waves found would increase rapidly, while for L too small no such waves would exist.

Each of the waves found is continued by "freezing" L at $L = 22$ and by "freeing" the wave speed c in its place. The parameter a_0 remains free as before. The results of these computations are shown projected onto the $a_0 - c$ plane in Figure 3. For example, the a_0-values at which the five waves with $L = 22$ were found in Figure 2a,b can be located in Figure 3 by drawing a horizontal line at $c = 0.05$. In Figure 4 a particular wave solution is shown at two different positions as it travels around the ring. Time integration indicates that this wave is asymptotically stable.

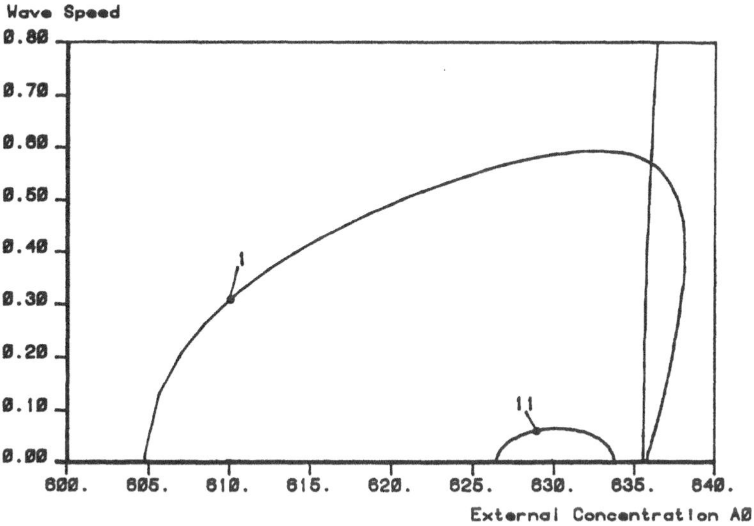

Figure 3 Loci of Traveling Waves on the Ring ($L = 22$).

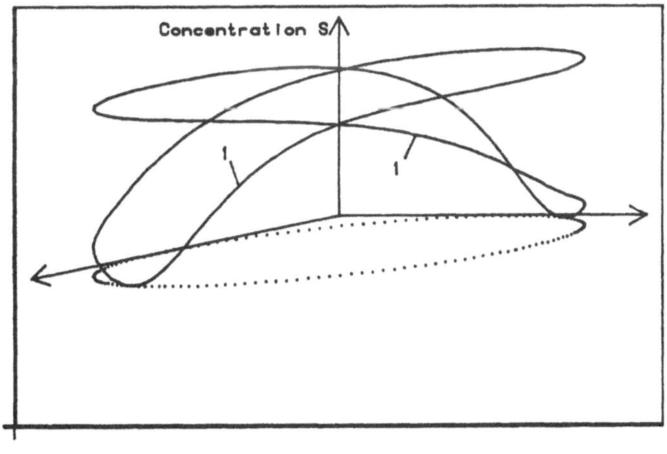

Figure 4 A Traveling Wave on the Ring ($a_0 = 610$, $c = 0.3086$).

5. Stationary Waves on a Ring. In addition to branches of traveling wave solutions to (1.1) on the ring, there are also several branches of ***stationary waves***, i.e., patterned solutions. Like the traveling waves these are non-unique, because an arbitrary phase shift (rotation around the ring) can be applied to them. But, unlike the traveling waves, they remain in the same position given initially (assuming stability).

To determine particular stationary waves to be used subsequently as starting points for the computation of entire branches of such waves, we have used the following two procedures:

(i) There are five distinct "traveling" waves in Figure 3 with wave speed $c = 0$, and these can serve as starting point for computing branches of stationary waves. More accurately, the curves in Figure 3 can be mirrored with respect to the a_0 axis to yield curves of traveling waves with negative wavespeed (This reflection principle can be verified using (3.1)). Thus the curves in Figure 3 actually cross the $c = 0$ line, and these crossings turn out to be bifurcation points also. Solutions on the bifurcating branches have zero wave speed, i.e., they are stationary waves.

(ii) All traveling waves from Figure 3 with $a_0 = 610, 620, 630, 636$, and 638 were integrated in time using the full partial differential equations (1.1). Many are unstable and ultimately settle into some other mode of behavior. In particular many unstable traveling waves evolve into a stable stationary wave. For example, a traveling wave at $a_0 = 630$ (labeled 11 in Figure 3) is unstable, and after time integration approaches the stationary wave shown in Figure 5.

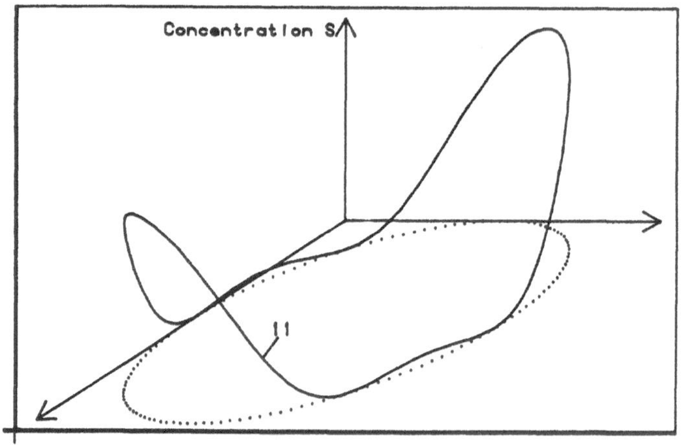

Figure 5 A Stationary Wave (Pattern) on the Ring ($a_0 = 630$, $c = 0$).

All stationary wave branches thus found are shown superimposed on the previously found uniform states and traveling waves in Figure 6. Of course we can no longer use the wave speed c as vertical axis as was done in Figure 3. Instead we use $\max_{z\in[0,L]} a(z)$ to represent the various solutions.

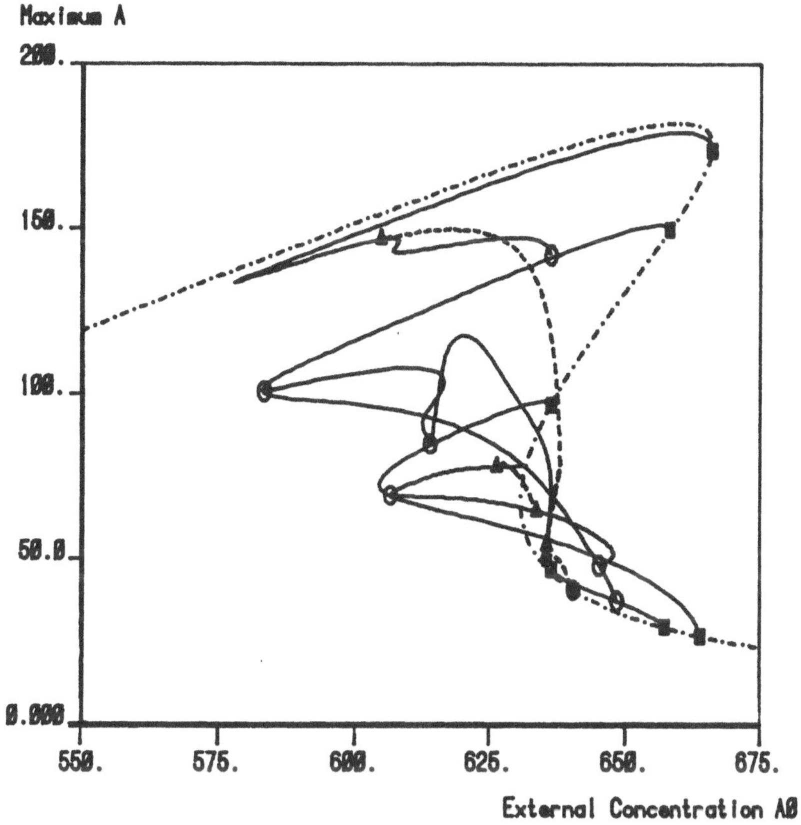

Figure 6 **The Complete Bifurcation Diagram.**

In the diagram we have used solid curves (———) to represent the stationary waves, while traveling waves (– – –) and uniform states (· – · – ·) are shown with the previously used curve types. The various bifurcations are indicated as follows: stationary waves from uniform states (■), stationary waves from traveling waves (▲), stationary waves from stationary waves (○), traveling waves from uniform states (●).

To illustrate the rich variety of solutions, we show several distinct stationary waves in Figure 7. All of these are for the same value of the bifurcation parameter ($a_0 = 610$), and time integration indicates that all are stable. Moreover they coexist with the stable traveling wave already shown in Figure 4. In addition there is also the spatially uniform state which is also stable at $a_0 = 610$.

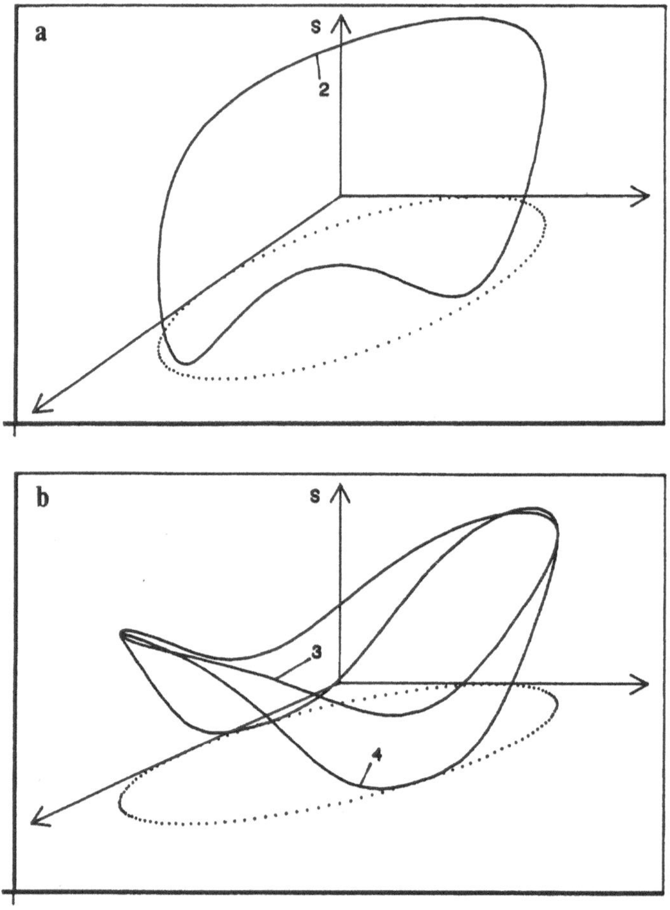

Figure 7a,b Stable Stationary Waves on the Ring ($a_0 = 610$, $c = 0$).

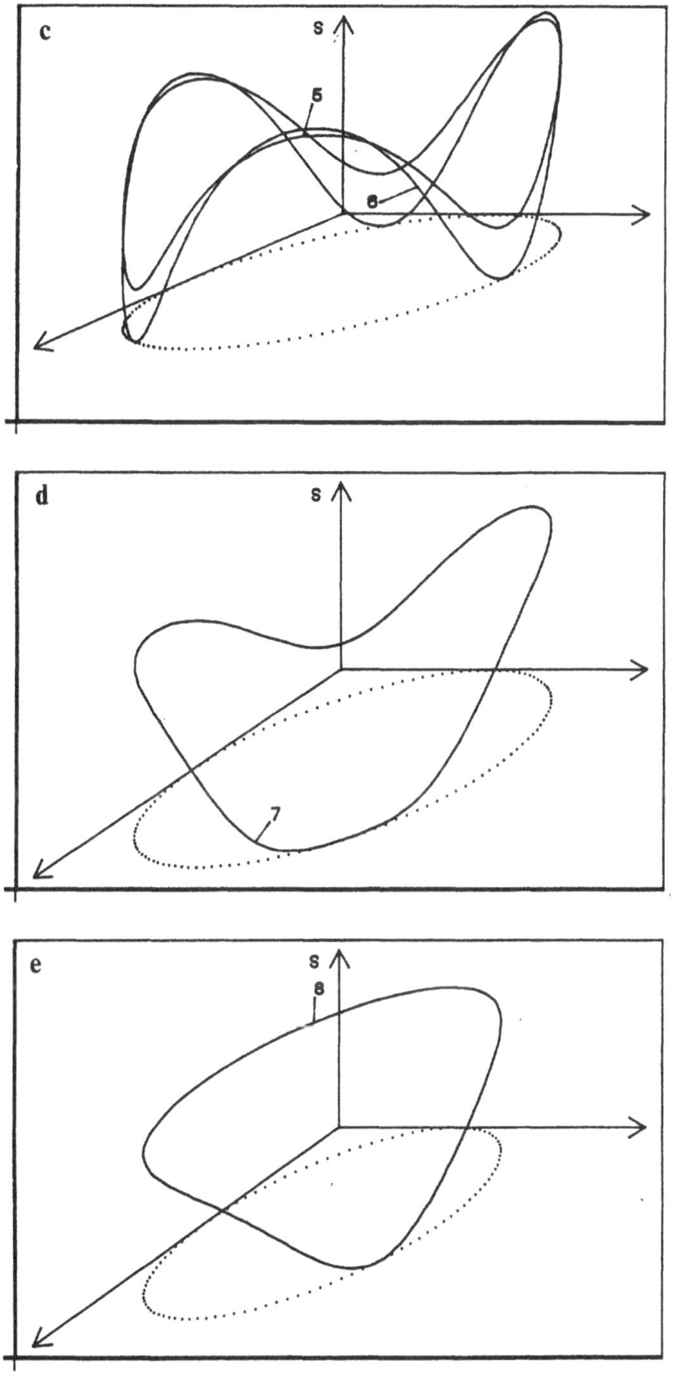

Figure 7c,d,e **Stable Stationary Waves on the Ring ($a_0 = 610$, $c = 0$).**

6. Numerical Methods. All computations were done using an extended version of AUTO [1] specially adapted for wave computations. Since traveling waves in (1.2) are periodic solutions of the reduced systems (3.2), we can make use of the basic routines for computing stable and unstable periodic solutions incorporated in the software. The branches of traveling waves on the ring correspond to 2-parameter curves of orbits of fixed period, the parameters being a_0 and c, and the "period" $L = 22$. This type of computation is also standard in AUTO.

Stationary waves are computed in exactly the same fashion as traveling waves. More specifically, we treat the wave speed c as unknown, even though it equals zero along stationary wave branches. This allows an extra equation to be imposed, viz. the same phase condition (cf. Section 4 in [1]) used for traveling waves. Note that a phase condition is also necessary for stationary waves on the ring, because of their rotational invariance.

A time integration scheme for parabolic systems of the form (1.2) is also included in the extended software. In particular it allows the spatial periodicity condition required in our ring geometry. Taking a step in time in this evolution computation is posed as a boundary value problem. This may result in some loss of efficiency, but it also makes available the high order discretization in space and the adaptive step size and mesh selection of AUTO. Thus our time integration scheme uses an adaptive mesh in *both* time *and* space.

In the extended version the system (3.2) is generated automatically: one need only specify the vector function $f(u, \lambda)$ in (1.2) (and optionally its partial derivatives) and the diagonal diffusion matrix D. The same user supplied "*driver*" can in fact be used in all stages of the analysis, including the time integration of (1.2). In the latter case also, the specific system required for taking a time step is generated automatically. One need only set a "switch" to indicate whether one seeks a steady state analysis (3.4), periodic solutions of (2.2), traveling or stationary waves using (3.2), or a time integration of (1.2). Furthermore the output data are compatible, so that, for example, one can start the computation of a branch of traveling or stationary waves using output data generated by a time integration, and conversely.

Also, the most computer time consuming part of the code (namely the *condensation of parameters*, cf. Section 3 in [1]) was rewritten in order to take full advantage of vector computer architectures. In fact, the entire program was run on an FPS 164 array processor from an IBM 4341 host computer. With this set up the computation of any branch of solutions required at most a few minutes.

Related Work. Several of the the underlying ideas concerning the existence of traveling waves can be found in [6]. For a recent review of literature on pattern formation see [5]. Numerical methods are described in [1] and are based on general techniques in [3]. A numerical analysis of wave phenomena in a nonlinear reaction diffusion equation is also given in [7].

Acknowledgement. The first author wishes to express his gratititude to Nguyen Thanh Long (Montréal) for the many hours he put into developing the interactive graphics program with 3D capability which can be used with AUTO.

References

[1] E. J. Doedel, J. P. Kernevez, Software for continuation problems in ordinary differential equations with applications, California Institute of Technology, 1985.

[2] J. P. Keener, Infinite period bifurcation and global bifurcation branches, SIAM J. Appl. Math. 41, 1981, 127-144.

[3] H. B. Keller, Numerical solution of bifurcation and nonlinear eigenvalue problems, in: Applications of Bifurcation Theory, P. H. Rabinowitz, ed., Academic Press 1977, 359-384.

[4] J. P. Kernevez, Enzyme Mathematics, North-Holland Press, 1980.

[5] S. A. Levin, L. A. Segel, Pattern generation in space and aspect, SIAM Review 27, No. 1, 1985, 45-67.

[6] H. G. Othmer, Nonlinear wave propagation in reacting systems, J. Math. Biol. 2, 1975, 133-163.

[7] P. Raschman, M. Kubíček, M. Marek, Waves in distributed chemical systems: Experiments and computations, in: New Approaches to Nonlinear Problems in Dynamics, P. H. Holmes, ed., SIAM 1980, 271-288.

On a Nonlinear Hyperbolic Equation Describing Transmission Lines, Cell Movement, and Branching Random Walks

Steven R. Dunbar[1] and Hans G. Othmer[2]

Department of Mathematics

University of Utah

Salt Lake City, Utah, USA

June 26, 1985

This is a preliminary and expository report on a nonlinear hyperbolic equation that arises from a variety of distinct phenomena. We derive the equation

$$\epsilon^2 v_{tt} + (1 + g(v))v_t = k^2 v_{xx} + f(v) \tag{1}$$

as the equation for the voltage along a transmission line with nonlinear shunt conductance and a series inductance along the length of the line, from simple models of movement and reproduction in tissue cells and one celled organisms, and from a mathematical treatment of a branching random walk. In addition, with the proper scaling and choice of f and g this nonlinear hyperbolic equation can be viewed as the equation that describes a continuum of coupled van der Pol oscillators. In equation (1) the value of ϵ^2 need not be small, but the choice of the notation ϵ^2 suggests analogies with other well known nonlinear partial differential equations, and we will mention some of these analogies below. The purpose of this report is to briefly explain and motivate all of these derivations and to present some basic results about the solutions of this equation. In addition, the purpose is to show how the probabilistic interpretation of the equation arising out of the branching random walk helps in the understanding and motivation of the results. Detailed proofs of the new results will be presented elsewhere.

The organization of the paper is as follows. In the first section we derive equation (1) from a model of a nonlinear transmission line by an application of

[1]Supported in part by NSF Grant DMS–8301840

[2]Supported in part by NSF Grant DMS–8301840 and NIH Grant GM–29123

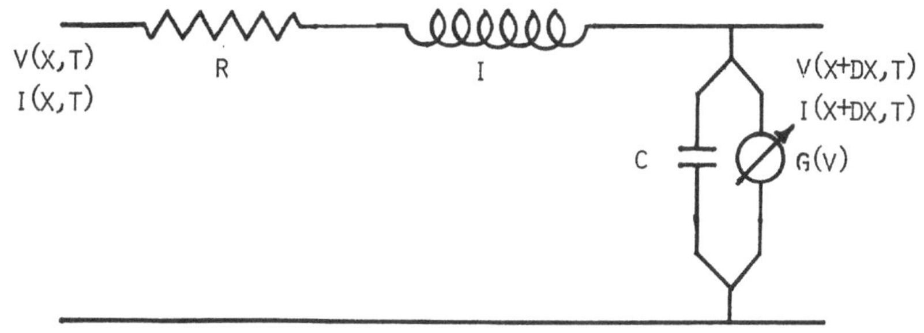

Figure 1: Nonlinear Transmission Line Circuit Equivalent

Kirchoff's laws. In the second section we present some basic facts about the movement of cells and present a model for this movement which leads to equation (1). In the third section we consider a branching Poisson random walk on the real line, and show that the function that defines the rightmost penetration of the process also satisfies equation (1). In the fourth section, we state some theorems asserting the existence of traveling wave solutions of certain speeds for equation (1). These results show that the minimal speed is an asymptotic speed of propagation in a certain sense. In the final section, we use the probabilistic interpretation to produce a heuristic argument that shows the possibility of persistent shocks in the solutions.

1. Nonlinear Transmission Lines

Consider a nonlinear transmission line with distributed inductance and resistance along the line, and distributed shunt capacitance and distributed nonlinear shunt conductance. The equivalent circuit appears in Figure 1. In the figure R is the series resistance along the length of the line, measured in ohms per meter and I is the distributed series inductance measured in henrys per meter. The distributed shunt capacitance is C, measured in farads per meter. The nonlinear shunt conductance is a function $G(v)$ of the voltage v through the shunt and is measured in mho-volts per meter. The voltage at a point x on the line at time t is denoted by $v(x,t)$.

By making use of the figure, one can write down Kirchoff's laws for voltage and current at points x and $x + dx$. Then letting $dx \to 0$ as in the derivation of the classical telegrapher's equation from mathematical physics [13], one arrives at the equation

$$LC\frac{\partial^2 v}{\partial t^2} + \frac{\partial}{\partial t}(LG(v) + RCv) = \frac{\partial^2 v}{\partial x^2} - RG(V).$$

By dividing through by RC and setting

$$\epsilon^2 = L/R, \quad g(v) = LG'(v)/RC, \quad k^2 = 1/RC, \quad f(v) = -G(v)/C,$$

we arrive at a normalized form of the equation, namely

$$\epsilon^2 v_{tt} + (1 + g(v))v_t = k^2 v_{xx} + f(v)$$

where as usual, subscripts denote partial derivatives. The derivation is general and does not place any requirements on the nonlinearity $G(v)$, other than that it be at least differentiable. The precise form of the nonlinearities $f(v)$ and $g(v)$ depend on the particular application being considered. An example in which they are of the form $av - bv^3$ and its derivative, respectively, will be considered later. The other derivations of the equation, and the available circuit components motivate this specification of the nonlinear shunt conductance.

The motivation for studying this particular transmission line comes from generalizing the Hodgkin-Huxley and Fitzhugh-Nagumo models of the nerve axon to explicitly include induction along the length of the nerve axon, [11] [15]. Although it is probably the case that in a normal neuron inductive effects are negligible, [15], it is possible that in extreme cases the effects may be noticeable. Nevertheless, using the ideas and circuit elements explained in [10] it is possible to construct a discrete version of this distributed transmission line. This provides a physical reason for the study of this equation on the one hand, and the actual circuit provides a convenient analog computer for the investigation of the equation on the other hand. We are currently investigating this circuit and details of the construction, and the results obtained will published elsewhere.

2. Cell Movement

It is common to assume that the movements of organisms can be modeled by a random walk [14]. However, within the last 15 years there have been several experimental studies that demonstrated the existence of significant correlations in the motions of several different types of cells. Investigations of the movements of a variety of tissue cells and one-celled organisms have shown that the movement of these cells is actually a sequence of "runs" and "turns". That is, the cell proceeds in a linear movement of some duration, typically of a random length of time at some characteristic speed, and then the cell changes direction. The change of direction is not uniformly distributed on the sphere in the case of three dimensional motion, or the circle in the case of two dimensional motion, but is "biased" toward the direction of the previous run. This bias may occur for a variety of reasons, among them taxis in the direction of a favorable substance. When such "persistence" exists the direction of a run is positively correlated with the direction of the previous run, and for this reason the random walk executed by the cell is often called a correlated random walk. This differs from the simplest random walk in which the direction chosen at each step is uniformly distributed on the sphere or the circle, and in which the steps are of unit length.

For instance, to search for food or to escape an unfavorable environment, *E. coli* alternates two basic behavioral modes, one a more or less linear motion called a run, and the other a highly erratic motion called tumbling, the purpose of which is to reorient the cell. Run times are typically longer than the time spent tumbling, and when bacteria move in a favorable direction, the run times are increased further. During a run the bacteria move at approximately constant speed in the most recently chosen direction. New directions are generated during tumbles, and when bacteria move in a unfavorable direction the run length decreases and the relative frequency of tumbling increases. The distribution of new directions is not perfectly uniform on the unit sphere, but has a slight bias in the direction of the preceding run. The effect of alternating these two modes and in particular, of increasing the run length when moving in a favorable direction, is that a bacterium executes a three dimensional random walk with drift in a favorable direction when observed on a sufficiently long time scale [1]. Gail and Boone [3] showed that mouse fibroblasts exhibit persistence in their direction of motion when they are observed over successive time intervals of 2.5 hours or less, but this correlation is not observed over longer time intervals. Hall [6] studied the motion of *Dictyostelium discoideum* amoeba and detected correlations in the angle of turn between successive steps, but no correlation in the length of successive steps. Hall and Peterson report similar findings by for human granulocytes [7].

In addition to providing biological background for the mathematical model developed in this section, these studies provide a biological motivation for considering the branching Poisson random walk treated in the following section.

A simple model for this sort of movement can be made by considering a "velocity jump process". Let $p(x,v,t)$ be the density of cells or particles at position $x \in R^3$ moving with velocity $v \in R^3$ at time t. Let $n(x,t) = \int p(x,v,t)\,dv$ be the number density of cells at x, whatever their velocity. Let the rate of change of p due to reaction or reproduction be given by $r(n)d(v)$, so that new cells with a velocity distribution $d(v)$ are born at a rate depending only on the number density. Let F denote the external force per cell acting on cells. We assume that the velocity of a cell is changing in a random way, just as the cells stop and turn according to the non-uniform turning distribution. The random velocity changes are assumed to be governed by a Poisson process of intensity λ, which means that the rate at which cells enter the phase space volume centered at (x,v) is given by

$$-\lambda p + \lambda \int T(v,v')p(x,v',t)\,dv'.$$

The redistribution kernel $T(v,v')$ gives the probability of a change in velocity from v' to v; thus $T(v,v') \geq 0$ and is normalized so that

$$\int T(v,v')\,dv = 1.$$

These conditions merely express the fact that no cells are lost during the process of changing velocity. With these hypotheses it follows that the evolution of p is governed by the partial differential equation

$$\frac{\partial p}{\partial t} + \nabla_x \cdot (vp) + \nabla_v \cdot (Fp) = r(n)d(v) - \lambda p + \lambda \int T(v,v')p(x,v',t)\,dv'.$$

For the present we assume that $F = 0$.

For most purposes one does not need complete information about p but only information about the first few moments of p, say the number density $n(x,t)$ introduced earlier and the mean velocity. Thus, if we integrate the equation with respect to v we obtain

$$\frac{\partial n}{\partial t} + \nabla_x \cdot (nu) = r(n)$$

where $nu = \int pv\,dv$. Similarly, multiplying by v and integrating over v gives an equation for the first v moment of p

$$\frac{\partial (nu)}{\partial t} + \nabla \cdot \int pvv\,dv = \lambda \int_{v'} \int_{v} T(v,v')vp(x,v',t)\,dv\,dv' - \lambda nu + \mu r(n)$$

where μ is the mean of the distribution $d(v)$ of the newborn cells. For simplicity we now restrict ourselves to one space dimension, although the theory can be developed in several space dimensions with some additional complexities. We assume that the speed v_0 of a cell is constant and we take the redistribution kernel $T(v, v') = \delta(v + v')$; thus only direction reversals are permitted. We also take the distribution of velocities of newborn cells to be

$$d(v) = \begin{cases} +v_0 & \text{with probability } \frac{1}{2} \\ -v_0 & \text{with probability } \frac{1}{2}, \end{cases}$$

which implies that $\mu = 0$. Then $v = \pm v_0$, $n(x,t) = (p^+ + p^-)$, and $nu = v_0(p^+ - p^-)$, where $p^\pm = p(x, \pm v_0, t)$ etc. Furthermore,

$$\nabla \cdot \int pvv\, dv = v_0^2 \frac{\partial}{\partial x}(p^+ + p^-).$$

The double integral term reduces to $-\lambda v_0(p^+ - p^-)$ and the balance equations become

$$\frac{\partial}{\partial t}(p^+ + p^-) + v_0 \frac{\partial}{\partial x}(p^+ - p^-) = r(n)$$

$$v_0 \frac{\partial}{\partial t}(p^+ - p^-) + v_0^2 \frac{\partial}{\partial x}(p^+ + p^-) = -2\lambda v_0(p^+ - p^-).$$

These two first order partial differential equations are easily reduced to the following single second order differential equation for n:

$$\frac{\partial^2 n}{\partial t^2} + (2\lambda - r'(n))\frac{\partial n}{\partial t} = v_0^2 \frac{\partial^2 n}{\partial x^2} + v_0 r(n).$$

This is clearly of the form of equation (1).

3. Branching Poisson Walks

In this section we discuss a probabilistic derivation of the nonlinear hyperbolic equation. To do this we must first discuss simple Poisson random walks as treated in [4], [8], [9].

Imagine a particle which is moving along the real line at constant speed 1, and suppose that the starting direction of the particle at time 0 is randomly chosen to be left with probability 1/2 and right with probability 1/2. Let the starting velocity be denoted by ϵ_0; then

$$\epsilon_0 = \begin{cases} +1 & \text{with probability } \frac{1}{2} \\ -1 & \text{with probability } \frac{1}{2}. \end{cases}$$

Suppose that the particle moves a random amount of time T_R, and then reverses its direction of motion. We assume that the distribution of the time T_R to reversal is exponential with parameter a, that is ,

$$Pr(T_R \in ds) = ae^{-as}ds.$$

The particle moves in this new direction a random amount of time, distributed with the same probability density, and then reverses direction. The number of reversals $N(t)$ on the time interval $[0,t]$ is a Poisson process with parameter a so that

$$Pr(N(t) = j) = e^{-at}(at)^j/j!.$$

The velocity at time t is $v(t) = \epsilon_0(-1)^{N(t)}$ and the position at time t is $x(t) = \epsilon_0 \int_0^t (-1)^{N(s)}\,ds$. Then one may think of $T(t) = \int_0^t (-1)^{N(s)}\,ds$ as a "randomized time" or a "scattered time". We call the resulting random walk on the real line a Poisson random walk.

The following theorem is due to Kac [8], and Kaplan [9].

Theorem 1 *Let v be a twice continuously differentiable function on the interval $(-t,t)$. Define $u(t) = E(v(\epsilon_0 T(t)))$, where $T(t) = \int_0^t (-1)^{N(s)}\,ds$ and $E(\cdot)$ is the expectation with respect to the Poisson process $N(t)$. Then*

$$\begin{aligned} u''(t) + 2au'(t) &= E(v''(T(t))) \\ \lim_{t\to 0} u(t) &= v(0) \\ \lim_{t\to 0} u'(t) &= 0. \end{aligned}$$

While the significance of this theorem may not be immediately apparent, it leads to the following interesting connection between the wave equation and the telegrapher's equation.

Theorem 2 *Suppose that $v(x,t)$ is a twice continuously differentiable solution of the wave equation*

$$v_{tt} = k_0^2 v_{xx}.$$

Then $u(x,t) = E(v(x,T(t)))$ is a solution of the telegrapher's equation

$$u_{tt} + 2au_t = k_0^2 u_{xx}.$$

Roughly speaking, this says that the expectation or average value of a solution of the wave equation evaluated at the randomized time is a solution operator for the telegrapher's equation, at least within the class of sufficiently smooth solutions.

Following [12] we apply Theorem 2 to a modified Poisson random walk that includes branching. Consider a particle confined to the real line that moves according to the following rules.

- The particle speed k_0 is constant and the initial direction of the particle is ϵ_0 where ϵ_0 is as defined above.
- The particle reverses directions at random times T_R where

$$Pr(T_R \in ds) = ae^{-as}ds.$$

- The particle "dies" and splits into two "daughter" particles at random times T_S where

$$Pr(T_S \in ds) = be^{-bs}ds.$$

- The daughter particles evolve according to similar rules.
  - Each daughter particle moves with speed k_0 and the starting direction of each daughter particle is ϵ_0, just as for the original particle.
  - Each daughter particle reverses direction at random times T_R distributed just as for the original particle.
  - Each daughter particle splits into further daughter particles at random times T_S distributed just as for the original particle.
- All random processes are independent of one another.

Then at time $t > 0$ there are a random number of particles, say $n(t)$ where $Pr(n(t) = j) = e^{-bt}(1 - e^{-bt})^{j-1}$, [2]. At a splitting the two daughter particles may both go left with probability 1/4, or they may go in opposite directions with probability 1/2, or they may both go right with probability 1/4. Let the positions of the $n(t)$ particles be $X_1(t), X_2(t), \ldots, X_{n(t)}(t)$ and let

$$v(x,t) = \prod_{j=1}^{n(t)} f(x + X_j(t)),$$

where

$$f(s) = \begin{cases} 1 & s \geq 0 \\ 0 & s < 0 \end{cases}$$

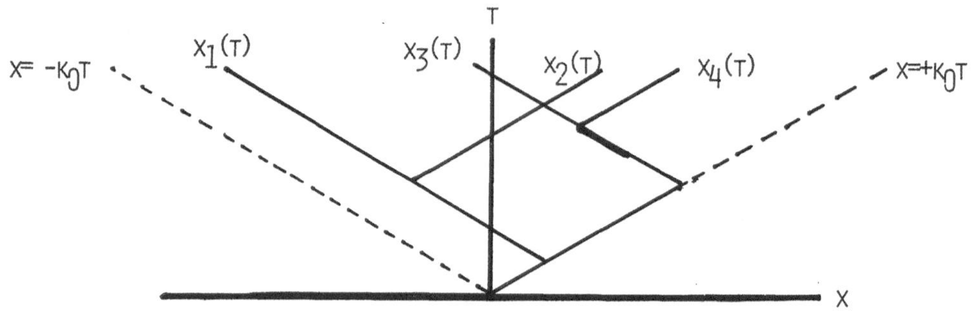

Figure 2: A Branching Random Walk

is the Heaviside step function.

It is easy to check that

$$v(x,t) = \begin{cases} 1 & x \geq - \min_{1 \leq j \leq n(t)} X_j(t) \\ 0 & x < - \min_{1 \leq j \leq n(t)} X_j(t). \end{cases}$$

Thus $v(x,t)$ is an indicator function measuring the maximal penetration of the process at time t. Define

$$u(x,t) \equiv E(v(x,t))$$

where the expectation is taken over the branching random walk process. Then by the spatial symmetry of the process

$$u(x,t) = Pr(x > \max_{1 \leq j \leq n(t)} X_j(t)).$$

Then $u(x,t)$ measures the cumulative probability of the penetration of the process to the position x by the time t. Obviously $u(x,t)$ is a monotone function, increasing from 0 at $x = -k_0 t$ to 1 at $x = k_0 t$. Since we expect a greater probability of locating the rightmost particle further to the right as the system of moving particles evolves, the most rapid variation of $u(x,t)$ is localized somewhat to the left of the line $x = k_0 t$, and this region of greatest increase should be moving to the right as t increases.

It is easy to derive an equation for $u(x,t)$ using a renewal argument, as McKean does for the case of branching Brownian motion. In fact, a little thought shows that

$$u(x,t) = Pr(T_S > t)E(f(x + X_1(t)) + \int_0^t Pr(T_S \in ds)E(u^2(x + X_1(s), t - s)\, ds.$$

The explanation of this "expectation by conditioning" is straightforward and goes as follows. The first summand is the probability that there have been no splittings up to time t multiplied by the expression for $u(x,t)$, which is simply the "product" over the single term $f(x + X_1(t))$. For the second summand we need the expectation of

$$\prod_{j=1}^{n(t)} f(x + X_j(t)), \tag{2}$$

given that the first splitting occurred at time s. The product factors into two subproducts, the first subproduct being over the factors corresponding to the descendants of one of the daughter particles of the first splitting, and the second subproduct being over the descendants of the other daughter particle. Then, in effect, given that the first splitting occurred at s, there are two independent branching Poisson random walk processes starting at time s and position $X_1(s)$. It follows that the expectation of the product (2) is the product of the expectation of each independent branch from the time s of the first splitting. Multiplying the results leads to the term

$$u^2(x + X_1(s), t - s),$$

and integration over all possible values of the first splitting time yields the second term.

Note that both of the expectation terms are just expectations of functions evaluated at the argument $x + k_0\epsilon_0 T(t)$, where $T(t)$ is the randomized time given by the movement of the original particle. But a function of the argument $x + k_0 t$ is a solution of the wave equation $v_{tt} = k_0^2 v_{xx}$. Therefore we apply the theorem of Kaplan and Kac and each of the expectation terms will satisfy a telegrapher's equation. (Actually, since the initial functions are not necessarily smooth, a limiting argument using smooth data is necessary). Therefore we can expect $u(x,t)$ to satisfy a hyperbolic equation, and after some straightforward manipulations we obtain the following equation for $u(x,t)$:

$$u_{tt} + ((2a + 2b) - 4bu)u_t = k_0^2 u_{xx} + (-b^2 - 2ab)u + (b^2 + 2ab)u^2. \tag{3}$$

Notice that this is of the same basic form as the nonlinear hyperbolic equation under consideration. Below, we will make some normalizations to bring this explicitly to the form of equation (1)

A generalization suggested by McKean [12] is possible. Suppose that each splitting produces a random number $j \geq 2$ of daughter particles with probability

b_j, where $\sum_{j=2}^{\infty} b_j = 1$. If we define the moment generating function $F(s) = b_2 s^2 + b_3 s^3 + \ldots$, then u satisfies the equation

$$u_{tt} + ((2a+2b) - 2bF'(u))u_t = k_0^2 u_{xx} + (-b^2 - 2ab)u + (b^2 + 2ab)F(u) \qquad (4)$$

For the special choices $a = 2b$ and $b_3 = 1$, in which case the process always produces 3 daughter particles, (4) reduces to an equation which is formally identical to an equation for a continuum of coupled van der Pol oscillators, namely

$$u_{tt} + 6b(1-u^2)u_t + 5b^2(u - u^3) = k_0^2 u_{xx}.$$

However, the interesting effects of the self excited oscillations that might occur for $u > 1$ do not arise in the probabilistic context since we always require $u \leq 1$.

In equation (3) it is convenient to make the transformation $v = 1 - u$ and to divide through the equation by $2a - 2b$ (here we suppose that $a > b$). Then if we define

$$\begin{aligned} \epsilon^2 &= 1/(2a-2b), & m &= 4b/(2a-2b), \\ k^2 &= k_0^2/(2a-2b), & l &= (b^2+2ab)/(2a-2b), \\ f(s) &= ls - ls^2, & g(s) &= ms \end{aligned}$$

the equation becomes

$$\epsilon^2 v_{tt} + (1 + g(v))v_t = k^2 v_{xx} + f(v). \qquad (5)$$

In equation (4) we can also make the transformation $v = 1 - u$. Gathering terms, dividing through the equation by $2a - 2(\mu - 1)b$, where μ is the mean number of particles produced at each splitting and where we assume that $a > (\mu - 1)b$, and finally making definitions similar to those above we again obtain equation (5).

4. Traveling Waves

The physical processes underlying the three distinct approaches to the same nonlinear hyperbolic equation suggest that there may be traveling wave solutions of the equation. Moreover, the equation has been written in a normalized form to emphasize the analogy with Fisher's equation which results when $\epsilon^2 = 0$ and $g = 0$, although it should be remembered that the coefficient ϵ^2 in front of the term v_{tt} need not be small and is only used suggestively for the analogy. In all three instances the governing equation is (5) with $g \in C^2[0,1]$, $1 + g(u) > 0$, and f such that $f \in C^2[0,1]$, $f(0) = f(1) = 0$, $f'(0)f'(1) < 0$, $f(s) > 0$ for $s \in (0,1)$.

Using some changes of variables, and well-known results about wave speeds for the traveling wave solutions of Fisher's equation, Hadeler [5] has recently proved the following theorem.

Theorem 3 *Suppose $1+g(s)>0$. Then equation (5) has smooth traveling wave solutions $u(x,t)=U(x-ct)$ for all speeds c such that*

$$c \in \left[\sqrt{\frac{4k^2 f'(0)}{1+4\epsilon^2 f'(0)}}, \frac{k}{\epsilon}\right).$$

Notice that unlike the case of Fisher's equation, here there is an upper bound on the possible wave speeds, which is a result of the hyperbolic nature of the equation. From the probabilistic interpretation, this means that with probability 1 the position x of the rightmost particle satisfies $x \leq kt/\epsilon$, with equality if there have not been any direction reversals up to time t. Also note that the minimal traveling wave speed is less than the speed that one would obtain from Fisher's equation, because of the presence of the term $1+4\epsilon^2 f'(0)$ in the denominator. Another comment is in order concerning the requirement that $1+g(s)>0$. In the case of the branching Poisson random walk with two daughters at every splitting, this requirement is equivalent to the condition that $a>b$, whereas in the branching Poisson random walk with a random number of daughters at each splitting, this is equivalent to requiring that $a>(\mu-1)b$. Roughly speaking, this says that in order to satisfy the requirements of the theorem the rate of reversal must be sufficiently large compared with the rate of branching.

Given the existence of the traveling wave solutions, one may prove the following theorem about the asymptotic speed of disturbances in the initial value problem.

Theorem 4 *Let $v(x,t)$ be the solution to the initial value problem on the real line*

$$\begin{aligned} \epsilon^2 v_{tt} + (1+g(v))v_t &= k^2 v_{xx} + f(v) \\ v(x,0) &= v_0(x) \\ v_t(x,0) &= 0 \end{aligned}$$

where the initial condition $v_0(x)$ is non-negative and has support which is bounded on the right. Let x_r be the right endpoint of the support of $v_0(x)$. Then for any $\bar{c}$ such that

$$\bar{c} \in \left(\sqrt{\frac{4k^2 f'(0)}{1+4\epsilon^2 f'(0)}}, \frac{k}{\epsilon}\right),$$

we have that $v(x_r+\bar{c}t,t) \to 0$ as $t \to \infty$.

A similar result holds on the left. It is easy to show that the solution tends pointwise to 1 because of the growth or reaction term $f(v)$. Therefore this theorem says that the bulk of the nonzero part of the solution asymptotically lies to the left of the line $x = x_r + c_* t$, where c_* is the minimal wavespeed given by

$$c_* = \sqrt{\frac{4k^2 f'(0)}{1 + 4\epsilon^2 f'(0)}}.$$

This theorem too has a probabilistic interpretation. Recall that $u = 1 - v$. Because the rate of reversal is greater than the rate of splitting, we do not expect to find the original particle or any of its descendants on the line $x = x_r + kt/\epsilon$ (or $x = x_l - kt/\epsilon$). This is because they have, with high probability, reversed their direction and moved to the left (respectively to the right). Having once moved off to the left a particle can never return to the line $x = x_r + kt/\epsilon$. Of course, the particle may reverse direction again but then it will move parallel and to the left of the line $x = x_r + kt/\epsilon$ (cf. Figure 2). Thus, we find that the spatial location having the highest probability density for the rightmost particle is localized to the left of the line $x = x_r + kt/\epsilon$. Thus the function $u(x,t)$, which is a cumulative probability distribution function for the random variable which gives the location of the rightmost particle, has its greatest increase somewhat to the left of the line $x = x_r + kt/\epsilon$ and the cumulative probability is essentially 1 to the right of this position. The theorem gives an asymptotic description of just how far to the left of the line $x = x_r + kt/\epsilon$ this region occurs.

5. Shock Effects

One of the important features of hyperbolic equations is the presence of shocks or discontinuities in the solutions. In some parameter regions equation (1) can develop and support such shocks, and the purpose of the present section is to give some heuristic arguments showing how these shocks develop.

Consider the equation (5), with $1 + g(v) > 0$ and with initial condition the step function

$$v_0(x) = \begin{cases} 1 & x \le 0 \\ 0 & x > 0. \end{cases}$$

Such a situation arises from the branching Poisson random walk process when the original particle starts at the origin (recall that $v = 1 - u$) and a is large compared with b, say $a > b$ in the case of splitting into two particles, and $a >$

$(\mu-1)b$ in the case of splitting into a random number of particles. Then there is always a shock, or discontinuity in the solution, arising from the unit step or discontinuity in the initial condition. However this shock, which propagates along the line $x = x_r + kt/\epsilon$, is exponentially diminishing like e^{-at}, corresponding to the probability that a particle (whether the original or one of its descendants) has not reversed its direction and moved leftward by time t. That the shock should be diminishing is also plausible from the first order time derivative term, which contributes a dissipative, smoothing effect.

It is easy to check that as $a \downarrow b$ or $a \downarrow (\mu-1)b$, according to the case at hand, we have

$$\sqrt{\frac{4k^2 f'(0)}{1+4\epsilon^2 f'(0)}} \to \frac{k}{\epsilon}.$$

That is,, the minimal speed of propagation tends to the maximal speed of propagation. Therefore the region of most rapid variation of u is concentrated at the spatial position $x = kt/\epsilon$. Furthermore, if $a < b$ or $a < (\mu-1)b$ according as the splitting is into two or a random number of daughter particles, then it is no longer true that $1+g(u) > 0$ and the two previous theorems about the existence of continuous traveling wave solutions no longer apply. Indeed, if $a < b$ or $a < (\mu-1)b$ in equation (4), and if the formula for the minimal wavespeed were still valid, the minimal wavespeed would be greater than the speeds of propagation allowed by the hyperbolic equation. This suggests that a shock will form at the spatial position $x = kt/\epsilon$.

The argument that a shock will develop when $1+g(u) < 0$ for some values of u can be made plausible from a probabilistic point of view as well. Recall that in the probabilistic interpretation the condition that $1+g(u) < 0$ for some values of u is equivalent to saying that the reversal rate a is slow compared with the splitting rate b, e.g. $a < b$ or $a < (\mu-1)b$. Therefore we may expect that as the original particle proceeds along its initial course, that is along the line $x = kt/\epsilon$, then with high probability it undergoes splitting before it reverses its direction. Now with high probability (i.e. probability 3/4) at least one of the daughters continues in the same direction, that is, the location of the daughter particle is along the line $x = kt/\epsilon$. Then this daughter undergoes splitting before it reverses direction and so on. Roughly speaking, because the splitting rate is sufficiently large compared with the reversal rate , there are particles "piling up"on the line $x = kt/\epsilon$, and so there is a non-zero probability of locating the rightmost particle at the point $x = kt/\epsilon$. This non-zero probability mass at $x + kt/\epsilon$ corresponds to a jump discontinuity in the cumulative probability distribution function $u(x,t)$ which is the solution to the equation (4). That is, there is a shock in the solution.

The above argument can be made rigorous and explicit. The details of the proof will be published elsewhere.

Acknowledgments The authors would like to thank Rex Saffer for many helpful discussions on the electrical circuit described in section 1.

Bibliography

[1] Berg, H.; *Random Walks in Biology*, Princeton:Princeton University Press (1983)

[2] Feller, W., *An Introduction to the Probability Theory and its Applications, Vol. I*; Wiley:New York, London, Sydney, (1968) p. 450.

[3] Gail, M., and Boone, C.; The locomotion of mouse fibroblasts in tissue culture, Biophys. J. **10** (1970), 980-993.

[4] Goldstein, S.; On diffusion by discontinuous movements, and on the telegraph equation, Quart. J. Mech. Appl. Math. **IV** (1951), 129-156

[5] Hadeler, K.; Hyperbolic traveling fronts, manuscript, March 1985

[6] Hall, R.; Amoeboid movement as a correlated walk, J. Math. Biol. **4**, (1977) 327-335.

[7] Hall, R., and Peterson, S.; Trajectories of human granulocytes, Biophys. J. **25** (1979), 497-509.

[8] Kac, M.; A stochastic model related to the telegrapher's equation, Rocky Mountain J. Math. **4** (1974), 497-509.

[9] Kaplan, S.; Differential equations in which the Poisson process plays a role, Bull. Amer. Math. Soc. **70** (1964), 264-268.

[10] Keener, J.; Analog circuitry for the van der Pol and FitzHugh-Nagumo equations, IEEE Transactions on Systems, Man and Cybernetics, Vol. SMC-13, (1983) 1010-1014.

[11] Lieberstein, M.; *Mathematical Physiology: Blood Flow and Electrically Active Cells*, Elsevier: New York, London, Amsterdam, (1973).

[12] McKean, H.; Application of Brownian motion to the equation of Kolmogorov-Petrovskii-Piscunov, Comm. Pure Appl. Math. **28** (1975), 323-331.

[13] Miller, K.; *Partial Differential Equations in Engineering Problems*, Prentice-Hall, Englewood Cliffs, New Jersey (1970).

[14] Okubo, A.; *Diffusion and Ecological Problems: Mathematical Models*, Biomathematics 10, Springer-Verlag: Berlin, Heidelberg, New York (1980).

[15] Scott, R.; The electrophysics of a nerve fiber, Rev. Mod. Phys. **47**, (1975), 487-533.

Lecture Notes in Biomathematics

Vol. 1: P. Waltman, Deterministic Threshold Models in the Theory of Epidemics. V, 101 pages. 1974.

Vol. 2: Mathematical Problems in Biology, Victoria Conference 1973. Edited by P. van den Driessche. VI, 280 pages. 1974.

Vol. 3: D. Ludwig, Stochastic Population Theories. VI, 108 pages. 1974.

Vol. 4: Physics and Mathematics of the Nervous System. Edited by M. Conrad, W. Güttinger, and M. Dal Cin. XI, 584 pages. 1974.

Vol. 5: Mathematical Analysis of Decision Problems in Ecology. Proceedings 1973. Edited by A. Charnes and W. R. Lynn. VIII, 421 pages. 1975.

Vol. 6: H. T. Banks, Modeling and Control in the Biomedical Sciences. V. 114 pages. 1975.

Vol. 7: M. C. Mackey, Ion Transport through Biological Membranes, An Integrated Theoretical Approach. IX, 240 pages. 1975.

Vol. 8: C. DeLisi, Antigen Antibody Interactions. IV, 142 pages. 1976.

Vol. 9: N. Dubin, A Stochastic Model for Immunological Feedback in Carcinogenesis: Analysis and Approximations. XIII, 163 pages. 1976.

Vol. 10: J. J. Tyson, The Belousov-Zhabotinskii Reaktion. IX, 128 pages. 1976.

Vol. 11: Mathematical Models in Medicine. Workshop 1976. Edited by J. Berger, W. Bühler, R. Repges, and P. Tautu. XII, 281 pages. 1976.

Vol. 12: A. V. Holden, Models of the Stochastic Activity of Neurones. VII, 368 pages. 1976.

Vol. 13: Mathematical Models in Biological Discovery. Edited by D. L. Solomon and C. Walter. VI, 240 pages. 1977.

Vol. 14: L. M. Ricciardi, Diffusion Processes and Related Topics in Biology. VI, 200 pages. 1977.

Vol. 15: Th. Nagylaki, Selection in One- and Two-Locus Systems. VIII, 208 pages. 1977.

Vol. 16: G. Sampath, S. K. Srinivasan, Stochastic Models for Spike Trains of Single Neurons. VIII, 188 pages. 1977.

Vol. 17: T. Maruyama, Stochastic Problems in Population Genetics. VIII, 245 pages. 1977.

Vol. 18: Mathematics and the Life Sciences. Proceedings 1975. Edited by D. E. Matthews. VII, 385 pages. 1977.

Vol. 19: Measuring Selection in Natural Populations. Edited by F. B. Christiansen and T. M. Fenchel. XXXI, 564 pages. 1977.

Vol. 20: J. M. Cushing, Integrodifferential Equations and Delay Models in Population Dynamics. VI, 196 pages. 1977.

Vol. 21: Theoretical Approaches to Complex Systems. Proceedings 1977. Edited by R. Heim and G. Palm. VI, 244 pages. 1978.

Vol. 22: F. M. Scudo and J. R. Ziegler, The Golden Age of Theoretical Ecology: 1923–1940. XII, 490 pages. 1978.

Vol. 23: Geometrical Probability and Biological Structures: Buffon's 200th Anniversary. Proceedings 1977. Edited by R. E. Miles and J. Serra. XII, 338 pages. 1978.

Vol. 24: F. L. Bookstein, The Measurement of Biological Shape and Shape Change. VIII, 191 pages. 1978.

Vol. 25: P. Yodzis, Competition for Space and the Structure of Ecological Communities. VI, 191 pages. 1978.

Vol. 26: M. B Katz, Questions of Uniqueness and Resolution in Reconstruction from Projections. IX, 175 pages. 1978.

Vol. 27: N. MacDonald, Time Lags in Biological Models. VII, 112 pages. 1978.

Vol. 28: P. C. Fife, Mathematical Aspects of Reacting and Diffusing Systems. IV, 185 pages. 1979.

Vol. 29: Kinetic Logic – A Boolean Approach to the Analysis of Complex Regulatory Systems. Proceedings, 1977. Edited by R. Thomas. XIII, 507 pages. 1979.

Vol. 30: M. Eisen, Mathematical Models in Cell Biology and Cancer Chemotherapy. IX, 431 pages. 1979.

Vol. 31: E. Akin, The Geometry of Population Genetics. IV, 205 pages. 1979.

Vol. 32: Systems Theory in Immunology. Proceedings, 1978. Edited by G. Bruni et al. XI, 273 pages. 1979.

Vol. 33: Mathematical Modelling in Biology and Ecology. Proceedings, 1979. Edited by W. M. Getz. VIII, 355 pages. 1980.

Vol. 34: R. Collins, T. J. van der Werff, Mathematical Models of the Dynamics of the Human Eye. VII, 99 pages. 1980.

Vol. 35: U. an der Heiden, Analysis of Neural Networks. X, 159 pages. 1980.

Vol. 36: A. Wörz-Busekros, Algebras in Genetics. VI, 237 pages. 1980.

Vol. 37: T. Ohta, Evolution and Variation of Multigene Families. VIII, 131 pages. 1980.

Vol. 38: Biological Growth and Spread: Mathematical Theories and Applications. Proceedings, 1979. Edited by W. Jäger, H. Rost and P. Tautu. XI, 511 pages. 1980.

Vol. 39: Vito Volterra Symposium on Mathematical Models in Biology. Proceedings, 1979. Edited by C. Barigozzi. VI, 417 pages. 1980.

Vol. 40: Renewable Resource Management. Proceedings, 1980. Edited by T. Vincent and J. Skowronski. XII, 236 pages. 1981.

Vol. 41: Modèles Mathématiques en Biologie. Proceedings, 1978. Edited by C. Chevalet and A. Micali. XIV, 219 pages. 1981.

Vol. 42: G. W. Swan, Optimization of Human Cancer Radiotherapy. VIII, 282 pages. 1981.

Vol. 43: Mathematical Modeling on the Hearing Process. Proceedings, 1980. Edited by M. H. Holmes and L. A. Rubenfeld. V, 104 pages. 1981.

Vol. 44: Recognition of Pattern and Form. Proceedings, 1979. Edited by D. G. Albrecht. III, 226 pages. 1982.

Vol. 45: Competition and Cooperation in Neutral Nets. Proceedings, 1982. Edited by S. Amari and M. A. Arbib. XIV, 441 pages. 1982.

Vol. 46: E. Walter, Identifiability of State Space Models with applications to transformation systems. VIII, 202 pages. 1982.

Vol. 47: E. Frehland, Stochastic Transport Processes in Discrete Biological Systems. VIII, 169 pages. 1982.

Vol. 48: Tracer Kinetics and Physiologic Modeling. Proceedings, 1983. Edited by R. M. Lambrecht and A. Rescigno. VIII, 509 pages. 1983.

Vol. 49: Rhythms in Biology and Other Fields of Application. Proceedings, 1981. Edited by M. Cosnard, J. Demongeot and A. Le Breton. VII, 400 pages. 1983.

Vol. 50: D. H. Anderson, Compartmental Modeling and Tracer Kinetics. VII, 302 pages. 1983.

Vol. 51: Oscillations in Mathematical Biology. Proceedings, 1982. Edited by J. P. E. Hodgson. VI, 196 pages. 1983.

Vol. 52: Population Biology. Proceedings, 1982. Edited by H. I. Freedman and C. Strobeck. XVII, 440 pages. 1983.

Vol. 53: Evolutionary Dynamics of Genetic Diversity. Proceedings, 1983. Edited by G. S. Mani. VII, 312 pages. 1984.

Vol. 54: Mathematical Ecology. Proceedings, 1982. Edited by S. A. Levin and T. G. Hallam. XII, 513 pages. 1984.

Vol. 55: Modelling of Patterns in Space and Time. Proceedings, 1983. Edited by W. Jäger and J. D. Murray. VIII, 405 pages. 1984.

Vol. 56: H. W. Hethcote, J. A. Yorke, Gonorrhea Transmission Dynamics and Control. IX, 105 pages. 1984.

Vol. 57: Mathematics in Biology and Medicine. Proceedings, 1983. Edited by V. Capasso, E. Grosso and S. L. Paveri-Fontana. XVIII, 524 pages. 1985.

ctd. on inside back cover

Lecture Notes in Biomathematics

Vol. 58: C.A. Macken, A.S. Perelson, Branching Processes Applied to Cell Surface Aggregation Phenomena. VIII, 122 pages. 1985.

Vol. 59: I. Nåsell, Hybrid Models of Tropical Infections. VI, 206 pages. 1985.

Vol. 60: Population Genetics in Forestry. Proceedings, 1984. Edited by H.-R. Gregorius. VI, 287 pages 1985.

Vol. 61: Resource Management. Proceedings, 1984. Edited by M. Mangel. V, 138 pages 1985.

Vol. 62: B. J. West, An Essay on the Importance of Being Nonlinear. VIII, 204 pages. 1985.

Vol. 63: Carme Torras i Genís, Temporal-Pattern Learning in Neural Models. VII, 227 pages, 1985.

Vol. 64: Peripheral Auditory Mechanisms. Proceedings, 1985. Edited by J.B. Allen, J.L. Hall, A. Hubbard, S. Neely and A. Tubis. VII, 400 pages. 1986.

Vol. 65: Immunology and Epidemiology. Proceedings, 1985. Edited by G.W. Hoffmann and T. Hraba. VIII, 242 pages. 1986.

Vol. 66: Nonlinear Oscillations in Biology and Chemistry. Proceedings, 1985. Edited by H.G. Othmer. VI, 289 pages. 1986.